An Introductory Course of
MATHEMATICAL ANALYSIS

T0296161

An Introductory Course of
MATHEMATICAL ANALYSIS

BY

CHARLES WALMSLEY, M.A.,

ASSISTANT LECTURER IN THE UNIVERSITY OF MANCHESTER
AND LATE ASSISTANT LECTURER IN THE UNIVERSITY
COLLEGE OF WALES, ABERYSTWYTH; FORMERLY
SCHOLAR OF KING'S COLLEGE, CAMBRIDGE

WITH A PREFACE BY

W. H. YOUNG, Sc.D., F.R.S.

CAMBRIDGE
AT THE UNIVERSITY PRESS
1926

CAMBRIDGE
UNIVERSITY PRESS

University Printing House, Cambridge CB2 8BS, United Kingdom

Cambridge University Press is part of the University of Cambridge.

It furthers the University's mission by disseminating knowledge in the pursuit of education, learning and research at the highest international levels of excellence.

www.cambridge.org
Information on this title: www.cambridge.org/9781316509739

First published 1926
First paperback edition 2015

A catalogue record for this publication is available from the British Library

ISBN 978-1-316-50973-9 Paperback

PREFACE

By
Dr W. H. YOUNG, F.R.S.

THIS book had its origin in, and is one of, the tangible results of my reorganisation of the Mathematical Department at Aberystwyth in the University of Wales, during my four years tenure of office there. The problem I had to solve was a difficult one. It was one of the crying needs of the Departments of Physics and Chemistry that their students should by the end of their first year have become acquainted with the concepts and processes of Differentiation and Integration. A considerable portion of these students, however, were, on their arrival at the College, entirely ignorant of Trigonometry, the knowledge of which is presupposed in treatises, even elementary, on the Calculus. In drawing up a Syllabus for the non-geometrical portion of the First Year Course, I was thus led to arrange that the notion of a trigonometrical function appeared at a comparatively late stage and in an analytical form; a procedure, for the rest, in accordance with what has been felt as desirable by more than one distinguished mathematician, in view of the very restricted range of the geometrical definition *per se* of a trigonometrical function. Moreover the well-known interest felt by the Celtic race for abstract ideas, as well as the circumstance that the First Year Class necessarily contained students who had in view the possibility of taking Honours in Mathematics at a later stage, justified a rather more theoretical treatment than has been usual in this country, not only in the above connection but throughout the course.

The Syllabus, as printed in the Programme of the University College of Wales for the year 1922—1923, is as follows:

"FIRST YEAR COURSE. *Elementary Algebraical Analysis:*—The fundamental laws of algebra; the real number; equalities and equations; inequalities and inequations; indices; nature and use of logarithms; simple infinite series; rational functions; formation of simple transcendental functions by means of power series; the idea of a limit; the idea of continuity; differentiation of rational functions and of the simpler functions represented by power series; the idea of a definite integral; its intuitive properties; identification of the geometrical and analytical definitions of the trigonometrical functions; simple applications of the notions of differentiation and integration."

The prominence given to *inequalities*, and more especially to *inequations*, in this Syllabus is in accordance with Continental practice. Their comparative neglect in English teaching has seriously handicapped in the past the researcher, and has interfered with the efficiency of the ordinary mathematical student. *Logarithms* were intended to be introduced almost simultaneously with *indices*, reappearing, of course, under the head of *simple transcendental functions represented by power series*. In the remainder of the course the *idea of limit* is fundamental, the method of approach being understood to be that by *monotone sequences** and making free use of the notions of *upper and lower bounds*, on which I have laid great stress and of which I have made systematic use in my papers.

The task of actually lecturing, under my direction, on the new Syllabus, I entrusted to one of my assistants, Mr Walmsley. In view of the somewhat revolutionary character of the changes introduced into the curriculum and with the object of securing that the matter communicated to the students was what was intended, partly also because I thought a great want would thereby be supplied, I suggested to Mr Walmsley that the Course should be written out and subsequently form the basis of a joint book by the two of us. Unfortunately, only the first few manuscript pages were submitted to me, so that the present volume appears under his name alone and on his sole responsibility.

I have only to add that the success that attended the reform at Aberystwyth warrants a confident hope that the present volume, attempting as it does to embody the main features of the Schedule, will find its place, and that an important one, among the books habitually used at school and college.

* Mr Walmsley has not followed my usage of restricting the use of the word *sequence* to such *successions* as have an unique limit; but where the succession is monotone the two notions, of course, coincide.

W. H. YOUNG.

COLLONGE, LA CONVERSION,
July, 1926.

AUTHOR'S PREFACE

I HAVE to express my indebtedness to various works of the following authors: G. H. Hardy, T. J. I'A. Bromwich, E. W. Hobson, W. H. Young, C. J. de la Vallée Poussin, and E. Goursat.

I have to thank Mr Arthur Berry, Vice-Provost of King's College, Cambridge, for many valuable improvements, and Mr H. W. Unthank, of the University College, Exeter, for reading the proofs.

C. W.

EYAM, DERBYSHIRE,
   August, 1926.

CONTENTS

CHAPTER III

CHAPTER IV

CHAPTER I

NUMBER

§ 1. FUNDAMENTAL LAWS OF ALGEBRA

1. ALGEBRA may be described as the general science of arithmetic In arithmetic the processes of addition, subtraction, multiplication, division and other derived operations arise and are applied in connection with particular numbers. It is seen that the methods employed, being applicable to all particular numbers encountered, with certain well-defined exceptions, must be capable of a general formulation. It is this general formulation which is the primary object of algebra.

In order to express the truths of arithmetic in the general (algebraic) form it is necessary to employ some kind of *symbolism*. The choice of a symbolism is to some extent arbitrary, but not without effect on the development of the science; the lack of an appropriate symbolism having caused in some instances a delay of hundreds of years in the progress of different branches of mathematics. In the present matter of the algebraic statement of the acts of arithmetic the use of the *letters of the alphabet* to represent the unidentified numbers under consideration is singularly appropriate and has been universal since the sixteenth century. Its fertility is sufficiently apparent in the use of formulae in elementary algebra, and in all branches of science, for it to be unnecessary here to enlarge upon it. It may be noted in passing however that the letters occurring in algebraic theorems do not always represent numbers with the same scope of generality. Thus in some theorems the letters (or some of them) may represent numbers belonging to the widest class contemplated,—that of the real or complex numbers,—including under this head, not only the whole numbers 1, 2, 3, etc., but also such entities as -1, $\frac{1}{2}$, $\sqrt{2}$, $\sqrt{-1}$, which are defined below; or they may be restricted to represent numbers of only a particular class, such as the whole numbers, or proper fractions, etc. On the other hand it is always the method of algebra to state theorems with the greatest possible generality, and the fewer the restrictions placed on the variables (i.e. letters) occurring in a theorem the more important (in general) will the theorem be.

2. Fundamental laws. The theorems on which algebra is built, —*the fundamental laws of algebra,*—are based on our intuitive ideas of *counting,* and therefore in the first instance are stated only for the ordinary *whole numbers* 1, 2, 3, etc. The laws are:—

(I) *The associative law for addition,* viz.: The terms of a sum of three numbers may be added together in any way preserving the original order without altering the sum; or, symbolically, if a, b, c are any three whole numbers then $a + (b + c) = (a + b) + c$.

(II) *The associative law for multiplication,* viz.: The terms of a product of three terms may be multiplied in any way preserving the original order without altering the product, or

$$a \times (b \times c) = (a \times b) \times c,$$

where a, b, c have the same significance as in (I).

(III) *The commutative law for addition,* viz.: The terms in a sum of two numbers may be added together in either order without altering the sum, or $a + b = b + a$.

(IV) *The commutative law for multiplication,* viz.: The terms in a product may be multiplied in either order without altering the product, or $a \times b = b \times a$.

(V) *The distributive law,* viz.: The product of a sum of two numbers by any third number is the sum of the products of the separate terms of the sum by the common multiplier, or

$$(a + b) \times c = (a \times c) + (b \times c).$$

In the statement of these laws we must be quite clear as to the meaning of the terms and symbols used. The simplest definition of the sum $a + b$ is to consider it as the number finally obtained if, having counted up to the number a, we continue the process of counting until an additional stock of b objects is exhausted. Multiplication is repeated addition. Subtraction and division will be the inverses of addition and multiplication respectively.

Probably no argument which could be put forward would make the student more convinced of the truth of (I) and (III) than he already is,—though such argument might prove useful in helping the student to appreciate the difficulties inherent in the foundations of all science. We will therefore be content to assume these laws as axioms. We shall have a few further remarks to make in this connection shortly. We can however convince ourselves of the truth of (II), (IV) and (V) by arguments similar to the following, dealing with the commutative law for multiplication (IV): From the definition of multiplication $a \times b$ is the number of objects in an imagined array consisting of b rows each containing a objects*; by intuition, or by extending the laws (I) and (III) to cover the case of any number of terms, we see that the number of objects in our array is the same in whatever order we count the objects; in particular therefore this number is the same as that obtained by considering the array of objects as a columns of b objects, i.e. $b \times a$; and the theorem is proved.

The fundamental laws can be extended to deal with sums, etc., of any number of terms; these extensions follow from the laws themselves and will provide an excellent exercise for the student. In most algebraic deductions from the fundamental laws it will be found most desirable to employ such simple extensions of the laws instead of the original laws themselves.

3. The fundamental laws are at the base of all algebraical analysis, and in fact bear to algebra much the same relation as do the axioms of Euclid to ordinary elementary geometry. Taking, as suggested above, *counting* as the basis of our definitions, we cannot avoid belief in the truth of the laws;—our intuitive ideas on counting would be at variance with a denial of the fundamental laws of addition; and, having accepted these intuitions, expressed in the form of these laws or in any other equivalent form, all the laws,—and thence briefly all algebra,—follow as a logical consequence. But it is interesting to see whether any contrary laws are logically possible, that is, whether, either by distrusting our intuitions or by adopting other more or less arbitrary definitions of "addition," etc., we could without contradiction build up a system of "algebra" on the supposition that our fundamental laws (or some of them) were untrue. In geometry it is possible to deny the Euclidean system of axioms and build up perfectly logical non-Euclidean geometries,—in which for example it is no longer impossible

* The symbol $a \times b$ is to be read as "a multiplied by b"; and in general in any such arithmetical symbol the operations occurring are to be performed successively from the left to the right, except when brackets or special conventions otherwise direct.

for two straight lines to enclose a space. In algebra likewise the answer to this question is in the affirmative. There are in fact, for example, *non-commutative* algebras, i.e. algebras in which the symbols $a \times b$ and $b \times a$ (say) are not equivalent. In such systems the symbols used cannot represent the same entities as in the ordinary algebra of whole numbers unless we are prepared to deny our intuitions of counting. But by using the symbols to represent different operations and combinations of operations of a more general character than our elementary addition, multiplication, etc. (a process to which we have to resort even in ordinary algebra once we pass on to consider the addition, etc., of non-integral numbers), quite definite non-commutative algebras can be obtained.

A trivial example of such a system would be obtained if a were taken to represent a motion of a point through a distance a in an easterly direction on the spherical earth, b a motion through a distance b in a northerly direction, and the multiplication of a by b the motion a followed by the motion b. The results of the two combined motions $a \times b$ and $b \times a$ would then differ by an amount depending on the latitudes of the points considered.

In this course, however, we shall not be concerned with non-commutative or other algebras in which the fundamental laws do not hold. But we shall use these laws as a guide to help us to introduce into algebra and arithmetic entities, other than whole numbers, which will also satisfy these laws and therefore be capable of being dealt with in the same way as the whole numbers.

4. Subtraction and division. If we introduce the notions inverse to those of addition and multiplication, viz., subtraction and division, the fundamental laws are capable of extension to a certain extent. Thus the distributive law (V) will remain true if the sum of the numbers be replaced by the difference, or the product with the third number by the quotient, or both sum and product be replaced by difference and quotient simultaneously, thus e.g. $(a - b) \div c = (a \div c) - (b \div c)$*. The associative and commutative laws cease to hold when addition or multiplication is replaced by subtraction or division; it is known for example that $a - b$ is not the same as $b - a$ and that $a \div (b \div c)$ is not the same as $(a \div b) \div c$. But it is easy to modify the two associative laws so that they will

* It is tacitly assumed here that the operations involved are possible. Thus $a > b$ (i.e. a is greater than b), so that b can be subtracted from a; and so on.

remain true in these cases by simply introducing appropriate *rules of signs*, e.g. by replacing $a-(b+c)$ by $a-b-c$ or $a\div(b\times c)$ by $a\div b\div c$ or $a-(b-c)$ by $a-b+c$*—*provided* the operations considered do not lead at any stage to any impossible operation, i.e. provided, for example, we are not led to the operation of subtracting a number from one not greater than itself nor to the operation of dividing a number by a number of which the first is not a multiple. It will be seen incidentally that any attempt to modify similarly the two commutative laws would, except in trivial cases, essentially lead to some such impossible operation.

EXAMPLES I.

1. Prove the extensions of the fundamental laws to the operations of subtraction and division mentioned above.

2. Prove by imagining m^2 units arranged in a square and adding up in suitable orders that

$$\text{(i)} \quad 1+3+5+\ldots+(2m-1)=m^2,$$

and $\quad\quad$ (ii) $\quad 1+2+3+\ldots+m=\tfrac{1}{2}(m^2+m).$

3. Prove by direct reference to the fundamental laws that

$$\text{(i)} \quad (a+b)^2=a^2+2ab+b^2,$$
$$\text{(ii)} \quad (a+b)^3=a^3+3a^2b+3ab^2+b^3,$$
$$\text{(iii)} \quad (a+b)(a-b)=a^2-b^2,$$

a and b being any whole numbers (with the restriction that in (iii) a must exceed b); the definitions of the square and cube of a number being supposed known.

4. Prove that if $a>b$ and $c>d$ then

$$(a-b)\times(c-d)=(ac+bd)-(ad+bc).$$

5. Prove that if $a>b$ and $c>d$ then

$$(a-b)(c-d)=ac-ad-bc+bd,$$

provided also $ac>ad+bc$.

[If ac is not greater than $ad+bc$ the expression $ac-ad-bc+bd$ is meaningless as a whole number.]

§ 2. RATIONAL NUMBERS

5. The inverse operations, subtraction and division, cannot, as we have seen, be applied to all whole numbers indiscriminately. For example (as we are at present dealing exclusively with whole

---

* To prove for example that $a+(b-c)=a+b-c$ we argue: From the definition of subtraction, if $b-c=x$, $b=x+c$; whence $(a+x)+c=a+(x+c)=a+b$ and therefore $a+(b-c)=a+x=(a+b)-c$. Or to prove $a-(b+c)=a-b-c$, let $a-(b+c)=x$, so that $a=x+(b+c)=(x+c)+b$ whence $a-b=x+c$ whence again $a-b-c=x$.

numbers) 6 cannot be subtracted from 4, because there is no whole number which when added to 6 will give the result 4; and again 7 cannot be divided by 3 because there is no whole number which when multiplied by 3 will give the result 7. The problem confronts us therefore, whether these impossibilities can be surmounted in any way; i.e. whether we can invent new entities, which we may wish to call "numbers," and new operations with these "numbers," which we may call "addition," etc., so that such hitherto impossible operations will be possible in the sense that, considered as an operation in our new "arithmetic," it gives a definite result which is a "number" of the new type. Whether or not such a new arbitrary "arithmetic" can be of any use for practical applications is not primarily the concern of pure mathematics; but as a matter of fact we shall find that our new arithmetic will have useful applications,—in particular to the important problem of measurement.

6. Fractional-number-pairs. Let us define first the *fractional numbers* or *fractions*. We observe first that to some pairs of numbers (e.g. 8 and 2) correspond, by the process of division, single numbers (4), whereas to other pairs (e.g. 7 and 3) there are no such corresponding numbers. For the first kind of pairs of numbers,—such as (8 and 2), or $\frac{8}{2}$,—which represent, or correspond to, whole numbers, we have the following laws of addition, etc.:

If p, q, r, s are whole numbers such that p and r are respectively divisible by q and s, then:

$$\frac{p}{q} + \frac{r}{s} = \frac{ps+qr}{qs}, \qquad \frac{p}{q} \times \frac{r}{s} = \frac{pr}{qs};$$

for (to prove the first relation)

$$\left(\frac{p}{q} + \frac{r}{s}\right) \times qs = \left(\frac{p}{q} \times qs\right) + \left(\frac{r}{s} \times qs\right) \text{ by the distributive law,}$$

$$= \left(\frac{p}{q} \times q\right) \times s + \left(\frac{r}{s} \times s\right) \times q \text{ by the associative law,}$$

$$= (p \times s) + (r \times q) \text{ by the definition of division,}$$

and therefore $\qquad \dfrac{p}{q} + \dfrac{r}{s} = \dfrac{ps+qr}{qs}$

by the definition of division,—all the numbers concerned being whole numbers.

Or, using the notation (p, q) instead of $\frac{p}{q}$, we have the following laws for the addition and multiplication of pairs of numbers of the type considered:

$$(p, q) + (r, s) = (ps + qr, qs)\}$$
$$(p, q) \times (r, s) = (pr, qs) \;\;\}\;\;\;\;\;\;\;\;\;\;\;\;\;\;(1).$$

In the statement of these laws there is no need for the restriction that the numbers p, r should be divisible by q, s as supposed, i.e. should be such that the various pairs of numbers correspond to (or represent) whole numbers;—*the pairs of numbers*

$$(p, q), \;\; (r, s), \;\; (ps + qr, qs), \;\; (pr, qs)$$

exist equally whether the whole numbers p, r are divisible by the numbers q, s or not. Let us then relax this restriction and use these laws as the *definitions* of "addition" and "multiplication" of "*fractional-number-pairs*,"—or, as we will say for the sake of brevity, *fractional numbers, or fractions**.

With these definitions we see easily that the fundamental laws which we have found for addition, etc., of whole numbers, are still formally true. Thus, if A and B are two fractional-number-pairs, say $A = (p, q)$ and $B = (r, s)$, then

$$A + B = (p, q) + (r, s)$$
$$= (ps + qr, qs),$$

and
$$B + A = (r, s) + (p, q)$$
$$= (rq + sp, sq)$$
$$= (ps + qr, qs),$$

by the commutative laws for whole numbers. Hence $A + B = B + A$, or the commutative law for addition holds. Similarly the other fundamental laws can be proved to be still true for our "fractional numbers," i.e. if we call a "fractional-number-pair" a *number* and if we call the operations on these fractional-number-pairs defined in equations (1) *addition* and *multiplication*, then the fundamental laws can be applied to these numbers and operations of addition, etc., verbally unchanged. Consequently all general statements about arithmetical operations on whole numbers deducible from the fundamental laws are equally true of these arbitrary operations

* Or positive rational numbers.

on these arbitrary fractional numbers; for example, in particular, if A and B are any two fractional numbers, then

$$(A + B)^2 = A^2 + 2AB + B^2.$$

7. Order. Our arithmetic is not yet the arithmetic of fractional numbers which we desire. The definitions so far given do not enable us to assign any *order* to the fractional numbers, whereas the notion of order is certainly essential in connection with whole numbers. If we wish to use our fractional numbers as numbers in any complete sense it is therefore necessary to give a definition determining which of two fractional numbers is the "greater" and which the "less."

We say that *the fraction (p, q) is greater than, equal to, or less than, the fraction (r, s) according as the whole number ps is greater than, equal to, or less than, the whole number qr;* or, in symbols,

$$(p, q) >, =, \text{ or } < (r, s) \Big\}$$
according as $\qquad ps >, =, \text{ or } < qr \qquad \Big\}$ ....(2).

In this definition we assume a knowledge of the notions of greater than, etc., as applied to whole numbers.

We notice that from the relations (2) we have

$$(ps, qs) = (p, q) \dots\dots\dots\dots\dots\dots(2a).$$

This relation,—which corresponds with the characteristic property of a quotient $\dfrac{p}{q}$ that $\dfrac{ps}{qs} = \dfrac{p}{q}$,—is interesting as being a statement of *equality* between two fractional-number-pairs which are plainly *different* pairs of numbers. The use of the notion and sign of equality in this somewhat arbitrary sense is logically important and interesting, but we will not dwell on it. It is sufficient for us to realise that with this *use* of the sign of equality, whatever it may *mean*, no contradiction can arise.

8. Number-pairs corresponding to whole numbers. We notice the following particular facts about the new arithmetic concerning the *special class* of fractional-number-pairs whose second number is 1:

$$(p, 1) + (q, 1) = (p + q, 1) \text{ and } (p, 1) \times (q, 1) = (pq, 1).$$

It follows that this particular class of number-pairs can be treated for addition and multiplication as a class by itself, for such operations lead only to number-pairs of this kind, having 1 for the second number. In fact a little reflection will shew that in dealing with such special number-pairs we can operate with them just as with the ordinary whole numbers p and q and the ordinary arithmetical operations of addition and multiplication; that in fact the new arithmetic of these special number-pairs is identical with the old arithmetic;—i.e. that *every result in either arithmetic is capable of interpretation as a result in the other.* If therefore in our arithmetic of fractional-number-pairs we agree to replace any number-pair whose second number is 1 by the ordinary whole number which is the first number, i.e. to put

$$(p, 1) = p \quad \dots\dots\dots\dots\dots\dots\dots(3),$$

the results we shall get will be quite valid, that is to say any algebraic result we obtain for whole numbers will be true as a result in ordinary arithmetic and any result for our fractional-number-pairs will be a true result in our new arithmetic; if in any case the numbers concerned are of both kinds the apparently whole numbers (e.g. p) are to be replaced by the corresponding fractional-number-pair $[(p, 1)]$ and the result interpreted as a result in the new arithmetic. Also, applying the relation $(2a)$ to these fractional-number-pairs having the second number 1 and coupling it with our convention (3), we obtain the other fundamental characteristic of a fraction that

$$\frac{p}{q} \times q = p;$$

for $\qquad\qquad (p, q) \times (q, 1) = (pq, q) = (p, 1).$
From this it follows that

$$(p, q) = (p, 1)/(q, 1).$$

This last relation is a relation in the new arithmetic and the operation of division is the new arbitrary kind of division (the inverse of the new multiplication).

But we notice that if p is divisible by q (with the quotient r say), —in the ordinary sense,—then the expression $(p, 1)/(q, 1)$ can be replaced by p/q or r; i.e. if p is divisible by q the fractional-number-pair (p, q) represents the ordinary quotient p/q; or, a fractional-number-pair (p, q) for which p is divisible by q behaves as far as arithmetical facts are concerned just like the quotient $\frac{p}{q}$. In future we shall write (p, q) as $\frac{p}{q}$ or p/q, whether p is divisible by q or not.

9. The arithmetic of our fractional-number-pairs may therefore be said to include the arithmetic of such quotients. Our fractional-number-pairs and their "arithmetic" are seen to be the extension of the number system and arithmetic of which we have been in search. In our new system such an operation as $3 \div 8$ is possible.

The advantage of this extended arithmetic from the point of view of the original arithmetic is considerable. We may, in proving a certain result, even in ordinary arithmetic of whole numbers, find it more convenient to use fractional numbers in the intermediate steps of the proof. The proof will nevertheless be a valid proof of the theorem concerned, provided only the final stages of the argument can be interpreted (by the convention (3)) as a relation between whole numbers.

Thus, to take a fundamental, if simple, example: if a, b, and c are whole numbers such that b is a multiple of c and ac a multiple of b, we can argue

$$a \div (b/c) = (a/b) \times c = (ac)/b$$

by using the simple extension to division of the associative law; and in this the expression $(a/b) \times c$ will be meaningless as a whole number unless a is a multiple of b; but nevertheless the result $a \div (b/c) = (ac)/b$ is a true result holding between whole numbers and this proof is quite valid.

10. Subtractive-number-pairs. In the same way we can extend the meaning of subtraction to cases hitherto impossible by introducing *subtractive-number-pairs* * $\{p, q\}$ subject to the definitions:

$$\{p, q\} + \{r, s\} = \{p + r, q + s\};$$
$$\{p, q\} \times \{r, s\} = \{pr + qs, ps + qr\};$$
$$\{p, q\} >, =, \text{ or } < \{r, s\}$$

according as

$$p + s >, =, \text{ or } < q + r;$$
$$\{p + q, q\} = p\dagger;$$

where p, q, r, s represent any whole numbers or fractions as hitherto defined.

It is necessary here also to introduce a new number, *zero* or 0, defined as the subtractive-number-pair $\{p, p\}$.

---

* These are the positive and negative rational numbers.

† This relation could be replaced by $\{p+1, 1\} = p$ and so brought more into agreement with relation (3) of p. 9.

It is at once evident that, with these definitions,
$$\{p, q\}=p-q$$
if the operation on the right is possible, for, from the fourth relation above,
$$\{(p-q)+q, q\}=(p-q),$$
i.e. $\{p, q\}=p-q$ if $p-q$ is a whole number or a fraction.

We see also that the operation $p-q$ can *always* be performed if p and q are interpreted as subtractive-number-pairs; for
$$\{q+r, r\}+\{p, q\}=\{p+q+r, q+r\}$$
$$=\{p+r, r\},$$
and therefore $\quad \{p+r, r\}-\{q+r, r\}=\{p, q\},$
i.e. $\qquad\qquad\qquad p-q=\{p, q\}.$

The number 0, as defined, possesses the usual properties, such as those expressed by the relations
$$p+0=p, \quad p\times 0=0;$$
for
$$p+0=\{p+q, q\}+\{p, p\}$$
$$=\{p+q+p, p+q\}$$
$$=\{p+q, q\}$$
$$=p,$$
and
$$p\times 0=\{p+q, q\}\times\{p, p\}$$
$$=\{(p+q)p+qp, (p+q)p+qp\}$$
$$=0.$$

11. Rational numbers. It is easy to see that, with our system of definitions, the fundamental laws will continue to hold if these subtractive-number-pairs are used as numbers. We are justified in considering them as numbers. We call them the *rational numbers*.

We shall call a rational number *positive* or *negative* according as it is greater than or less than 0 in accordance with our definitions. We shall use the term *integer* to denote either a whole number or a negative whole number, i.e. a negative number expressible as $\{q, p+q\}$ where p and q are ordinary whole numbers. We shall write in future the negative rational number $\{q, p+q\}$ as $-p$, and, where needed to avoid ambiguity, we shall write $+p$ for the positive number p.

12. Zero. In our system of positive and negative rational numbers there is one arithmetical operation which is still impossible; —division by the number zero. For there is evidently no rational number x whatever such that $2/0 = x$ or $x \times 0 = 2$. We could overcome this limitation by adding to our system a new number, *infinity*, (denoted by ∞), defined to have the property that if x is any rational number (other than zero) $0 \times \infty = x$. But we should then have introduced further difficulties. The operations ∞ / ∞, $0 \times \infty$ as well as $0/0$ would be meaningless, (or at least have indefinite meanings), and evidently some vital alteration in our system of algebra would have to be made to render these operations definite. We therefore prefer to avoid the use of such a notion of infinity,—more especially as it is often convenient to use the term infinity in another connection. With this special exception of division by zero *our system of rational numbers is complete in that all operations of addition, subtraction, multiplication and division* applied not only to the original whole numbers, but to any rational numbers, *are possible and obey the fundamental laws of algebra.*

EXAMPLES II.

1. Prove from the definition of a fraction as a fractional-number-pair that
$$\frac{a}{b} \div \frac{c}{d} = \frac{ad}{bc}, \ a, b, c, d \text{ being whole numbers.}$$

2. Prove from the definition of a rational number as a subtractive-number-pair that
$$a+(-a)=0, \quad a-(-b)=a+b, \quad (-a)\times(-b)=a\times b, \quad (-a)\times b = -ab,$$
a and b being whole numbers or fractions.

3. Prove from the definitions of the text that if a and b are two rational numbers then $a >$, $=$, or $< b$ according as the difference $a-b$ is positive, zero, or negative.

4. It is a consequence of the fundamental laws that
$$a^2+b^2-2ab=a^2-2ab+b^2=(a-b)^2=(b-a)^2.$$
If a and b are whole numbers and $a < b$, $a^2-2ab+b^2$ and $(a-b)^2$ are both meaningless considered as ordinary arithmetical combinations of whole numbers, but a^2+b^2-2ab and $(b-a)^2$ nevertheless represent whole numbers and the relation $a^2+b^2-2ab=(b-a)^2$ is true. Compare this with the remarks and example on p. 10 (Par. 9).

5. The equality of Ex. 5, p. 5, $(a-b)(c-d)=ac-ad-bc+bd$ can now be stated without any restrictions on the magnitudes of the letters; and if $a > b$, and $c > d$, both sides of the equality can in all cases be interpreted as whole numbers.

6. Prove $\quad (a-b)^3=a^3-3a^2b+3ab^2-b^3.$

7. Shew that if it is possible in any way to define the sums and products, etc., of negative numbers so as to conform to the fundamental laws, then the rule of signs, $(-a)(-b)=ab$, must be satisfied.

[The negative numbers are supposed to be defined as subtractive-number-pairs for which the notions of addition, etc., have not yet been defined.]

§ 3. The Problem of Measurement. Irrational Numbers

13. The system of rational numbers which has now been established can be applied to the problem of measurement, and the operations of addition, etc., can then be given a simple geometrical (or physical) interpretation. The practical problem of the measurement of any physical quantity such as a speed, an electric current, a temperature, a weight, a time interval, or a length may be typified by the single problem of the measurement of a length along a straight line. Most often in actual fact these measurements are made by means of readings along a scale and in all cases the measurement may be reduced to this method. We may consider the problem of measurement then as that of the representation of lengths along a straight line by means of numbers.

14. Rational numbers represented by points on a straight line. Let $X'OX$ (Fig. 1) be an unlimited straight line, and O a fixed point on it. Take some unit of measurement (say one inch) and mark off distances of 1 unit, 2 units, 3 units, etc., from O in one

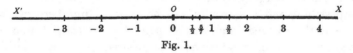

Fig. 1.

direction, say to the right, and also similar distances in the other direction from O. We have thus a geometrical representation of the positive and negative integers which is evidently appropriate, for we see that we can represent the operations of addition and subtraction simply as motions of translation to the right or left respectively. Thus the point numbered -3 is the point obtained when we follow the motion of say 2 units to the right from the origin O by a motion of 5 units to the left; i.e. the point numbered -3 is the point which corresponds to the operation of subtracting the number 5 from the number 2 in this representation of subtraction as a motion to the left. We need not stress this point; it is evident that this representation is appropriate for all operations of addition and subtraction of positive and negative integers. We have a point on the line corresponding to (or *numbered*) any positive or negative integer.

We can mark also points on the line corresponding to all the (positive and negative) *rational numbers* (e.g. 1/2, 2/3, − 5/6 etc.); the point which will correspond to the number 5/7, for example, will be obtained by dividing the portion of the line between the point O and the point 1 into 7 equal portions and taking the fifth of the points of division,—or by taking the first of the 7 points of division dividing the portion of the line from O to the point marked 5 into 7 equal parts; and these two methods will evidently give the same point to be numbered 5/7. We have now a point numbered by every rational number, and our representation of addition and subtraction as motions of translation is evidently still applicable; the straight line is now available to represent all rational numbers and all operations of addition and subtraction of such rational numbers.

15. The position we have reached in regard to rational numbers may be summarised thus:

We can perform all the arithmetical operations of addition, multiplication, subtraction and division for all rational numbers whatever, (with the single exception that division by the number zero is impossible), and the results will be rational numbers; we can represent the rational numbers as points on a straight line and give geometrical interpretations, if so desired, of the arithmetical operations. It appears moreover at first sight that the line will be entirely covered by the rational numbers and that therefore the system of rational numbers is complete in two directions,—(1) as a system allowing all arithmetical operations, and (2) as affording a complete representation of the straight line and measurements along the straight line;—or briefly the system appears adequate for all arithmetical problems and all problems of measurement.

16. Inadequacy of rational numbers for problem of measurement. It is true that the four elementary arithmetical operations (with the exception of division by zero) are always possible with numbers of this system, and it is true that between any two rational points on the line (however close together) any number of other rational points can be interpolated (thus between 1/2 and 4/7 there is certainly at least one rational number with any number greater than 14 as denominator and therefore any number of such points can be so interpolated). So far the above presumption is true, but it is not true that the system is adequate for the solution of the problem of measurement along a straight line; as will be seen in a moment, *there are points on the line which correspond to no rational number whatever.* Moreover if we consider certain extended arithmetical operations such as extracting the square root (even of positive integers) we shall find the system inadequate in this respect also. We can easily prove, for example, that *there is no rational number whatever whose square is* 2, thus :—If there were such a rational number it could be expressed as a fraction in its lowest terms as p/q say, p and q being whole numbers having no common divisor, and we should then have

$$(p/q)^2 = 2, \text{ i.e. } p^2 = 2q^2 \quad \dots\dots\dots\dots\dots\dots(1),$$

but q^2 is a whole number and therefore p^2 is an even number, whence it follows that p must also be even (because if p were odd p^2 would be odd also); hence p^2 must be a multiple of 4 whence equation (1) shews that q^2 must be a multiple of 2 whence as before q must be a multiple of 2; i.e. p and q must both be divisible by 2, which is contrary to the hypothesis that p/q was a fraction in its lowest terms; whence it follows that there can be no rational number (such as p/q) whose square is 2.

But, in geometry (e.g. Euclid, Book II) a geometrical construction for the side of a square equal in area to a given rectangle is given. We can therefore, for example, construct geometrically a straight line of a length such that the square on it is 2 square units in area; i.e. we can mark on our straight line a point corresponding to what we should naturally call $\sqrt{2}$. Since there is no rational number $\sqrt{2}$ this point cannot be one of the rational points already numbered. Though, in a sense, the rational points cover the line infinitely densely yet there are *other* points on the line. Or, expressed differently, there are distances which cannot be measured by numbers of the system of rational numbers; i.e. *the rational numbers are inadequate for the problem of measurement.*

17. In order to make the system of numbers adequate to measure all conceivable lengths, the so-called *irrational numbers* must be introduced. If we were content to let the foundations of analysis rest on our intuitions concerning the nature of a straight line (instead of basing it entirely on the arithmetical concept of whole number) we could say that to every point on the straight line of Fig. 1 there corresponds a number, rational or irrational. The addition of such numbers could then moreover easily be represented as motions to the right, and similarly for subtraction. But, while willing to accept any help in the way of suggestions which we can get from our geometrical intuitions, it is of fundamental importance that we should give strictly arithmetical definitions of all the notions used in analysis,—and in particular therefore of the irrational numbers.

18. A property of the straight line. Let us first investigate the problem geometrically. We have seen that the rational points on the line of Fig. 1 (p. 13) are arranged infinitely densely on the line. It follows that if P is any point on the line which is not a rational point, rational points can be found on either side of P, as close to P as we may desire (short of actual coincidence with P). There are therefore rational numbers which are "approximations" to the number corresponding to P (if there is to be such a number), the degree of approximation being as close as we like. Thus, for example, we readily agree, because

$$1^2 < 1\cdot4^2 < 1\cdot41^2 < 1\cdot414^2 < 2 < 1\cdot415^2 < 1\cdot42^2 < 1\cdot5^2 < 2^2,$$

that the irrational number $\sqrt{2}$ must "lie between" the numbers 1, 1·4, 1·41, 1·414, on the one hand and 1·415, 1·42, 1·5, 2 on the other. This idea cannot be used directly as an arithmetic method of *definition* of the irrational number $\sqrt{2}$, because in it we have presupposed the existence (and properties) of this irrational number $\sqrt{2}$; but we notice that the point P (or the point $\sqrt{2}$) *divides the rational points of the line into two classes* such that (1) every rational point lies in one or other of the classes, and (2) every point of one class (the lower class) lies to the left of every point of the other class (the upper class). This simple property of a point on a line,—of dividing the line into two parts,—gives us the clue to the strict arithmetical definition of irrational numbers.

19. Dedekind's definition of a real number. *If the whole set of rational numbers is divided into two classes, L and R, such that (1) every rational number whatever is in one or other of the two classes and (2) every number of the one class, L, is less than every*

number of the other class, R, then the classification or division cor-
responds to, or defines, a real number. There is no logical objection
to saying that the real number *is* the classification, or even one of
the classes of rational numbers concerned (say *L*), but the student
will probably find it more satisfactory to say that the real number
corresponds to the classification and is defined by it. The essential
thing to notice is that we have defined the new system of real
numbers by means of the previously defined rational numbers; not
as pairs of such numbers (as in the case of the definition of the
rational numbers by means of the whole numbers, § 2 above) but
as classes (or pairs of classes) of such rational numbers. It is an
essentially arithmetical definition.

20. As an example of the use of this definition we may note the
example of $\sqrt{2}$ just mentioned. We know that, if *x* typifies any
positive rational number such that $x^2 < 2$ or any negative (or zero)
rational number and *y* any positive rational number such that
$y^2 > 2$, then every rational number whatever will belong to the class
of *x*'s or to the class of *y*'s; and every number of the *x* class is less
than every number of the *y* class; i.e. this division of the rational
numbers into these two classes has the two essential properties
necessary (and sufficient) for the definition of a real number. The
real number so defined is called $\sqrt{2}$. It is easily seen that the point
of division on the line which will divide the rational numbers
marked on the line into these two classes will be the point which
would be obtained as $\sqrt{2}$ by any geometrical or other method.

Or again, take the classification into *x*'s and *y*'s such that $x \leqslant 1$,
$y > 1$. Here again the two fundamental properties are satisfied
and a real number is therefore defined. The number evidently
corresponds to the rational point 1. In such a case it is convenient
to identify the real number with the corresponding rational number
(in this case 1).

If a real number does not correspond to a rational number in
this way it is said to be *irrational*.

If the real number concerned corresponds to a rational number the definition
may be expressed in alternative forms; e.g. the real number corresponding to
the rational number 1 may be defined as the classification into (1) all rational
numbers $\leqslant 1$ and rational numbers > 1, or (2) all rational numbers < 1 and
all rational numbers $\geqslant 1$. This suggests a modification of the definition, by

omitting the number to be defined (e.g. 1) from both classes. These differences
of definition are evidently not vital and we shall in this course use whichever
method of classification is most convenient for the matter in hand.

21. Arithmetical properties of real numbers. We have still
to define the arithmetical operations of addition etc. on these real
numbers. Reference once more to the straight line of Fig. 1 will
make clear how these definitions have to be made. Expressed
strictly arithmetically they will be as follows.

Denoting by $(x \,|\, y)$ the Dedekindian classification into classes
typified by x and y, then if $(x \,|\, y)$ is the classification which defines the
real number α and $(x' \,|\, y')$ is that which defines the real number β,
$\alpha + \beta$ *is the number defined by the classification* $(x + x' \,|\, y + y')$.

That the classification $(x + x' \,|\, y + y')$ does fulfil the two funda-
mental conditions of p. 16, and so in all cases actually defines a real
number, is easily seen; and it is at the same time easily seen that
the commutative law for addition, $\alpha + \beta = \beta + \alpha$, is true.

To define the product of two real numbers it is simplest first
to define *positive* and *negative* real numbers and then consider the
different cases separately. The real number α defined by the classi-
fication $(x \,|\, y)$ is said to be *positive* if some of the numbers x (of the
lower class) are positive; it is *negative* if some of the numbers y
are negative*. Now if α and β are two positive real numbers
defined by the classifications $(x \,|\, y)$ and $(x' \,|\, y')$ *the product* $\alpha\beta$ *is
the number defined by the classification* $(xx' \,|\, yy')$, the lower class
being completed by the addition of all numbers which $\leqslant 0$.

This definition is easily extended to the other possibilities which
may arise; and it is easily seen that the fundamental laws affecting
multiplication are still true.

Subtraction and division are defined as the inverses of addition
and multiplication.

The notions of inequality as applied to two irrational numbers
α, β will agree with the case of rational numbers already dealt with
if we agree to call α *greater or less than* β *according as* $\alpha - \beta$ *is
positive or negative*.

22. Adequacy of real numbers. We do not develop in detail
this Dedekindian theory of irrational numbers, the general lines of

---

* See p. 11 above. The student should verify that these definitions correspond
with the plain facts of Fig. 1 and agree with the definitions for rational numbers.

which are now sufficiently clear. The details may be left to the student. The essential point is that *we have a purely arithmetical definition of real numbers which not only corresponds to our ordinary notion of distance on a straight line and so solves the problem of measurement, but also ensures that all the fundamental laws* (p. 2) *are satisfied by irrational numbers, which can accordingly be manipulated by ordinary algebraical rules in the same way as integral and rational numbers.* As in the extension from whole numbers to rational numbers in § 2 above, the new "arithmetic" of real numbers includes our old arithmetic of rational numbers as a special case.

It may be noted that the introduction of real (irrational) numbers has made certain operations, such as extracting the square root of 2, possible; but not all such operations have been rendered possible by this extension. For example it is still impossible to extract the square root of -2, i.e. there is no real number whose square is -2. The number system can be extended further— by the introduction of *imaginary* or *complex numbers*—so as to make all such operations possible; but this extension lies outside the main scope of this course*.

The system of all real numbers is called the *arithmetic continuum* of real numbers. Henceforth in this course all numbers with which we shall be concerned will be real numbers unless the contrary is stated or implied.

It will occur to the student that we may have Dedekindian classifications of the real numbers (rational and irrational). But every Dedekindian classification of the system of real numbers corresponds to one and only one such classification of the system of rational numbers, i.e. to a real number. The introduction of such classifications therefore does not introduce a new class of numbers. It will often be desirable in this course to use such classifications of the real numbers instead of the corresponding classifications of the rational numbers.

EXAMPLES III

1. Prove that there is no rational number whose square is 3.

2. Prove that $\sqrt[3]{2}$ is irrational.

3. Give Dedekind's definition of $\sqrt[3]{2}$ and, by finding numbers of the two classes of rational numbers by which the number $\sqrt[3]{2}$ is defined, find rational approximations to $\sqrt[3]{2}$ correct to within ·1.

4. Prove from the definition of addition and multiplication of real numbers,

---

* A short account of complex numbers is given in the Appendix.

defined as Dedekindian classifications of the rational numbers, that the associative, commutative and distributive laws, viz.

$$a+(b+c)=(a+b)+c, \quad a\times(b\times c)=(a\times b)\times c, \quad a+b=b+a, \quad a\times b=b\times a,$$

and
$$(a+b)c=ac+bc,$$

hold when the numbers concerned (a, b, c) are any real numbers.

5. Prove that the number $\sqrt{2}$, as defined on p. 17, is such that $\sqrt{2}\times\sqrt{2}$ (as defined on p. 18) is equal to 2.

6. Prove that between any two real numbers there lie both rational and irrational numbers.

§ 4. ALGEBRAIC CONSEQUENCES OF THE FUNDAMENTAL LAWS. EQUALITIES AND EQUATIONS

23. Equalities. Mathematics has been described as tautology. Without entering upon a discussion as to the full meaning or limitations of this description it may be useful to keep it in mind while considering *equalities*. Two numbers are *equal* simply if they are the same number. The notion of equality thus appears at first sight as somewhat unnecessary, but this is far from being the case. The fact is that in algebra we do not know what actual numbers our letters represent; it being in fact essential to algebra that we should not know. It is possible that different calculations performed on the same set of numbers may lead to the same number as result, no matter what the particular numbers may have been. Our notion of equality applied in such a case (e.g. in the relation $(a + b)^2 = a^2 + 2ab + b^2$) thus implies rather more than the mere identity of two numbers; it implies the equivalence of the two different calculations concerned,—an equivalence which may quite possibly be of interest and provide a definite addition to our knowledge (as we know in fact to be repeatedly the case).

Formally an *equality* (or *identity*) is the statement of the identity of two numbers represented in general by two different sets of symbols. We have already met equalities of wide scope in the fundamental laws, e.g. $a + b = b + a$, a and b being any two real numbers. We collect here for reference other standard equalities which we shall find of constant use, all of which can be proved by application of the fundamental laws.

(i) $(a \pm b)^2 = a^2 \pm 2ab + b^2,$

(ii) $(a+b)(a-b) = a^2 - b^2,$

(iii) $(a \pm b)^3 = a^3 \pm 3a^2b + 3ab^2 \pm b^3,$

(iv) $(a \pm b)(a^2 \mp ab + b^2) = a^3 \pm b^3,$

(v) $a^3 + b^3 + c^3 - 3abc = (a + b + c)(a^2 + b^2 + c^2 - bc - ca - ab),$

(vi) $(x \pm a)(x \pm b) = x^2 \pm x(a + b) + ab,$

(vii) $(x - a)(x - b)(x - c) = x^3 - x^2(a + b + c)$
$$+ x(bc + ca + ab) - abc,$$

(viii) $(1 + x + x^2 + x^3 + \dots + x^{n-1})(1 - x) = 1 - x^n,$

(ix) $a + (a + b) + (a + 2b) + (a + 3b) + \dots$
$$+ (a + \overline{n - 1}b) = \frac{n}{2}(2a + \overline{n - 1}b).$$

In these equalities a, b, c, x represent any real numbers without restriction and n any positive integer; and in (i), (iii), (iv) and (vi) the ambiguous signs $\pm$, $\mp$ are used "respectively," i.e. either the upper signs only or the lower signs only must be used throughout the equality.

24. Induction. A special method of proof which is of considerable use and power is that known as *mathematical induction*. The method is useful in proving general results, such as (viii) and (ix), concerning positive integers where the result is easily seen to hold for the first few integers, 1, 2, etc. It will be understood by a consideration of its application to the proof of equality (viii) above.

We know from (ii) that
$$(1 + x)(1 - x) = 1 - x^2;$$
therefore $(1 + x + x^2)(1 - x) = (1 + x)(1 - x) + x^2(1 - x)$
$$= 1 - x^2 + x^2 - x^3$$
$$= 1 - x^3,$$
and therefore again
$$(1 + x + x^2 + x^3)(1 - x) = (1 + x + x^2)(1 - x) + x^3(1 - x)$$
$$= 1 - x^3 + x^3 - x^4$$
$$= 1 - x^4,$$
and so on; the argument by which we prove the truth of our desired result for $n = 4$ from the known result for $n = 3$ is evidently general. To make the argument sound we argue:

If the equality (viii), viz.
$$(1 + x + x^2 + \dots + x^{n-1})(1 - x) = 1 - x^n,$$

is true for *some* value of n then

$$(1 + x + x^2 + \ldots + x^n)(1 - x) = (1 + x + x^2 + \ldots + x^{n-1})(1 - x) + x^n - x^{n+1}$$
$$= 1 - x^n + x^n - x^{n+1} = 1 - x^{n+1},$$

i.e. the result of the same form is true for the next greater integral value of n; but we have seen that the result *is* true when $n = 2$, therefore it is true when $n = 3$, and again when $n = 4$, and when $n = 5$, and so on; the result is therefore true for all positive integral values of n (the case when $n = 1$ being trivial).

25. Binomial theorem. Another equality of great importance is the *binomial theorem*. We shall have occasion later to establish this theorem in its general form. We here state and prove the special case known as *the binomial theorem for a positive integral index*:

If x is any real number and n any positive integer then

$$(1 + x)^n = 1 + nx + \frac{n(n-1)}{1 \cdot 2} \cdot x^2 + \ldots$$
$$+ \frac{n(n-1)\ldots(n-r+1)}{1 \cdot 2 \ldots r} \cdot x^r + \ldots + x^n.$$

The expression on the right contains $n + 1$ terms of which the term $\dfrac{n(n-1)\ldots(n-r+1)}{1 \cdot 2 \ldots r} x^r$, which is the $(r+1)$th (r being any integer from 1 to n), may be called the typical term.

We prove the theorem by induction, thus:

If

$$(1 + x)^n = 1 + nx + \ldots + \frac{n(n-1)(n-2)\ldots(n-r+1)}{1 \cdot 2 \cdot 3 \ldots r} \cdot x^r + \ldots + x^n$$

then

$$(1 + x)^{n+1} = (1 + x)^n (1 + x)$$
$$= 1 + nx + \frac{n(n-1)}{1 \cdot 2} \cdot x^2 + \ldots$$
$$+ \frac{n(n-1)\ldots(n-r+1)}{1 \cdot 2 \ldots r} \cdot x^r + \ldots + x^n$$
$$+ x + nx^2 + \ldots$$
$$+ \frac{n(n-1)\ldots(n-r+2)}{1 \cdot 2 \ldots (r-1)} \cdot x^r + \ldots + nx^n + x^{n+1}$$
$$= 1 + (n+1)x + \frac{(n+1)n}{1 \cdot 2} \cdot x^2 + \ldots$$
$$+ \frac{(n+1)n(n-1)\ldots(n+1-r+1)}{1 \cdot 2 \cdot 3 \ldots r} \cdot x^r + \ldots + x^{n+1},$$

because

$$\frac{n(n-1)\ldots(n-r+1)}{1.2\ldots r} + \frac{n(n-1)\ldots(n-r+2)}{1.2\ldots(r-1)}$$

$$= \frac{n(n-1)\ldots(n-r+2)}{1.2\ldots r}(n-r+1+r)$$

$$= \frac{(n+1)n(n-1)\ldots(n+1-r+1)}{1.2.3\ldots r},$$

for all integral values of r from 1 to n.

Therefore if the equality holds for some given positive integral value of n it holds also for the next greater integral value of n, and therefore similarly for all greater values of n.

But the theorem is true when $n=2$ because we know that

$$(1+x)^2 = 1 + 2x + x^2;$$

therefore the theorem is true for all integral values of n greater than or equal to 2 ; and the theorem is proved (it being obviously true also for $n=1$).

26. Equations. In all *equalities* the numbers concerned may, broadly speaking, be any numbers whatever, without affecting the fact of the identity of the two numbers represented;—there may be broad restrictions placed on some of the numbers, but, provided the numbers belong to certain specified classes the equality always holds.

Equations on the contrary are not statements of the equality of two numbers except for certain exceptional values of the numbers concerned. In fact the sign " = " in an equation is essentially hypothetical; it does not state a universal fact as in an equality. The equation $x^2 + 2x + 1 = 4$ for example does not state the fact that $x^2 + 2x + 1 = 4$, for we know that $x^2 + 2x + 1 = (x+1)^2$ for all real values of x and therefore $x^2 + 2x + 1$ can never be equal to 4 except in the two special cases when x represents one of the two numbers 1 and -3; the equation is in effect meaningless by itself and has a meaning only in reference to some such context as " if x is a number such that $x^2 + 2x + 1 = 4$." There is always an " answer " or solution (possible or impossible) implied in an equation. To avoid confusion therefore it might be preferable to use two different signs for the " = " in equations and equalities, and in fact when attention is directed to this difference the sign " ≡ " is often used in equalities and the sign " = " reserved for equations ; but there is really no vital need for this duplication of symbols. After all, the meanings of the " = " in equalities and equations are similar ; moreover a mere glance will in general suffice to decide whether a statement containing the sign " = " is an equality or an equation,—no one could, for example, for a moment consider the equation $x + 1 = 2$ to be an equality; and finally the use of the same sign for the two ideas has definite advantages in that in dealing with equations the use of equalities is facilitated by such use. This method of using the same symbol and word to represent two different but similar ideas, when this can be done without confusion and with advantage, is characteristic of mathematics. It is the method we have adopted above in introducing the notions of non-integral "numbers" and operations of "addition" etc. with such numbers.

27. Reversible and irreversible operations. To solve an equation involving an unknown, say x, we have to determine all possible values of x, if any, which satisfy the equation. In simple cases we aim at simplifying the equation by ordinary algebraical manipulation until we have reduced it to a form in which it is obvious that x can only have a certain value, or one of certain several values. For example, in the case of a simple equation, by collecting on the left-hand side all terms containing x and on the right-hand side all terms not containing x, we reduce the equation to the form $ax = b$, whence, by division (if $a \neq 0$), x is necessarily b/a.

We have hitherto applied algebraic operations only to numbers ; and, in thus extending the operations to equations, we have to take special care in some points. In the example mentioned it is easily seen, by reversing the process used, that $x = b/a$ is actually a solution. There may however be cases in which a value of x obtained by some such process does not satisfy the equation. Thus if $\sqrt{x} = -1$ (where $\sqrt{x}$ is understood to mean—as always in this course—the positive number whose square is x), by squaring we get $x = 1$. What we have proved is that no number other than 1 can satisfy the equation ; but $x = 1$ is obviously not a solution. If we try to reverse the steps of our original argument we cannot pass from the last equation $x = 1$ to the preceding equation $\sqrt{x} = -1$.

Generally if, by algebraic manipulation, we reduce an equation to a simpler form, from which we can infer that x has only certain values, any one of these values of x will satisfy the original equation if all the operations used are reversible, but not necessarily if any of the operations is irreversible. The two irreversible operations most relevant to the solution of equations are that of squaring, just considered, and that of multiplying by zero.

28. Polynomial equations. Standard processes of solving simple and quadratic equations are given in all books on elementary algebra. These equations are the simplest forms of an important class of equation expressed in the form

$$a_n x^n + a_{n-1} x^{n-1} + \ldots + a_1 x + a_0 = 0 \ldots\ldots\ldots(1),$$

where n is an integer and $a_0, \ldots a_n$ are independent of x. The expression on the left-hand side of this equation is called a *polynomial*; and n is called the *degree* (or order) of the polynomial and of the equation.

For $n = 3$ or 4, standard processes of solution, of a complicated kind, are given in books on higher algebra or the theory of equations.

An equation of the above type can be reduced to one of lower degree if one root is known or can be found by any means, by using the *factor theorem*, viz:

The polynomial $a_n x^n + a_{n-1} x^{n-1} + \ldots + a_0$ has $x - \alpha$ as a factor if, and only if, $x = \alpha$ is a root of the equation (1) obtained by equating this polynomial to zero.

To prove this theorem we observe that the expression

$$(a_n x^n + a_{n-1} x^{n-1} + \ldots + a_1 x + a_0) - (a_n a^n + a_{n-1} a^{n-1} + \ldots + a_1 a + a_0)\ldots(2)$$

is necessarily divisible by $x - a$ because it can be written

$$a_n (x^n - a^n) + a_{n-1} (x^{n-1} - a^{n-1}) + \ldots + a_1 (x - a)$$

and each of the terms of this last expression is divisible by $x - a$. It follows that the expression in the first bracket of (2) is divisible by $x - a$ if and only if the expression in the second bracket is zero, i.e. if and only if $x = a$ is a root of equation (1).

The factor theorem shews at once that if $x = \alpha$ is a root of the equation (1) the polynomial is divisible by $x - \alpha$. The quotient will be of degree one lower than equation (1), and the roots of (1) other than α are the roots of the equation obtained by equating this new polynomial to zero.

With equations of degree higher than the second the processes of solution are often impracticable. In such cases it is often simple to obtain *approximate solutions* of the equation correct to any desired degree of accuracy. We shall see later (Ch. III) that if the expression $a_n x^n + \ldots + a_0$ is positive for one value of x and negative for another value of x, then it must be zero for some intermediate value of x, i.e. there is a root of equation (1) between these two values of x. By continued application of this principle the two values of x between which the root lies can be brought closer and closer together and thence the root sought for obtained more and more accurately.

29. Equations involving more than one unknown also occur. If the number of equations is equal to the number of unknowns, the equations being *simultaneous*, the solution is generally reduced to that of equations each containing only one unknown. If the equations are all simple (or of the first degree), i.e. of the form

$$ax + by + cz = d$$

(taking the case of three unknowns x, y, z), the solution, if it exists, can always be obtained simply or can be written down at once by means of determinants, as shewn in advanced books on algebra. There is no other important class of simultaneous equations which can be solved simply.

EXAMPLES IV.

1. Establish the equalities enunciated in the text (pp. 20, 21).

2. Prove the divisibility of $a^n - b^n$ by $a - b$; and of $a^n + b^n$ by $a + b$ if n is odd.

3. Prove that if $a_1, a_2, \ldots a_n$ and $b_1, b_2, \ldots b_n$ are two sets of n real numbers and if

$$s_1 = a_1, \quad s_2 = a_1 + a_2, \quad \ldots \quad s_n = a_1 + a_2 + \ldots + a_n,$$

then

$$a_1 b_1 + a_2 b_2 + \ldots + a_n b_n = s_1 (b_1 - b_2) + s_2 (b_2 - b_3) + \ldots + s_{n-1} (b_{n-1} - b_n) + s_n b_n.$$

[This is sometimes called Abel's equality.]

4. Deduce equality (ix) of p. 21 from Ex. 2, p. 5.

5. Shew that if $a_1, a_2, \ldots a_n$ are the n roots of the equation

$$a_n x^n + a_{n-1} x^{n-1} + \ldots + a_1 x + a_0 = 0$$

of the nth degree, then

$$a_1 + a_2 + \ldots + a_n = -\frac{a_{n-1}}{a_n}$$

and that

$$a_1 a_2 \ldots a_n = (-1)^n \frac{a_0}{a_n}.$$

6. Prove by factorising the quadratic equation $ax^2 + bx + c = 0$ that the roots are $(-b \pm \sqrt{b^2 - 4ac})/2a$.

7. Solve the equations

$$\text{(i)} \quad 2x^3 - 13x^2 - 10x + 21 = 0,$$
$$\text{(ii)} \quad (x-1)(x^2 - 2x - 2) = (x-2)(x^2-1).$$

8. Solve approximately the equation $x^3 - x^2 + 1 = 0$, finding the root which lies between -1 and 0 correct to one place of decimals.

9. Solve the simultaneous equations

$$x^2 = y, \quad y^2 + x - 2y = 0.$$

§ 5. INEQUALITIES AND INEQUATIONS

30. Inequalities. The notions of *inequality*, viz. *greater* and *less*, which are intuitive as applied to whole numbers and have been defined above for rational and real numbers (pp. 8, 10, 18), determine an order in the system of real numbers and are of fundamental importance. In this section we shall consider those deductions from these notions which will be of use to us.

We recall the definition that of two real numbers, a and b, a is greater than, equal to, or less than b (i.e. $a > b$, $a = b$, or $a < b$) according as the real number $a - b$ is positive, zero, or negative.

We here introduce the additional symbols "$\geqslant$" and "$\leqslant$" to denote "greater than or equal to" and "less than or equal to" respectively.

The statement of the fact of inequality of two numbers is termed an *inequality*. Inequalities are of great use and importance, particularly in higher analysis. From the point of view of technique the main difference between inequalities and equalities lies in the care which has to be taken to distinguish between positive and negative numbers in inequalities. Thus, e.g., if a is positive, $2a > a$, whereas if a is negative, $2a < a$. There are certain inequalities holding for positive and negative numbers alike; thus $a + 1 > a$ and $(a + b)^2 \geqslant 4ab$, and again if $a < b$ and $b < c$ then $a < c$; but quite ordinary operations applied to inequalities will lead to contrary results according as the numbers are positive or negative. We proceed to study the laws which inequalities follow and to give a few standard examples.

31. Laws of inequalities. Firstly, whatever real numbers the letters a, b, A, B represent, we have* :

---

* In these laws of inequalities the sign "$<$" may be replaced by "$>$" (or vice versa) throughout. Similar laws with "$\geqslant$" and "$\leqslant$" can also be at once framed.

(I) If $a < A$ then $a + b < A + b$.

(II) If $a < A$ then $a - b < A - b$.

(III) If $a < A$ and $b < B$ then $a + b < A + B$.

(IV) If $a < A$ and $b > B$ then $a - b < A - B$.

These results shew in particular that from given inequalities other inequalities can be deduced either by adding the same number to each side or by adding the corresponding sides of two similar* inequalities, a process which may conveniently be called adding the inequalities.

If we wish to use the operation of multiplication it is necessary to restrict the multiplier to be positive. We have in fact:

(V) If b is positive and $a < A$ then $ab < Ab$.

This covers also the case of division.

If the multiplier b were negative, $ab > Ab$.

Finally if we wish to multiply together two inequalities we have further to restrict all the numbers concerned to be positive. In this case:

(VI) If a, A, b, B are positive and $a < A$, $b < B$ then $ab < AB$.

Proofs of these laws are hardly necessary.

The first four are immediate deductions from the definition of addition (and subtraction). The remaining two are hardly less obvious. (V) for example is proved by putting $a = A - d$ where $d > 0$, whence $ab = (A - d) b = Ab - db < Ab$ because $db > 0$; (VI) follows from (V).

Operations of subtraction (e.g. II and IV) can be included under addition by means of the useful further inequality: If $a < A$ then $-a > -A$.

To sum up: Any inequality may be operated on without reversion of any sign of inequality

(i) by the addition of *any* number or of any inequality similar to the first inequality; or

(ii) by the multiplication by any *positive* number; or

(iii) by the multiplication by any similar inequality provided that the terms of the inequalities are all *positive*.

Another allowable operation, a special case of the multiplication of two positive inequalities, is that of squaring. Extracting the square root, is also valid provided the positive value of the square root is taken in each case.

---

* Inequalities such as $a > b$, $c > d$ are similar; the inequalities $a > b$, $c < d$ are not similar.

32. Standard inequalities. Many standard inequalities can best be proved by the use of the differential calculus. We give here a list of those standard inequalities which are useful and can be easily proved in an elementary manner.

(i) If a and b are any two positive real numbers
$$(a + b)/2 \geqslant \sqrt{(ab)};$$
or *the arithmetic mean of two positive numbers is greater than or equal to the geometric mean.*

(ii) If x is any real number greater than -1 and n is any positive integer, then
$$(1 + x)^n > 1 + nx,$$
unless $x = 0$ or $n = 1$, in which cases
$$(1 + x)^n = 1 + nx.$$

(iii) If a and b are any two (unequal) positive real numbers and n is any positive integer greater than 1, $\dfrac{a^n - b^n}{a - b}$ lies between* na^{n-1} and nb^{n-1}.

(This inequality is true for any real value of n. See Ex. 2, p. 91 below.)

(iv) If a and b are any two unequal positive real numbers and m and n any two positive integers, $\dfrac{a^m - b^m}{a^n - b^n}$ lies between $\dfrac{m}{n} a^{m-n}$ and $\dfrac{m}{n} b^{m-n}$.

(v) *Abel's lemma.*
If $a_1, a_2, \ldots a_n$; $b_1, b_2, \ldots b_n$ are two sets of numbers such that
$$b_1 \geqslant b_2 \geqslant b_3 \geqslant \ldots \geqslant b_n \geqslant 0$$
and if M and m are the greatest and least respectively of the set of numbers
$$a_1, a_1+a_2, a_1+a_2+a_3, \ldots, a_1+a_2+\ldots+a_n,$$
then $\quad mb_1 \leqslant a_1b_1 + a_2b_2 + \ldots + a_nb_n \leqslant Mb_1,$
i.e. the sum of the n products $a_1b_1, a_2b_2, \ldots, a_nb_n$ lies between mb_1 and Mb_1 or equals one (or both) of these numbers.

(vi) If $a_1, a_2, \ldots, a_n$ are n positive numbers, then the arithmetic mean, viz. $\dfrac{a_1+a_2+\ldots+a_n}{n}$, $\geqslant$ the geometric mean, viz. $\sqrt[n]{(a_1 a_2 \ldots a_n)}$.

33. Proofs. We append proofs in view of the importance of the inequalities and the instructiveness of the proofs:

* The word *between* is used throughout this course in the strict (exclusive) sense. Thus x lies between a and b only if $a < x < b$ or $b < x < a$, the values $x = a$ and $x = b$ being excluded.

(i) We have $(a+b)^2 - 4ab = a^2 + 2ab + b^2 - 4ab$
$$= a^2 - 2ab + b^2$$
$$= (a-b)^2$$
$$\geqslant 0 \text{ whatever } a \text{ and } b \text{ may be.}$$

Therefore $(a+b)^2 \geqslant 4ab$ whatever a and b, and hence
$$a+b \geqslant \sqrt{(4ab)}$$
provided $a+b$ is positive. Hence
$$(a+b)/2 \geqslant \sqrt{(ab)}$$
if a and b are both positive, it being necessary and sufficient for the " reality " of the geometric mean $\sqrt{(ab)}$ that a and b should have the same sign, and for the positiveness of $a+b$ that that sign should be $+$. The sign " $\geqslant$ " is used and not the sign " $>$ " because it is possible (if $a = b$) for the two sides to be equal.

(ii) By induction:

If $(1+x)^n > 1 + nx$ for some particular value of n, it follows, by multiplying by $1+x$, that
$$(1+x)^{n+1} = (1+x)^n (1+x)$$
$$> (1+nx)(1+x),$$
the multiplier $1+x$ being positive under the conditions stated;
i.e. $\qquad (1+x)^{n+1} > 1 + (n+1)x + nx^2,$
which evidently $\qquad > 1 + (n+1)x.$

But it is plain that $(1+x)^2 > 1 + 2x$, except when $x = 0$, the two sides of the inequality being then equal; therefore $(1+x)^3 > 1 + 3x$; and again $(1+x)^4 > 1 + 4x$, and so on, giving in general the result stated, with the exceptional case when $x = 0$. The other exceptional case of equality, when $n = 1$, is evident.

(iii) We have from equality (viii) of p. 21 (or by direct division)
$$\frac{a^n - b^n}{a-b} = a^{n-1} + a^{n-2}b + a^{n-3}b^2 + \dots + b^{n-1},$$
there being n terms in the expression last written*.

Supposing, for definiteness, that $a > b$ (a and b being both positive by hypothesis) we have evidently, if $n > 1$,
$$a^{n-1} > a^{n-2}b > a^{n-3}b^2 > \dots > b^{n-1},$$

---

* If the student has any difficulty in appreciating a *general* proof applicable at once to any value of n, he should first work through the proof with one or two definite values for n, e.g. $n=2$, or 3. A similar remark applies also to the proof of inequality (iv), where the student may find it advisable to work through the proof with say the values 5 and 3 for m and n.

whence it follows that the sum of these terms is less than the sum of n terms each equal to a^{n-1} and greater than the sum of n terms each equal to b^{n-1}; whence

$$nb^{n-1} < \frac{a^n - b^n}{a-b} < na^{n-1} \text{ if } a > b > 0;$$

and similarly the same result holds with the signs of inequality reversed if $0 < a < b$.

It is easy to see further that the result holds also if a and b are both negative, though the proof would need some modification.

(iv) To prove, more generally, that $\dfrac{a^m - b^m}{a^n - b^n}$ lies between $\dfrac{m}{n} a^{m-n}$ and $\dfrac{m}{n} b^{m-n}$ let us suppose, for definiteness, that $a > b > 0$ and $m > n$.

By dividing both numerator and denominator of the given fraction by the common factor $a - b$, the fraction is seen to be equal to

$$\frac{a^{m-1} + a^{m-2}b + \ldots + ab^{m-2} + b^{m-1}}{a^{n-1} + a^{n-2}b + \ldots + ab^{n-2} + b^{n-1}},$$

which equals

$$\frac{a^{m-n}(a^{n-1} + a^{n-2}b + \ldots + b^{n-1}) + a^{m-n-1}b^n + \ldots + b^{m-1}}{a^{n-1} + a^{n-2}b + \ldots + b^{n-1}},$$

which equals

$$a^{m-n} + b^n \frac{a^{m-n-1} + a^{m-n-2}b + \ldots + b^{m-n-1}}{a^{n-1} + a^{n-2}b + \ldots + b^{n-1}},$$

which is less than $\quad a^{m-n} + b^n \dfrac{(m-n)a^{m-n-1}}{nb^{n-1}},$

which equals $\quad a^{m-n} + \dfrac{m-n}{n} a^{m-n-1}b,$

which is less than $\quad a^{m-n}\left(1 + \dfrac{m-n}{n}\right),$

i.e. is less than $\dfrac{m}{n} a^{m-n}$, and one part of the theorem is proved.

Similarly the same fraction may be expressed as

$$b^{m-n} + a^n \frac{b^{m-n-1} + \ldots + a^{m-n-1}}{b^{n-1} + \ldots + a^{n-1}},$$

whence it follows that under the same conditions as before the given fraction is greater than $\dfrac{m}{n} b^{m-n}$.

The theorem is completely proved in the case supposed; the other cases can be easily deduced or proved similarly.

(v) To prove Abel's lemma let us denote the sums

$$a_1, \quad a_1+a_2, \quad a_1+a_2+a_3, \quad ..., \quad a_1+a_2+...+a_n$$

by $s_1, \quad s_2, \quad s_3, \quad ..., \quad s_n.$

Then we have at once

$$a_1b_1+a_2b_2+...+a_nb_n = s_1b_1+(s_2-s_1)\,b_2+...+(s_n-s_{n-1})\,b_n$$
$$= s_1\,(b_1-b_2)+s_2\,(b_2-b_3)+...+s_{n-1}\,(b_{n-1}-b_n)+s_nb_n.$$

But, M being the greatest of the numbers $s_1, s_2, ..., s_n$, we have

$$s_1 \leqslant M, \quad s_2 \leqslant M, \quad ..., \quad s_n \leqslant M\,;$$

and, since by hypothesis

$$b_1 \geqslant b_2, \quad b_2 \geqslant b_3, \quad ..., \quad b_{n-1} \geqslant b_n, \quad b_n \geqslant 0,$$

and therefore b_1-b_2, b_2-b_3, ..., $b_{n-1}-b_n$, and b_n are all positive (or zero), it follows that

$$s_1\,(b_1-b_2) \leqslant M\,(b_1-b_2), \quad s_2\,(b_2-b_3) \leqslant M\,(b_2-b_3), \quad ...,$$
$$s_{n-1}\,(b_{n-1}-b_n) \leqslant M\,(b_{n-1}-b_n) \text{ and } s_nb_n \leqslant Mb_n$$

(from our fundamental inequalities), whence we have, by adding,

$$s_1\,(b_1-b_2)+...+s_{n-1}\,(b_{n-1}-b_n)+s_nb_n$$
$$\leqslant M\,(b_1-b_2)+M\,(b_2-b_3)+...+M\,(b_{n-1}-b_n)+Mb_n = Mb_1*.$$

Hence the sum $a_1b_1+a_2b_2+...+a_nb_n \leqslant Mb_1\,;$

and similarly we have the same sum $\geqslant mb_1$. The theorem is proved.

In this theorem, if none of the numbers $a_1, a_2, ..., a_n$ is negative, the numbers m and M are respectively s_1 (i.e. a_1) and s_n.

It should be specially noted that, though the numbers $a_1, a_2, ..., a_n$ in the theorem are unrestricted, the numbers $b_1, b_2, ..., b_n$ are restricted; as written in order they must not increase at any stage and must none of them be negative.

The theorem may evidently be modified by replacing m and M by any numbers respectively less than and greater than all the sums $s_1, s_2, ..., s_n$.

(vi) To prove (vi) we have to prove $a_1a_2...a_n \leqslant \left(\dfrac{a_1+a_2+...+a_n}{n}\right)^n$.

Write M for the arithmetic mean, i.e. $M = \dfrac{a_1+a_2+...+a_n}{n}$, and consider the product $a_1a_2...a_n$.

If all the numbers $a_1, a_2, ..., a_n$ are equal, they are all equal to M, and $a_1a_2...a_n = M^n$.

---

* By this use of the sign "=" we mean that the expression immediately preceding the sign is equal to the expression following the sign; the relation between the expression $s_1\,(b_1-b_2)+...+s_{n-1}\,(b_{n-1}-b_n)+s_nb_n$ and Mb_1 is of course one of inequality. This convention conflicts with the practice of some books.

If the numbers are not all equal, there must be at least one of them which is less than M and at least one greater than M; for, if, e.g., all the numbers $\geqslant M$ and were not all $= M$, we should have

$$a_1 + a_2 + \ldots + a_n > M + M + \ldots + M$$
$$= n \cdot M,$$

and M would $> M$, which is impossible.

Suppose $a_1 < M$ and $a_2 > M$.

If the two terms a_1 and a_2 in the product $a_1 a_2 \ldots a_n$ were replaced by M and $a_1 + a_2 - M$, the product would be increased; because

$$a_1 a_2 \ldots a_n - M(a_1 + a_2 - M) a_3 \ldots a_n = a_3 \ldots a_n (M - a_1)(M - a_2) < 0$$

(because $a_3 \ldots a_n > 0$, $M - a_1 > 0$, and $M - a_2 < 0$).

Calling, for convenience at this stage,

$$M, \quad a_1 + a_2 - M, \quad a_3, \quad \ldots, \quad a_n$$

respectively $a_1', \qquad a_2', \qquad a_3', \quad \ldots, \quad a_n',$

we have proved that $a_1 a_2 \ldots a_n < a_1' a_2' \ldots a_n',$

where $a_1' = M$.

If the numbers a_1', a_2', a_3', $\ldots$, a_n' are not all equal (and $= M$), the same argument will apply to shew that

$$a_1' a_2' a_3' \ldots a_n' < a_1'' a_2'' a_3'' \ldots a_n'',$$

where $a_1'' = M$ and $a_2'' = M$.

This process can be repeated until all the numbers in the product are equal and equal to M. The product then $= M^n$.

Hence, if the numbers a_1, a_2, $\ldots$, a_n are not all equal,

$$a_1 a_2 \ldots a_n < \left(\frac{a_1 + a_2 + \ldots + a_n}{n} \right)^n,$$

while, if the numbers are equal,

$$a_1 a_2 \ldots a_n = \left(\frac{a_1 + a_2 + \ldots + a_n}{n} \right)^n.$$

The inequality (vi) is established.

34. Inequations. As the equating of an algebraic expression to some number or other algebraic expression leads to an equation, so the statement of the inequality of two expressions leads to an *inequation*, or contingent inequality, satisfied for certain values of the unknown concerned. The problem of solving such inequations therefore arises.

The solution of inequations can be carried out on lines suggested by those followed in the solution of equations. Inequations can in fact be operated upon as equations (with the qualification that multiplication or division by a negative number will necessitate

the reversion of the sign of inequality and therefore is not a
directly valid operation) and so brought into standard forms. If
the left-hand side of the inequation, when the right-hand side has
been reduced to zero, can be factorized, the complete solution
follows at once. For example the inequation

$$(x-2)(x+1)(x-3) < 0$$

has evidently as its complete solution all the values of x for which
either one or three of the factors is, or are, negative, i.e. $x < -1$
or $2 < x < 3$. In fact in general the solution of the inequation

$$(x - a_1)(x - a_2) \ldots (x - a_n) < 0,$$

in which $\qquad\qquad a_1 < a_2 < a_3 < \ldots < a_n,$

if n is even is

$$a_1 < x < a_2, \ a_3 < x < a_4, \ \ldots, \ a_{n-1} < x < a_n,$$

and if n is odd is

$$x < a_1, \ a_2 < x < a_3, \ \ldots, \ a_{n-1} < x < a_n;$$

and evidently the sign "$<$" could be altered to "$\leqslant$," "$>$," or "$\geqslant$"
throughout the inequation and solution.

It may happen that a quadratic expression (e.g. $x^2 - 2x + 2$) has
no real factors. By the introduction of complex numbers* the
equation $x^2 - 2x + 2 = 0$ can be solved and thus complex factors
($x - 1 - \sqrt{-1}$ and $x - 1 + \sqrt{-1}$) found for the expression, but
such factors can be of no help in solving the inequation

$$x^2 - 2x + 2 < 0$$

for neither of the factors can be said to be positive or negative or
greater than or less than any number,—these notions being applicable only to real numbers.

However, if $x^2 + px + q$ is a quadratic expression,

$$x^2 + px + q = \left(x + \frac{p}{2}\right)^2 + q - \frac{p^2}{4}$$

and will therefore have two real factors or not according as this is
of the form $x^2 - a^2$ or not (a being a real number); i.e. according
as $q - \dfrac{p^2}{4} < 0$ or not.

If $p^2 < 4q$ the expression will not have real factors, but then the

---

* See Appendix.

expression $\left(x + \frac{p}{2}\right)^2 + q - \frac{p^2}{4}$ is necessarily positive for all values of x, because $\left(x + \frac{p}{2}\right)^2 \geqslant 0$ and $q - \frac{p^2}{4} > 0$. The inequation

$$x^2 + px + q < 0,$$

in this case, will have no real roots; the inequation

$$x^2 + px + q > 0$$

will have every real number as a root.

It will be noticed that the solutions of inequations do not consist of certain isolated values of the unknown, but (unlike solutions of equations) the solutions in general consist in the restricting of the unknown to lie within certain ranges.

It is important to remember that, in multiplying (or dividing) an inequality by any expression containing an unknown x, the sign of inequality is reversed or not according as the expression is negative or positive. Until an inequation is solved, however, the sign of such an expression is not known, and it is essential to take into account both the possibilities as to sign. For example, the inequation $\frac{x}{x+1} > 0$ is equivalent to $x > 0$ if $x + 1 > 0$ (giving the solution $x > 0$) but is equivalent to $x < 0$ if $x + 1 < 0$ (giving the solution $x < -1$); the roots of the inequation consist of all the numbers x for which either $x < -1$ or $x > 0$. (The number $x = -1$ is clearly inadmissible.)

35. Other types of equations and inequations, such as $x! > x^2$,—where $x!$, called *factorial x*, and occasionally written $\lfloor x$, means the product

$$x(x-1)(x-2)(x-3) \ldots 2 . 1,—$$

are often met with in analysis. The solution of such equations and inequations which cannot be written in the form

$$a_n x^n + a_{n-1} x^{n-1} + \ldots + a_1 x + a_0 \gtreqless 0$$

is however a matter requiring special consideration for each particular type and is best dealt with as required.

Simultaneous inequations in more than one unknown are not important.

EXAMPLES V.

1. Prove that if $a > b$ then $-a < -b$ and that if c is negative and $a > b$ then $ac < bc$.

2. By taking $a = 1 + x$ and $b = 1$ deduce inequality (ii) of the text (p. 29) from inequality (iii).

3. Prove
$$1 - x + x^2 > \frac{1}{1+x} > 1 - x$$
if $x > 0$.

4. Prove $a^3 + b^3 + c^3 \geqslant 3abc$ if $a + b + c \geqslant 0$.

5. Prove that, if m and n are positive integers such that $m > n$ and a is any positive real number, then $a^m > a^n$ if $a > 1$ and $a^m < a^n$ if $a < 1$.

6. Shew that if $a > b$ then $a^n > b^n$
  (i) for all positive integral values of n if also $|a| > |b|$,
and (ii) for all odd positive integral values of n only if $|a| < |b|$*.

7. Prove that if x is any real positive number less than unity and if ϵ is any positive number whatever (however small) then $x^n < \epsilon$ for all positive integral values of n greater than $\frac{x}{\epsilon} \frac{1-\epsilon}{1-x}$; and that if x is any real positive number greater than unity and K any positive number whatever (however great) then $x^n > K$ for all positive integral values of n greater than $\frac{K-1}{x-1}$.

$\left[\text{For the first part put } x = \dfrac{1}{1 + \dfrac{1-x}{x}} \text{ and apply inequality (ii) of the text to}\right.$

the denominator, whence $x^n < \dfrac{1}{1 + n\dfrac{1-x}{x}} \cdot \Bigg]$

8. Shew by using Abel's lemma that if $0 < x < 1$ and M and m are the greatest and least of the n numbers
$$a_1, \quad a_1 + a_2, \quad a_1 + a_2 + a_3, \quad \ldots, \quad a_1 + a_2 + \ldots + a_n,$$
then $\qquad m \leqslant a_1 + a_2 x + a_3 x^2 + \ldots + a_n x^{n-1} \leqslant M.$

9. Shew that $2^n > n$ for all positive integral values of n.

10. Shew that $(1+x)^n > n$ for all positive integral values of n greater than some value depending on x, x being any positive number.

[From inequality (ii) of the text an integer m can be chosen sufficiently great to make $(1+x)^m > 1 + \dfrac{1}{x}$. From the same inequality $(1+x)^p > \dfrac{m+p}{1+\dfrac{1}{x}}$ if p is any positive integer such that $1 + px > \dfrac{m+p}{1+\dfrac{1}{x}}$, i.e. if $p > \dfrac{m-1-\dfrac{1}{x}}{x}$; whence it

---

* For the definition of the modulus $|x|$ of a real number x, see p. 44 below.

follows that if n is any positive integer ($m+p$ say) greater than $m+\dfrac{m-1-\dfrac{1}{x}}{x}$ (where m is the integer chosen above)

$$(1+x)^n=(1+x)^{m+p}=(1+x)^m(1+x)^p$$
$$>\left(1+\frac{1}{x}\right)\frac{m+p}{1+\dfrac{1}{x}}$$
$$=m+p$$
$$=n$$

and the theorem is proved.

It can be seen in fact that $(1+x)^n>n$ for all positive integral values of n which exceed $1+\dfrac{1}{x^3}$; this does not exclude the possibility of the inequality holding for smaller values of n also.]

11. Prove that $px^2+qx+r>0$ for all real values of x if $q^2<4pr$ and $p>0$.

12. Solve the inequations:

(i) $x^2-48x+551<0$,

(ii) $x^4-3x^3-2x^2+12x-8>0$,

(iii) $\dfrac{x}{x+1}<2$,

(iv) $n!>n^2$,   (n being a positive integer) *.

[The complete solution of (iii) is $x>-1$ or $x<-2$. It may be illustrated graphically by drawing and comparing the two graphs $y=\dfrac{x}{x+1}$, $y=2$.]

13. Shew that $2^n<n!$ if $n>3$ and $n!<n^n$ if $n>1$.

14. Shew that, if a, b, c, ..., k are positive numbers less than 1,

(i) $(1+a)(1+b)(1+c)...(1+k)>1+(a+b+c+...+k)$,

and   (ii) $(1-a)(1-b)(1-c)...(1-k)>1-(a+b+c+...+k)$.

Deduce that, if, in addition, $a+b+c+...+k<1$,

(iii) $(1+a)(1+b)(1+c)...(1+k)<\dfrac{1}{1-(a+b+c+...+k)}$,

and   (iv) $(1-a)(1-b)(1-c)...(1-k)<\dfrac{1}{1+(a+b+c+...+k)}$.

15. Prove that if x, y, x', y' are any real numbers

$$\sqrt{(x^2+y^2)}-\sqrt{(x'^2+y'^2)}\leqslant\sqrt{\{(x+x')^2+(y+y')^2\}}\leqslant\sqrt{(x^2+y^2)}+\sqrt{(x'^2+y'^2)}.$$

[If $y=y'=0$ these relations reduce to the simple, but useful, relations

$$|x|-|x'|\leqslant|x\pm x'|\leqslant|x|+|x'|.]$$

* See p. 35 above.

§ 6. INFINITE SEQUENCES

36. Approximations to irrational numbers. In § 3 we saw
how to define irrational numbers arithmetically by means of
Dedekindian classifications of the system of rational numbers. The
definition of actual classifications corresponding to particular
numbers (e.g. $\sqrt{2}$, $\sqrt[3]{10}$) was left however to be determined separately in each separate case, and no general method of obtaining
rational numerical approximations to irrational numbers was given.
Since, however, rational numbers in the two Dedekindian classes
defining an irrational number can be found as close together as
may be desired*, it follows that rational *approximations* to any
irrational number can be found arbitrarily close to the irrational
number in question. For practical purposes such approximations,
if expressed as decimals, will be of more utility in the representation
of irrational numbers than the somewhat theoretical Dedekindian
classification. If we wish to have a set of rational approximations to
an irrational number of *arbitrary* accuracy (short of absolute accu-
racy†) we must have an *indefinite number* of such approximating
rational numbers; for, if we had only a definite number of rational
numbers (e.g. 1, 1·4, 1·41, 1·414) we could not thereby obtain a
number arbitrarily close to any irrational number (e.g. $\sqrt{2}$, which in
fact differs by more than ·0001 from any one of these four approximations). It will be necessary therefore to consider *indefinitely continued sequences of approximations* (e.g. the unending sequence of
numbers obtained at the successive stages in extracting the square
root of 2 by the ordinary arithmetical method, viz. 1, 1·4, 1·41, 1·414,
..., the dots indicating that the sequence of numbers is supposed
continued indefinitely according to some supposed law;—for by
continuing the sequence sufficiently far we can obtain a number
as close to $\sqrt{2}$ as we may wish).

In other words, *the effective representation of irrational numbers
demands the consideration of infinite sequences of numbers.*

This consideration,—which is a matter of capital importance
from several points of view, theoretical and practical,—will be the
concern of the present section and the one next following.

* This geometrically evident property of the system of rational numbers is not
difficult to prove.

† Absolute accuracy is of course impossible if the number concerned is really
irrational.

37. Upper and lower bounds of a set of numbers. We begin [*] by considering the trivial case of a finite set of numbers, e.g. the six numbers 1, 3, 2, 8, 5, 6.

Of these numbers one, viz. 8, is the greatest; any real number greater than 8 exceeds all numbers of the set, and any number less than 8 is less than at least one number of the set. The number 8 marks the upper boundary between the numbers of the set and all other real numbers; it is therefore called the *upper bound* of the set. Similarly 1 is the least number of the set and is called the *lower bound* of the set. This set is also said to be *bounded* above and below because numbers can be found greater than (and less than, respectively) all the numbers of the set.

38. Sequences. Let us see to what extent these definitions can be applied to indefinitely continued sets or *sequences*. A *sequence* is any set of numbers written in some definite order, but we shall interpret the term to *exclude* finite sequences; i.e. in future a sequence will mean an indefinitely continued sequence (sometimes called *unending* or *infinite*). For this purpose let us consider three particular examples:

(*a*) the sequence 1, 1/2, 1/3, 1/4, ...;

(*b*) the sequence $\frac{1}{2}$, $\frac{3}{4}$, $\frac{7}{8}$, $\frac{15}{16}$, ...;

(*c*) the sequence 1, 2, 3, 4, ...

The sequence (*a*) is similar to the above finite set in one respect: it has one number, 1, greater than all the other numbers of the sequence and it is *bounded above*. As before we call 1 the *upper bound* of the sequence. The sequence however differs from the finite set considered in that there is *no least number* of the sequence;— *whatever number of the sequence we take* (e.g. 1/20) *there are other numbers of the sequence less than it.* Nevertheless the numbers of the sequence are all positive numbers and their representative points on the line of Fig. 1 therefore all lie to the right of the origin O; so that the sequence is certainly *bounded below* in the sense that there are numbers less than all the numbers of the sequence. Is there a lower bound? That is,—is there a number which marks the lower boundary between the numbers of the

---

[*] The student is strongly advised to revert to the straight line of Fig. 1 repeatedly throughout the present § and to mark off for himself on such a straight line the numbers and sets dealt with.

sequence and all other numbers, i.e. a number such that all real numbers greater than it exceed at least one number of the sequence and all numbers less than it are less than all the numbers of the sequence ? Geometrically the answer is evident : the point O (i.e. the number 0) represents such a number. Analytically also this is verified, for if x is any real number greater than 0 we can find a number of the sequence less than x by simply going to the nth number of the sequence (i.e. $1/n$) where $n > 1/x$, such a value of n evidently necessarily existing whatever positive number (however small) x may be ; and at the same time, any number less than 0 is clearly less than all the numbers of the sequence. *Though the sequence has no least number, yet it has a lower bound* (viz. 0).

The sequence (b) is seen similarly to be bounded below ; the least number, viz. $\frac{1}{2}$, is the lower bound ; and the sequence has no greatest number, but it has an upper bound, viz. 1.

Take now the sequence (c). This sequence evidently has a least number, 1,—which will be, as before, the lower bound of the sequence ; but it has no upper bound, for whatever number we take, no matter how large, we can find a number of the sequence (as many as we like in fact) greater than it. The sequence in fact is *unbounded above*,—as is once more geometrically evident.

39. Definitions of upper bound, etc. We can now put down the following definitions, applicable to any sets or sequences of real numbers, finite or infinite :

A sequence of numbers is *bounded above* if some real number, K say, can be found greater than all the numbers of the sequence.

The sequence is *bounded below* if some real number, K say, can be found less than all the numbers of the sequence.

The sequence is *bounded* if bounded above and bounded below.

The *upper bound* of a sequence of numbers is the least real number which is at least as great as all the numbers of the sequence, if there is such a number.

The *lower bound* of a sequence of numbers is the greatest real number which is at least as small as all the numbers of the sequence, if there is such a number.

To complement these definitions we must now prove the following simple but fundamental theorem :

If a sequence of real numbers is bounded above, it necessarily has an upper bound, rational or irrational, and similarly, if bounded below, it has a lower bound.

The student will probably think a proof of this theorem unnecessary; the theorem is in fact geometrically evident, being merely an illustration of the continuity (or unbrokenness) of the straight line of Fig. 1. The need for a proof is however more apparent if one reflects that if irrational numbers were excluded from our consideration such a sequence as that instanced above obtained by carrying out indefinitely the process for extracting the square root of 2 (viz. 1, 1·4, 1·41, 1·414, ...), though bounded above, yet would have no upper bound (because there is no rational number having the properties of an upper bound of this sequence).

The proof is:

Let the numbers* of the sequence be

$$s_1, s_2, s_3, \ldots \qquad \ldots\ldots\ldots\ldots\ldots\ldots\ldots\ldots(1).$$

We know there are some real numbers (e.g. K say) which exceed all the numbers $s_1, s_2, s_3, \ldots$ of the sequence (1),—because of the hypothesis that the sequence is bounded above; and there are other real numbers (e.g. k say) which are less than some (i.e. at least one) of the numbers $s_1, s_2, s_3, \ldots$ of the sequence (1); this division of the system of real numbers into two classes (i) those numbers K which exceed all the numbers of the sequence (1) and (ii) those numbers k which are less than at least one of the numbers of the sequence (1) is a Dedekindian division† of the system of real numbers, because every number k is less than every number K, and every real number, with only one exception, is included in one or other of the two classes. The classification therefore defines a real number $(M = (k \mid K))$, which is such that all numbers greater than M belong to the K class and therefore exceed all the numbers $s_1, s_2, s_3, \ldots$ of the sequence, and all numbers less than M belong to the k class and are therefore less than at least one of the numbers $s_1, s_2, s_3, \ldots$ of the sequence; i.e. M is the upper bound of the sequence.

Similarly the existence of a lower bound (m) of a sequence which is bounded below may be proved.

* The numbers of a sequence will often be referred to as the *terms* of the sequence.
† See p. 16 above.

The existence of the lower and upper bounds of the sequences
(a) and (b) above (p. 39) are thus but particular instances of this
general theorem.

40. Monotone sequences. Unique limit. Definitions. Suppose now we have a sequence which is not only bounded above,
but also *steadily increasing*, i.e. the terms in order steadily become
greater and greater.

Take for example the sequence

$$·9, ·99, ·999, ·9999, \ldots \quad \ldots\ldots\ldots\ldots\ldots\ldots(2),$$

the successive terms being decimals composed of a steadily increasing number of 9's.

We see at once that the upper bound of this sequence is 1,
because evidently 1 (and therefore, *a fortiori*, every number greater
than 1) exceeds every number of the sequence (2), and any number
less than 1 is less than some numbers of the sequence; this last
fact being consequent on the fact that the difference between 1 and
the nth term of the sequence equals $1/10^n$, and, no matter how
close to 1 we take a number $x\ (< 1)$, we can find a power of 10
(10^n) which exceeds the number $\dfrac{1}{1-x}$*, i.e. we can find a number
of the sequence (2) which exceeds x. The number 1 therefore satisfies
the terms of the definition of the upper bound of the sequence.

It should be clearly realised that there is no doubt whatever
that this upper bound 1 is greater than all and not equal to any
of the terms of this sequence, and that any number whatever less
than 1 is actually less than some numbers of the sequence; and
that the upper bound is the number 1 and not some imagined
" next number just below 1."

In this sequence we notice one quite remarkable fact: the numbers
of the sequence form a succession of approximations to the upper
bound, of steadily greater and greater accuracy; and in fact, that as
we progress in the sequence the terms all tend to coincide towards
the upper bound 1. The relation between this fact and our problem of approximations of unlimited accuracy to irrational numbers
is apparent. Such sequences as tend together in this manner so as
to form a set of approximations of unlimited accuracy to some

* See Ex. 7, p. 36.

number (whether that number is the upper bound of the sequence or not) are said to be *convergent* and the number so approximated to is called the *unique limit* of the sequence*.

The special sequences discussed above belong to the important type of sequences which are called *monotone* or *monotonic*. The term monotone is applied to any sequence which is either *non-decreasing* or *non-increasing* according to the definition:

The sequence s_1, s_2, s_3, ... is non-decreasing if

$$s_1 \leqslant s_2 \leqslant s_3 \leqslant \dots ;$$

it is non-increasing if $s_1 \geqslant s_2 \geqslant s_3 \geqslant \dots.$

We have at once the following fundamental general theorem concerning monotone sequences.

A monotone bounded sequence necessarily has a unique limit, which is the upper bound of the sequence if the sequence is non-decreasing and the lower bound if the sequence is non-increasing.

Proof:

Let the sequence s_1, s_2, s_3, ... be non-decreasing; so that

$$s_1 \leqslant s_2 \leqslant s_3 \leqslant \dots,$$

and, because the sequence is bounded, it has an upper bound, L say, which is such that any number greater than L is greater than all the terms of the sequence and any number less than L is less than some term (or terms) of the sequence. It follows that, however small we take the positive number ϵ, there is at least one term s_n which exceeds $L - \epsilon$ (and $\leqslant L$) and therefore, since all the subsequent terms of the sequence $\geqslant s_n$, all the terms of the sequence, from the nth onwards, exceed $L - \epsilon$; i.e. the terms of the sequence tend towards the number L, i.e. the sequence has the unique limit L.

Similarly for a non-increasing sequence.

In future, to denote the fact that a sequence s_1, s_2, s_3, ... is non-increasing and has the lower bound and unique limit L we shall write s_1, s_2, s_3, ... $\searrow L$, or, shortly, $s_n \searrow L$†; and if the sequence is non-decreasing and has the upper bound and unique limit L, we shall write s_1, s_2, s_3, ... $\nearrow L$, or $s_n \nearrow L$†. We make use of the shortened form of statement that s_n *tends decreasingly* (or increasingly, as the case may be) *to* L.

---

* Definitions which are more strictly analytical are given on the next page.

† See footnote, next page.

41. Definitions for any sequence. We shall find that monotone sequences are much the most important type of sequence. We shall moreover be able to reduce the consideration of sequences in general to the consideration of such monotone sequences. A sequence which is not monotone may or may not be convergent, i.e. have a single definite limiting point towards which the numbers of the sequence tend. For general sequences we define a *limit* (or *limit point, limiting point, limiting value, limiting number*) of any sequence s_1, s_2, s_3, ... as *any number L, within an arbitrarily small neighbourhood of which* (i.e. within the range of values from $L - \delta$ to $L + \delta$, for any and every possible choice of the number δ) *there lie numbers of the sequence*; a number of the sequence itself not being a limit of the sequence unless it is repeated *indefinitely often* as a term of the sequence or there are *other* terms of the sequence within the arbitrarily small neighbourhood.

With this definition a sequence may have *any number of limits. If it is bounded and has only one such limit we say the sequence is convergent and the limit is called the unique limit.*

It is clear that a bounded monotone sequence has a unique limit according to this definition.

If the sequence s_1, s_2, s_3, ... is convergent and has the unique limit L, whether that sequence is monotone or not, we write s_1, s_2, s_3, ... $\rightarrow L$ or, shortly, $s_n \rightarrow L$; and we shall say s_n *tends to* L*. We write also alternatively $\lim_{n \to \infty} s_n = L$.

For convenience we define the *modulus* (or *absolute value, numerical value, positive value*), $|x|$, of a real number x to be the number x if x is positive (or zero) or $-x$ if x is negative.

In the above definition of a limit, a number s_n of the sequence will lie within an arbitrarily small neighbourhood ($L - \delta$ to $L + \delta$) of L if $|s_n - L| < \delta$. The number L will therefore be a limit of the sequence s_1, s_2, s_3, ... if and only if, for any and every possible choice of the positive number δ, there are terms of the sequence, e.g. s_n say, such that $|s_n - L| < \delta$; with the special proviso in the case when L is one of the terms of the sequence.

---

* When the shortened statements, $s_n \rightarrow L$, etc. are used, the phrase "as n increases indefinitely" or "as n tends to infinity" or "as $n \rightarrow \infty$" is often added when it is desired to make clear that the rank of the terms of the sequence is denoted by n.

We have also that the sequence s_1, s_2, s_3, ... will have a *unique limit* L if, and only if, corresponding to every possible choice of the positive number ϵ there is always a term s_n of the sequence such that $|s_n - L| < \epsilon$ and all terms, say s_m, subsequent to s_n in the sequence, also satisfy the inequality $|s_m - L| < \epsilon\,$*.

It is geometrically evident that if the sequence satisfies these conditions it has a unique limit according to our definition. We can prove this as follows:

Such a sequence plainly has the number L as a limit. Let L' be any number different from L. Take $\epsilon = \dfrac{|L-L'|}{2}$. There is a term s_n such that, for it and all terms s_m subsequent to it, $|s_m - L| < \epsilon$, and therefore

$$|s_m - L'| = |s_m - L + L - L'|$$
$$> ||L - L'| - \epsilon|$$
$$= \frac{|L-L'|}{2}.$$

Of the terms of the sequence preceding s_n (excluding those, if any, which equal L') there will be one or more nearer to L' than are the others, and therefore there is a positive number, Δ say, such that $|s_m - L'| > \Delta$ for all terms s_m preceding s_n (excluding those terms which $= L'$). Taking δ to be the lesser of the two numbers $\dfrac{|L-L'|}{2}$ and Δ, we have proved that there is a positive number δ for which $|s_m - L'|$ is not less than δ for any term of the sequence (with the possible exception of a certain number of them which may $= L'$). L' is therefore not a limit of the sequence. Thus L is the unique limit of the sequence.

It can also be proved conversely that if a sequence has a unique limit L, then the above conditions are satisfied.

For the purposes of this course the notion of *unique limit* is of much greater importance than that of limit in the wider sense. It is customary, and often convenient, to use the word *limit* in the sense of *unique* limit. Where there is no danger of confusion (e.g. where we have occasion to speak of *the* limit of a sequence) we shall use the word in this way. Wherever emphasis is needed we shall use the term *limiting number* for limit in the wider sense. It must not be forgotten that the notions of unique limit and limiting number have vital differences, nor, in particular, that only sequences which have unique limits are said to be convergent.

A consideration of a few examples in connection with the geo-

---

* This statement is applicable also to sequences of complex numbers. See Appendix.

metrical representation of Fig. 1 will soon convince one of the fundamental theorem that *any bounded sequence necessarily has at least one limiting number.* For a proof of this theorem,—the Bolzano-Weierstrass theorem,—the student is referred elsewhere[*].

42. Examples of sequences. The following examples of sequences will be instructive:

(a) $\frac{1}{2}, 1\frac{1}{2}, \frac{3}{4}, 1\frac{3}{4}, \frac{7}{8}, 1\frac{7}{8}, \frac{15}{16}, 1\frac{15}{16}, \dots,$

the nth term being $1 - 1/2^{\frac{n+1}{2}}$ or $2 - 1/2^{\frac{n}{2}}$ according as n is odd or even.

Here the sequence is bounded above and below; the upper bound $= 2$, the lower bound $= \frac{1}{2}$; the lower bound is one of the numbers of the sequence; the upper bound is not a number of the sequence but is a limiting number; there is no unique limit; there are two limiting numbers, 1 and 2.

(b) $\frac{1}{2}, 1\frac{1}{2}, \frac{3}{4}, 1\frac{1}{4}, \frac{7}{8}, 1\frac{1}{8}, \frac{15}{16}, 1\frac{1}{16}, \dots,$

the nth term being $1 - 1/2^{\frac{n+1}{2}}$ or $1 + 1/2^{\frac{n}{2}}$ according as n is odd or even.

The sequence is bounded above and below; the upper bound $= 1\frac{1}{2}$; the lower bound $= \frac{1}{2}$; neither the upper bound nor the lower bound is a limit; there is one unique limit, 1; the sequence is convergent.

(c) $\sin 1°, \sin 2°, \sin 3°, \dots$

The numbers of this sequence repeat themselves indefinitely, for
$$\sin(180° - 1°) = \sin 1°, \quad \sin(360° + 1°) = \sin 1°, \text{ etc.};$$
there are in fact only 181 distinct numbers in this sequence, viz.
$$0, \pm \sin 1°, \pm \sin 2°, \dots, \pm \sin 90°.$$

Each of these numbers is repeated indefinitely often and therefore is a limit (limiting number). This sequence therefore has 181 different limits, of which two, viz. 1 and -1, are the upper bound and the lower bound.

Note. It is essential to consider a number which is repeated indefinitely as a limit (as is done in our definition above); otherwise our fundamental theorems would cease to hold. As a matter

[*] See e.g. G. H. Hardy, *Course of Pure Mathematics* (3rd edition), or T. J. I'A. Bromwich, *Theory of Infinite Series* (2nd edition).

of fact it is easy to modify this particular sequence so as to retain the same 181 limits while ensuring that these limits are not repeated. The sequence whose nth term is $\sin(n + 1/n)$ will have the same 181 limits while none of the terms are so repeated.

43. General sequences by method of monotone sequences.

To shew how the consideration of *general sequences* may be made to depend on that of *monotone sequences* let us consider *any* sequence

$$s_1,\ s_2,\ s_3,\ \ldots \quad \ldots\ldots\ldots\ldots\ldots\ldots\ldots\ldots\ldots\ldots\ldots\ldots\ldots\ldots(3)$$

which is bounded above and below.

Let U_1 be the upper bound of the sequence $s_1,\ s_2,\ s_3,\ \ldots$
$\qquad U_2$ be the upper bound of the sequence $s_2,\ s_3,\ s_4,\ \ldots$
$\qquad U_3$ be the upper bound of the sequence $s_3,\ s_4,\ s_5,\ \ldots$
$\qquad\qquad$ etc.,

and $\quad L_1$ the lower bound of the sequence $s_1,\ s_2,\ s_3,\ \ldots$
$\qquad L_2$ the lower bound of the sequence $s_2,\ s_3,\ s_4,\ \ldots$
$\qquad L_3$ the lower bound of the sequence $s_3,\ s_4,\ s_5,\ \ldots$
$\qquad\qquad$ etc.

It is evident that in all cases

$$U_1 \geqslant U_2 \geqslant U_3 \geqslant \ldots > \text{some number } K$$

and $\qquad L_1 \leqslant L_2 \leqslant L_3 \leqslant \ldots < \text{some number } K'$.

The two sequences $\qquad U_1,\ U_2,\ U_3,\ \ldots$
$\qquad\qquad\qquad\qquad\qquad L_1,\ L_2,\ L_3,\ \ldots$

are monotone and bounded and therefore each has a unique limit (S and S' say);

i.e. $U_n \searrow S$ and $L_n \nearrow S'$.

It is easy to prove now that *the sequence* (3) *is convergent if and only if these two limits S, S' are equal, and, in the case of convergence, the unique limit* $= S = S'$.

For, in any case all the limits of the sequence (3) must lie between S and S' inclusive. Because if $S + \delta$ is any number which exceeds S (which is the lower bound of the sequence $U_1,\ U_2,\ U_3,\ \ldots$) $S + \dfrac{\delta}{2}$ will exceed some number U_n of the sequence $U_1,\ U_2,\ U_3,\ \ldots$ and will therefore exceed all the numbers $s_n,\ s_{n+1},\ s_{n+2},\ \ldots$ of the original sequence from the nth term onwards, so that, between $S + \dfrac{\delta}{2}$ and $S + \dfrac{3\delta}{2}$ none of the numbers $s_n,\ s_{n+1},\ s_{n+2},\ \ldots$ will lie, i.e. the number $S + \delta$ can be enclosed in an interval $\left(S + \dfrac{\delta}{2},\ S + \dfrac{3\delta}{2}\right)$ containing only a definite number $(< n)$ of the numbers $s_1,\ s_2,\ s_3,\ \ldots$ of the sequence. That is the number $S + \delta$ cannot be a limit of the sequence. Similarly any number $< S'$ cannot be a limit of the sequence, and we have proved that all the limits therefore lie between the two numbers S, S' inclusive.

Moreover the numbers S and S' certainly are limits of the sequence $s_1, s_2, s_3, \ldots$ for if, e.g., S were not a limit of this sequence, we could find an interval $S - \delta$, $S + \delta$ ($\delta > 0$) in which only a definite number of the numbers $s_1, s_2, s_3, \ldots$ lie; so that, for all values of n sufficiently great, s_n either $< S - \delta$ or $> S + \delta$ and then every corresponding U_n either $\leqslant S - \delta$ or $\geqslant S + \delta$, which is impossible because S is the unique limit of the sequence $U_1, U_2, U_3, \ldots$; therefore S must be a limit of the sequence $s_1, s_2, s_3, \ldots$. Similarly S' must be also a limit.

The theorem now follows at once. If $S = S'$ all the limits of the sequence $s_1, s_2, s_3, \ldots$ lie between S and S', and therefore all coincide with S; i.e. the sequence has a unique limit S.

On the other hand, if the sequence $s_1, s_2, s_3, \ldots$ has a unique limit, this limit must be S (and S'); and the theorem is proved.

We note incidentally in the proof of this theorem that in the general case the numbers S, S' are the greatest and least of the limits of the sequence $s_1, s_2, s_3, \ldots$. They are called the *upper* and *lower limits* of the sequence.

We have immediately *the necessary and sufficient condition of convergence:*

The sequence $s_1, s_2, s_3, \ldots$ is convergent if and only if the difference $|s_{n+p} - s_n|$ is less than any positive number whatever (ϵ) however small, for all positive integral values of p and all values of n, sufficiently great, i.e. given the number ϵ, a number n_0 can be found so great that for all values of n greater than n_0 $|s_{n+p} - s_n|$ is less than ϵ for all positive integral values of p.

For: if this condition is satisfied, since S, the greatest of the limits, is a limit of the sequence, there are values of n so great that the given conditions are satisfied and at the same time s_n differs from S by less than any arbitrarily small positive number ϵ; therefore, for all positive integral values of p, s_{n+p}, which differs from s_n by less than ϵ, and from S by $|s_{n+p} - s_n + s_n - S|$, must differ from S by less than $\epsilon + \epsilon$, i.e. 2ϵ; i.e. *all* terms s_{n+p} of the sequence sufficiently far on differ from S by less than an arbitrary small number, and the number S therefore is the unique limit of the sequence.

On the other hand, if the sequence has a unique limit (S say), both s_n and s_{n+p} must differ from S by less than any arbitrarily small positive number ϵ if n is sufficiently great; and therefore the difference between them, viz. $|s_{n+p} - s_n|$, must be arbitrarily small if n is sufficiently great.

The sufficiency and necessity of the condition are established. This theorem, the most general theorem in the theory of convergence of sequences, is often known as the *principle of convergence.*

The consideration however of general sequences lies beyond our scope and the above indications are given only to shew the important rôle which may be played by monotone sequences in the consideration of general sequences.

44. Special sequences. Let us now consider in detail some special sequences, with reference to their convergence and the evaluation of their unique limits (when convergent). The student is strongly advised to work through some of the easier examples at the end of this section in order to familiarise himself thoroughly with the ideas of convergent sequences; and in all cases the help of the geometrical representation of Fig. 1 should be used.

(i) The sequence whose nth term is x^n is convergent and has the unique limit 0 if the number x lies between -1 and 1 (i.e. $-1 < x < 1$); is convergent and has the unique limit 1 if $x = 1$; and otherwise is not convergent.

To prove this we see that if $x = 1$ all the terms of the sequence are 1, and therefore the sequence has as unique limit this indefinitely repeated number 1. Similarly, if $x = 0$, the sequence has the unique limit 0.

If $0 < x < 1$ the sequence is steadily decreasing and

$$x = \frac{1}{1 + \dfrac{1-x}{x}} = \frac{1}{1+y}$$

where $y \left(= \dfrac{1-x}{x} \right)$ is some positive number; therefore, by inequality (ii), p. 29,

$$x^n = \frac{1}{(1+y)^n} < \frac{1}{1+ny} < \frac{1}{ny},$$

which is less than any positive number, ϵ say, for all values of n greater than $1/(\epsilon y)$.

Therefore the lower bound and unique limit of the sequence in this case is 0; i.e. $x^n \searrow 0$ if $0 < x < 1$.

If $-1 < x < 0$, the terms of the sequence are alternately positive and negative, but their moduli as before tend decreasingly to zero; therefore $x^n \to 0$:

In the case when $x > 1$ we have

$$x^n = [1 + (x-1)]^n > 1 + n(x-1) > n(x-1),$$

which exceeds any number K whatever, however large, so soon as $n > K/(x-1)$, and the sequence is therefore unbounded above.

Similarly, if $x < -1$, the sequence is unbounded above and below, because the terms are alternately positive and negative and their moduli exceed any number K.

Finally, if $x = -1$, the terms are alternately 1 and -1. These two numbers are both limiting numbers, and the sequence is not convergent, though bounded both above and below.

(ii) The sequence x, $2x^2$, $3x^3$, $4x^4$, ... converges to the unique limit 0 if $-1 < x < 1$, but is otherwise unbounded.

Taking first x to be positive, the sequence will not be a steadily decreasing sequence unless $0 < x < \frac{1}{2}$, for the second term will exceed the first; but in any case, if $0 < x < 1$, *sooner or later* the terms of the sequence will steadily decrease; for $nx^n > (n+1)x^{n+1}$, so soon as $n/(n+1) > x$ (x being positive), i.e. so soon as $1 + 1/n < 1/x$, which will be the case so soon as $n > x/(1-x)$.

Therefore, from and after some term, the sequence is steadily decreasing. It is moreover bounded below, because all the numbers of the sequence are evidently positive; therefore it is convergent and its unique limit is greater than or equal to zero.

To shew that the limit is *equal* to zero we can use the inequality of Ex. 10, p. 36.

We have, if $0 < x < 1$,

$$x = \frac{1}{1+y},$$

where y is some positive number, viz. $(1-x)/x$.

Therefore $x^m = \dfrac{1}{(1+y)^m} < \dfrac{1}{m}$ by the inequality cited, if m is any positive integer sufficiently large; whence it will follow that $x^{2m} < \dfrac{1}{m^2}$,

and therefore $2mx^{2m} < 2m\,\dfrac{1}{m^2} = \dfrac{2}{m} \to 0$ as m increases indefinitely;

and also $(2m+1)x^{2m+1} < (2m+1)\,\dfrac{1}{m^2} = \dfrac{2}{m} + \dfrac{1}{m^2} \to 0$; and therefore, whether n is odd or even ($n = 2m+1$ or $2m$), $nx^n \to 0$.

When x is negative the terms are alternately positive and negative, but their moduli are the same as when x has the corresponding positive value; and therefore again $nx^n \to 0$.

When $x \geqslant 1$ or $\leqslant -1$, that the sequence is unbounded is plain; for the nth term is, in modulus, greater than or equal to n.

(iii) The sequence whose nth term is

$$1 + x + x^2 + \ldots + x^{n-1},$$

where x is any real number lying between -1 and $+1$, is convergent and has the limit $1/(1-x)$; if $x \leqslant -1$ or $\geqslant +1$ the sequence is not convergent.

Here, if $|x| < 1$, the nth term of the sequence

$$= s_n = 1 + x + x^2 + \ldots + x^{n-1} = \frac{1 - x^n}{1 - x} = \frac{1}{1 - x} - \frac{x^n}{1 - x}.$$

Since $\left|\dfrac{x^n}{1 - x}\right|$ will be less than any positive number ϵ, however small, if n is sufficiently large, we see that the terms s_n tend towards the unique limit $\dfrac{1}{1 - x}$; or, more precisely, within an arbitrarily small distance ϵ of $\dfrac{1}{1 - x}$, there lie all the terms s_n of the sequence farther on in the sequence than the mth term s_m, where $\left|\dfrac{x^m}{1 - x}\right| < \epsilon$; which will be the case if m is any integer exceeding $\dfrac{x}{\epsilon(1 - x)^2}$; and therefore the sequence is convergent and its unique limit is $\dfrac{1}{1 - x}$.

If $x \geqslant 1$ the sequence is steadily increasing but not bounded above; if $x < -1$ the sequence is unbounded above and below; whilst if $x = -1$ the terms of the sequence are alternately 1 and 0 and therefore, though bounded above and below, the sequence has not then a unique limit. Q.E.D.

45. The number e. Consider the sequence $(1 + 1)^1$, $(1 + 1/2)^2$, $(1 + 1/3)^3$, ... This sequence, which we shall find of great importance in the next chapter, will be seen to be steadily increasing and bounded, and therefore it must have a unique limit, its upper bound. This unique limit is in fact an irrational number—the irrational number e—which we shall evaluate as accurately as is here convenient.

We first prove that the sequence is steadily increasing, by applying the binomial theorem (p. 22 above) to the expression which is the nth term of the sequence, and comparing the expansion with the corresponding expansion for the $(n + 1)$th term.

We have:

$$\left(1 + \frac{1}{n}\right)^n = 1 + n\frac{1}{n} + \frac{n(n-1)}{1.2}\frac{1}{n^2} + \frac{n(n-1)(n-2)}{1.2.3}\frac{1}{n^3} + \ldots$$
$$+ \frac{n(n-1)\ldots(n-r+1)}{1.2\ldots r}\frac{1}{n^r} + \ldots + \frac{n(n-1)\ldots2.1}{1.2\ldots(n-1)n}\frac{1}{n^n}$$

(where r denotes any positive integer less than n)

$$= 1 + 1 + \frac{1}{2!}\left(1 - \frac{1}{n}\right) + \frac{1}{3!}\left(1 - \frac{1}{n}\right)\left(1 - \frac{2}{n}\right) + \ldots$$

$$+ \frac{1}{r!}\left(1 - \frac{1}{n}\right)\ldots\left(1 - \frac{r-1}{n}\right) + \ldots + \frac{1}{n!}\left(1 - \frac{1}{n}\right)\ldots\left(1 - \frac{n-1}{n}\right)$$

and

$$\left(1 + \frac{1}{n+1}\right)^{n+1} = 1 + 1 + \frac{1}{2!}\left(1 - \frac{1}{n+1}\right) + \ldots$$

$$+ \ldots + \frac{1}{r!}\left(1 - \frac{1}{n+1}\right)\ldots\left(1 - \frac{r-1}{n+1}\right) + \ldots$$

$$+ \frac{1}{n!}\left(1 - \frac{1}{n+1}\right)\ldots\left(1 - \frac{n-1}{n+1}\right) + \frac{1}{(n+1)^{n+1}}.$$

After the first two terms, every term in the second expansion is seen to exceed the corresponding term in the first, the terms in both expansions are all positive, and there is moreover an additional term $\left(\frac{1}{(n+1)^{n+1}}\right)$ in the second expansion; therefore the second expression is the greater, i.e. $\left(1 + \frac{1}{n}\right)^{n} < \left(1 + \frac{1}{n+1}\right)^{n+1}$ for all positive integral values of n; i.e.

$$\left(1 + \frac{1}{1}\right)^{1} < \left(1 + \frac{1}{2}\right)^{2} < \ldots < \left(1 + \frac{1}{n}\right)^{n} < \ldots,$$

i.e. *the sequence is steadily increasing.*

The sequence is *bounded* above because the nth term equals

$$\left(1 + \frac{1}{n}\right)^{n} = 1 + 1 + \frac{1}{2!}\left(1 - \frac{1}{n}\right) + \frac{1}{3!}\left(1 - \frac{1}{n}\right)\left(1 - \frac{2}{n}\right) + \ldots$$

$$+ \frac{1}{r!}\left(1 - \frac{1}{n}\right)\ldots\left(1 - \frac{r-1}{n}\right) + \ldots + \frac{1}{n!}\left(1 - \frac{1}{n}\right)\left(1 - \frac{2}{n}\right)\ldots\left(1 - \frac{n-1}{n}\right)$$

$$< 1 + 1 + \frac{1}{2!} + \frac{1}{3!} + \ldots + \frac{1}{r!} + \ldots + \frac{1}{n!}$$

$$= 1 + 1 + \frac{1}{1.2} + \frac{1}{1.2.3} + \ldots + \frac{1}{1.2\ldots r} + \ldots + \frac{1}{1.2\ldots n}$$

$$< 1 + 1 + \frac{1}{1.2} + \frac{1}{1.2.2} + \ldots + \frac{1}{1.2\ldots 2} + \ldots + \frac{1}{1.2\ldots 2}$$

(by replacing all the factors greater than 2 in the denominators by 2)

$$= 1 + 1 + \frac{1}{2} + \frac{1}{2^2} + \ldots + \frac{1}{2^{r-1}} + \ldots + \frac{1}{2^{n-1}}$$

$$= 1 + \frac{1 - \left(\frac{1}{2}\right)^n}{1 - \frac{1}{2}} = 1 + 2 - \frac{1}{2^{n-1}} = 3 - \frac{1}{2^{n-1}} < 3.$$

The sequence is therefore bounded, and its upper bound and unique limit is some number which $\leqslant 3$. The limit is approximated to indefinitely closely by the terms of the sequence; we have therefore only to evaluate terms of the sequence sufficiently far on to obtain the limit correct to any degree of accuracy; but we do not yet know how far in the sequence we shall have to go to obtain any desired degree of accuracy of approximation. It is essential in the evaluation of any limit that we should know within definite limits how accurate our approximation is.

In this case we see that

$$s_n = (1 + 1/n)^n < 1 + 1 + \frac{1}{2!} + \frac{1}{3!} + \ldots + \frac{1}{n!}$$

$$= 1 + 1 + \frac{1}{2!} + \ldots + \frac{1}{r!} + \left[\frac{1}{(r+1)!} + \ldots + \frac{1}{n!} \right]$$

$$= 1 + 1 + \frac{1}{2!} + \ldots + \frac{1}{r!} + \frac{1}{(r+1)!} \left[1 + \frac{1}{r+2} + \ldots + \frac{1}{(r+2) \ldots n} \right]$$

$$< 1 + 1 + \frac{1}{2!} + \ldots + \frac{1}{r!} + \frac{1}{(r+1)!} \left[1 + \frac{1}{r+2} \right.$$
$$\left. + \frac{1}{(r+2)^2} + \ldots + \frac{1}{(r+2)^{n-r-1}} \right]$$

$$= 1 + 1 + \frac{1}{2!} + \ldots + \frac{1}{r!} + \frac{1}{(r+1)!} \frac{1 - \dfrac{1}{(r+2)^{n-r}}}{1 - \dfrac{1}{r+2}}$$

$$< 1 + 1 + \frac{1}{2!} + \ldots + \frac{1}{r!} + \frac{1}{(r+1)!} \frac{r+2}{r+1}$$

for all integral values of n, r being any integer less than n.

By taking $r = 10$ say, we see that, for all values of n greater than 10,

$$s_n < 1 + 1 + \frac{1}{2!} + \ldots + \frac{1}{10!} + \frac{12}{11 . 11!},$$

which equals

$$2\cdot00000000$$
$$+\ \cdot50000000$$
$$+\ \cdot16666667$$
$$+\ \cdot04166667$$
$$+\ \cdot00833333$$
$$+\ \cdot00138889$$
$$+\ \cdot00019841$$
$$+\ \cdot00002480$$
$$+\ \cdot00000276$$
$$+\ \cdot00000028$$

i.e. $2\cdot71828181$
$$+\ \cdot00000003$$

i.e. $2\cdot71828184 < 2\cdot718282.$

The required limit will be less than (or at most equal to) this number.

Reverting to the expression for s_n, viz.

$$s_n = 1 + 1 + \frac{1}{2!}\Big(1 - \frac{1}{n}\Big) + \frac{1}{3!}\Big(1 - \frac{1}{n}\Big)\Big(1 - \frac{2}{n}\Big) + \dots$$
$$+ \frac{1}{r!}\Big(1 - \frac{1}{n}\Big)\dots\Big(1 - \frac{r-1}{n}\Big) + \dots + \frac{1}{n!}\Big(1 - \frac{1}{n}\Big)\dots\Big(1 - \frac{n-1}{n}\Big),$$

we see that, on the other hand,

$$s_n > 1 + 1 + \frac{1}{2!} + \frac{1}{3!} + \dots + \frac{1}{r!} + \dots + \frac{1}{n!}$$
$$- \bigg[\frac{1}{2!}\frac{1}{n} + \dots + \frac{1}{r!}\Big(\frac{1}{n} + \frac{2}{n} + \dots + \frac{r-1}{n}\Big) + \dots$$
$$+ \frac{1}{n!}\Big(\frac{1}{n} + \frac{2}{n} + \dots + \frac{n-1}{n}\Big)\bigg],$$

because $\Big(1 - \frac{1}{n}\Big)\Big(1 - \frac{2}{n}\Big)\dots\Big(1 - \frac{r-1}{n}\Big)$

$$= \Big(1 - \frac{1}{n} - \frac{2}{n} + \frac{2}{n^2}\Big)\Big(1 - \frac{3}{n}\Big)\dots\Big(1 - \frac{r-1}{n}\Big)$$

$$> \Big(1 - \frac{1}{n} - \frac{2}{n}\Big)\Big(1 - \frac{3}{n}\Big)\dots\Big(1 - \frac{r-1}{n}\Big)$$

$$> \Big(1 - \frac{1}{n} - \frac{2}{n} - \frac{3}{n}\Big)\Big(1 - \frac{4}{n}\Big)\dots\Big(1 - \frac{r-1}{n}\Big)$$

$$> \dots\dots\dots\dots\dots\dots\dots\dots\dots\dots\dots\dots\dots\dots\dots$$

$$> \Big(1 - \frac{1}{n} - \frac{2}{n} - \frac{3}{n} - \dots - \frac{r-1}{n}\Big)$$

for any positive integer r less than n.

Therefore

$$s_n > 1 + 1 + \frac{1}{2!} + \frac{1}{3!} + \dots + \frac{1}{r!} + \dots + \frac{1}{n!}$$

$$- \left[\frac{1}{2!}\frac{1}{n} + \frac{1}{3!}\frac{2.3}{2.n} + \dots + \frac{1}{r!}\frac{(r-1)r}{2.n} + \dots + \frac{1}{n!}\frac{(n-1)n}{2.n} \right]$$

because

$$1 + 2 + 3 + \dots + (r-1) = \frac{(r-1)r}{2}$$

(by equality (ix), p. 21, or Ex. 2, p. 5);

i.e. $s_n > 1 + 1 + \frac{1}{2!} + \frac{1}{3!} + \dots + \frac{1}{r!} + \dots + \frac{1}{n!}$

$$- \frac{1}{2n} \left[1 + 1 + \frac{1}{2!} + \dots + \frac{1}{(r-2)!} + \dots + \frac{1}{(n-2)!} \right]$$

$$> \left(1 + 1 + \frac{1}{2!} + \frac{1}{3!} + \dots + \frac{1}{n!} \right) - \frac{3}{2n},$$

because the expression in square brackets has been proved above to be always less than 3, whatever the value of the integer n.

It will follow in particular from this result that, whatever positive integer n may be, greater than 10,

$$s_n > 1 + 1 + \frac{1}{2!} + \dots + \frac{1}{10!} - \frac{3}{2n},$$

and, taking $n = 10^7$ say, and using the value of

$$1 + 1 + \frac{1}{2!} + \dots + \frac{1}{10!}$$

just calculated, we see that for all values of n sufficiently great ($\geqslant 10^7$),

$$s_n > 2{\cdot}7182818 - {\cdot}00000015 > 2{\cdot}7182816.$$

We have now proved that, for values of n sufficiently great, s_n lies between $2{\cdot}7182816$ and $2{\cdot}718282$. The unique limit of the sequence must lie between these two values, and its value therefore, correct to six places of decimals, is $2{\cdot}718282$.

We have, in the above process, seen how this limit e can be calculated to any desired degree of accuracy whatever. We have proved that

$$e < 1 + 1 + \frac{1}{2!} + \dots + \frac{1}{n!} + \frac{n+2}{(n+1)(n+1)!}$$

and
$$> 1 + 1 + \frac{1}{2!} + \dots + \frac{1}{n!} - \frac{3}{2m},$$

where n is any positive integer and m is any positive integer which exceeds n. From the second of these relations we can deduce that

$$e \geqslant 1 + 1 + \frac{1}{2!} + \dots + \frac{1}{n!};$$

because, if not, a positive number δ could be found such that

$$1 + 1 + \frac{1}{2!} + \dots + \frac{1}{n!} - \delta$$

would exceed e; but an integer m could be found such that

$$\delta > \frac{3}{2m} \left(\text{viz. } m \geqslant \frac{3}{2\delta}\right)$$

and the second relation would then be contradicted.

Hence we know that
$$e \geqslant 1 + 1 + \frac{1}{2!} + \frac{1}{3!} + \dots + \frac{1}{n!}$$

and
$$< 1 + 1 + \frac{1}{2!} + \frac{1}{3!} + \dots + \frac{1}{n!} + \frac{n+2}{(n+1)(n+1)!}.$$

If therefore we wish to calculate e with an error less than say $1/10^{10}$, we have only to find an integer n such that

$$\frac{n+2}{(n+1)(n+1)!} \leqslant \frac{1}{10^{10}},$$

and then e is given to the required degree of accuracy by

$$1 + 1 + \frac{1}{2!} + \frac{1}{3!} + \dots + \frac{1}{n!}.$$

To obtain this degree of accuracy $n = 13$ will suffice because

$$\frac{14 \cdot 14!}{15} > 10^{10}.$$

In the same way we can determine what value of n will suffice to give the value of e correct to any desired degree of accuracy whatever.

In this discussion we have proved that the sequence whose nth term is $1 + 1 + \frac{1}{2!} + \frac{1}{3!} + \dots + \frac{1}{n!}$ is convergent and has the limit e. In the language of the next section this proves that the infinite series $1 + 1 + \frac{1}{2!} + \frac{1}{3!} + \dots$ is convergent and has the "sum" e. This series is discussed independently in the next section.

Sequences may be defined in many ways. Some different methods will be found in the following examples. As we can see from the sequences considered above, the methods by which a sequence may be studied and its limit (if any) evaluated depend largely on the particular way in which the sequence is defined. A large part of the theory of infinite sequences and series is concerned with the special consideration of sequences defined in special ways. In the next section we shall give special consideration to sequences defined by *series*.

EXAMPLES VI*.

1. Discuss the following sequences, determining whether or not they are bounded and whether or not they are monotone; when bounded determine the upper and lower bounds and the limiting numbers of the sequences; when convergent determine the unique limits:

(i) $\qquad 1/2,\ 2/3,\ 3/4,\ 4/5,\ \dots\ n/(n+1),\ \dots$

[Bounded and steadily increasing; lower bound $=\frac{1}{2}$, upper bound $=1=$ the unique limit ; convergent.]

(ii) $\qquad 1,\ 1/3,\ 1/9,\ 1/27,\ \dots\ 1/3^{n-1},\ \dots$

[Bounded; steadily decreasing; upper bound $=1$, lower bound $=$ unique limit $=0$; convergent.]

(iii) $\qquad \frac{3}{1},\ \frac{5}{2},\ \frac{7}{3},\ \dots\ \dfrac{2n+1}{n},\ \dots$

[Bounded; steadily decreasing; upper bound $=3$, lower bound $=$ unique limit $=2$; convergent.]

(iv) $\qquad 2,\ 2^2,\ 2^3,\ 2^4,\ \dots\ 2^n,\ \dots$

[Bounded below, not bounded above; steadily increasing; lower bound $=2$; no upper bound, and no limit; not convergent.]

(v) $\qquad 2,\ 1,\ 1\frac{3}{4},\ 1\frac{1}{4},\ 1\frac{5}{8},\ 1\frac{3}{8},\ \dots,$

the nth term being $1\frac{1}{2}+(\frac{1}{2})^{\frac{n+1}{2}}$ or $1\frac{1}{2}-(\frac{1}{2})^{\frac{n}{2}}$ according as n is odd or even.

[Bounded; not monotone; upper bound $=2$, lower bound $=1$; convergent, unique limit $=1\frac{1}{2}$. Sequences of this type, which oscillate regularly, tending to a unique limit from above and below, are of frequent occurrence.]

(vi) $\qquad 1,\ 2,\ 1/2,\ 2\frac{1}{2},\ 1/4,\ 2\frac{3}{4},\ 1/8,\ 2\frac{7}{8},\ \dots,$

the nth term being $1/2^{\frac{n-1}{2}}$ or $3-1/2^{\frac{n}{2}-1}$ according as n is odd or even.

[Bounded, not monotone; lower bound $=0$, upper bound $=3$; two limits, 0 and 3; not convergent.]

* The student will find it instructive to sketch graphs to represent the sequences in these examples. Thus in 1 (i) the graph $y=n/(n+1)$ is plotted and is seen to tend to the limit 1.

(vii) $\qquad \sqrt[2]{2},\ \sqrt[3]{3},\ \sqrt[4]{4},\ \sqrt[5]{5},\ \dots\ \sqrt[n]{n},\ \dots.$

[Bounded; not monotone, but, after the third term steadily decreasing; upper bound = greatest term = $\sqrt[3]{3}$, lower bound = 1 = unique limit; convergent.

Proof: $\sqrt[n]{n} > \sqrt[n+1]{(n+1)}$ if $n^{n+1} > (n+1)^n$, i.e. if $n > (1+1/n)^n$. But

$$(1+1/n)^n < 3$$

for all integral values of n, and therefore $\sqrt[n]{n} > \sqrt[n+1]{(n+1)}$ if $n \geqslant 3$; it is easily seen that $\sqrt[3]{3} > \sqrt[2]{2}$, because $3^2 = 9 > 8 = 2^3$.

That the lower bound and unique limit is 1 can be deduced from Ex. 10, p. 36; for if $1+x$ be any number greater than 1, n can be found so large that $(1+x)^n > n$, i.e. $\sqrt[n]{n} < 1+x$, and therefore any number $1+x$ greater than 1 exceeds some term of the sequence; whence the lower bound $\leqslant 1$; but all the terms of the sequence > 1, and therefore the lower bound $= 1$.]

(viii) $\qquad\qquad 2,\ 1/2,\ 3,\ 1/3,\ 4,\ 1/4, \dots,$

the nth term being $(n+3)/2$ if n is odd and $2/(n+2)$ if n is even.

[Bounded below, not bounded above; lower bound = 0, which is the only limiting number; not convergent and no unique limit, because unbounded.]

(ix) $\qquad 1,\ 1/2,\ 1/3,\ 2/3,\ 1/4,\ 2/4,\ 3/4,\ 1/5,\ 2/5,\ 3/5,\ 4/5,\ \dots,$

the nth term being p/q where q is the integral part of $\frac{3}{2}+\frac{1}{2}\sqrt{(8n-15)}$ and $p = n - \dfrac{q^2}{2} + \dfrac{3q}{2} - 2$. The sequence is defined more naturally by the process suggested by the terms expressed,—i.e. the fractions are taken in order with steadily increasing denominators and numerators.

[This sequence (which the student may notice includes among its terms all proper fractions—i.e. all rational numbers between 0 and 1—and is for that reason of considerable interest in the theory of the arithmetic continuum and sets of points) is bounded above and below; upper bound = 1, lower bound = 0; every real number between 0 and 1 inclusive is a limit of this sequence; it is not convergent.]

2. Investigate the convergence (or otherwise) of the sequence

$$1-x,\ 1+x^2,\ 1-x^3,\ \dots\ 1+(-x)^n,\ \dots,$$

discussing the different cases which arise according to the value of the real number x.

3. Shew that the sequences whose nth terms are

$$1/n,\ 1/n^2,\ 1/(\sqrt{n}),\ 1/2^n,\ 1/10^n,\ 1/n!,\ n/n!,\ n^2/n!,\ 2^n/n!,\ n!/n^n$$

all converge to the unique limit zero.

4. Shew that the sequence $10,\ \sqrt[2]{10},\ \sqrt[3]{10},\ \sqrt[4]{10},\ \dots$ is steadily decreasing and has the lower bound and unique limit 1.

$$\left[\text{Consider } \frac{9}{\sqrt[n]{10}-1} \text{ as } \frac{(\sqrt[n]{10})^n - 1}{(\sqrt[n]{10})-1} \text{ and apply inequality (iii) of p. 29.} \right]$$

5. Shew that the sequence whose nth term is $\sin \dfrac{360°}{n+1}$ is bounded above and below, decreasing from the third term onwards, and convergent; and that the upper bound = the greatest term = 1, and the lower bound = the least term = the unique limit = 0.

6. Investigate, for different values of the real number x, the convergence or otherwise of the sequences whose nth terms are

$$x^n,\ nx^n,\ x^n/n,\ n!\,x^n,\ x^n/n!,\ (\sin x)/n,\ (nx)^n,\ (x/n)^n.$$

7. Prove that the sequence whose $(n-1)$th term is $\left(1-\dfrac{1}{n}\right)^n$ is steadily increasing and has $\dfrac{1}{e}$ for its upper bound and unique limit.

8. Prove successively the following general theorems on monotone and other sequences:

  (i)   If $s_1, s_2, s_3, \ldots \searrow s$ and $s_1', s_2', s_3', \ldots \searrow s'$, then
$$s_1+s_1',\ s_2+s_2',\ s_3+s_3',\ldots \searrow s+s'.$$

  (ii) If $s_1, s_2, s_3, \ldots \nearrow s$ and $s_1', s_2', s_3', \ldots \nearrow s'$, then
$$s_1+s_1',\ s_2+s_2',\ s_3+s_3',\ldots \nearrow s+s'.$$

 (iii) If $s_1, s_2, s_3, \ldots \to s$ and $s_1', s_2', s_3', \ldots \to s'$, then
$$s_1+s_1',\ s_2+s_2',\ s_3+s_3',\ldots \to s+s'.$$

[(iii) can be deduced from (i) and (ii) by the method of p. 47, using the relations

$$U(s_1, s_2, s_3, \ldots) + U(s_1', s_2', s_3', \ldots) \geqslant U(s_1+s_1',\ s_2+s_2',\ s_3+s_3', \ldots)$$
$$\geqslant L(s_1+s_1',\ s_2+s_2',\ s_3+s_3', \ldots) \geqslant L(s_1, s_2, s_3, \ldots) + L(s_1', s_2', s_3', \ldots),$$

where $U(s_1, s_2, s_3, \ldots)$, $L(s_1, s_2, s_3, \ldots)$ etc. denote the upper and lower bounds of the sequence $s_1, s_2, s_3, \ldots$ etc. It also follows directly from the condition for the existence of a unique limit on p. 45.]

9. It is not true in general that the upper and lower bounds of the sequence $s_1+s_1',\ s_2+s_2', \ldots$ are the sums of the upper and lower bounds of the separate sequences $s_1, s_2, \ldots$ and $s_1', s_2', \ldots$, but it is true if the separate sequences are both increasing or both decreasing.

10. Prove that if $s_n \to s$ and $s_n' \to s'$ then $s_n s_n' \to ss'$ and $s_n/s_n' \to s/s'$, provided in the latter case that s' (and the terms s_n') $\neq 0$.

11. Prove that in general if the terms of the sequence $s_1, s_2, s_3, \ldots$ alternately increase and decrease by steadily decreasing amounts (i.e. so that

$$s_1 < s_3 < s_5 < \ldots < \ldots < s_6 < s_4 < s_2)$$

and if the difference between any two successive terms tends to zero as the terms are taken indefinitely far on in the sequence, then the sequence is convergent and its unique limit lies between any two consecutive terms of the sequence.

12. Prove that the sequence

$$1+\tfrac{1}{2},\ \ 1+\frac{1}{2+\tfrac{1}{2}},\ \ 1+\cfrac{1}{2+\cfrac{1}{2+\tfrac{1}{2}}},\ \ \ldots$$

is convergent. Evaluate its limit, correct to two decimal places, and prove that it equals $\sqrt{2}$.

[We argue:

$$\sqrt{2}=1+(\sqrt{2}-1)=1+\frac{(\sqrt{2}-1)(\sqrt{2}+1)}{\sqrt{2}+1}=1+\frac{1}{\sqrt{2}+1}$$

$$=1+\frac{1}{2+(\sqrt{2}-1)}=1+\cfrac{1}{2+\cfrac{1}{2+(\sqrt{2}-1)}}, \text{ etc.,}$$

and it is easy to see that this expression lies between s_n and s_{n+1}, the nth and $(n+1)$th terms on the sequence. To complete the proof, the difference between s_n and s_{n-1} is shewn to tend to zero.]

13. Evaluate the limit of the sequence whose nth term is

$$4[1-1/3+1/5-1/7+\ldots+(-1)^{n+1}/(2n-1)],$$

correct to within ·1.

14. Shew that if $x_1, x_2, x_3, \ldots$ is any convergent sequence having the unique limit a, then, if k is any positive integer, the sequence $x_1{}^k, x_2{}^k, x_3{}^k, \ldots$ is convergent and has the unique limit a^k. (Continuity* of the power x^k.)

[This may be done from the definition of p. 44; or by the method of monotone sequences, thus:

If $U_1, U_2, U_3, \ldots$ are the upper bounds of

$$x_1, x_2, x_3, \ldots; \; x_2, x_3, \ldots; \; x_3, x_4, \ldots; \; \ldots$$

and $L_1, L_2, L_3, \ldots$ the corresponding lower bounds, we have the sequence

$$U_1, U_2, U_3, \ldots \searrow a,$$

and

$$L_1, L_2, L_3, \ldots \nearrow a,$$

whence easily

$$U_1{}^k, U_2{}^k, U_3{}^k, \ldots \searrow a^k,$$

and

$$L_1{}^k, L_2{}^k, L_3{}^k, \ldots \nearrow a^k.$$

But we know $L_n \leqslant x_n \leqslant U_n$ and therefore $L_n{}^k \leqslant x_n{}^k \leqslant U_n{}^k$. The result follows.]

15. Shew that if the sequence $x_1, x_2, x_3, \ldots \to a$, and k is any positive integer, then the sequence

$$\frac{x_1{}^k-a^k}{x_1-a}, \; \frac{x_2{}^k-a^k}{x_2-a}, \; \frac{x_3{}^k-a^k}{x_3-a}, \; \ldots \to ka^{k-1}.$$

(Differentiability* of x^k.)

16. Shew that if $h_1, h_2, h_3, \ldots$ is any sequence which tends to zero, then the sequence whose nth term is $(x+h_n)^2+3(x+h_n)-2 \to x^2+3x-2$, if x is any real number; and that the sequence whose nth term is

$$\frac{[(x+h_n)^2+3(x+h_n)-2]-[x^2+3x-2]}{h_n} \to 2x+3.$$

17. Prove that the number e is irrational.

[If e were rational it could be expressed as a fraction m/n; but e differs from $1+1+1/2!+\ldots+1/n!$ by less than $(n+2)/(n+1)(n+1)!$, and therefore the integer $m.n!/n$ would differ from the integer $n!(1+1+1/2!+\ldots+1/n!)$ by less than $(n+2)n!/(n+1)(n+1)!$, i.e. $(n+2)/(n+1)^2$,—a *proper* fraction. This is impossible.]

* See Chapter III below.

18. The upper bound of the set of numbers x of the lower class of a Dedekindian classification $(x\,|\,y)$ is the number defined $(x\,|\,y)$,—as is also the lower bound of the numbers y. E.g. $\sqrt{2}$ is the upper bound of those numbers x which are negative or are such that $x^2 < 2$.

[Such a set of numbers is not a sequence, but the definition given of upper and lower bounds will apply.]

§ 7. INFINITE SERIES

46. If we attempt to perform the division of 1 by $1 - x$ the process will continue indefinitely, the remainders being successively x, x^2, x^3,... and the quotient apparently the sum of the unending series $1 + x + x^2 + \ldots$. If, however, we ask ourselves what the meaning of this process can be, we are compelled to admit that all we know for certain is that at the nth stage of the division, when the terms in the quotient are $1 + x + x^2 + \ldots + x^{n-1}$, the remainder is x^n; or that

$$\frac{1}{1-x} = 1 + x + x^2 + \ldots + x^{n-1} + \frac{x^n}{1-x};$$

and that this is true at any and every stage. The idea of the process being "completed," giving as quotient the sum of the infinite series $1 + x + x^2 + \ldots$ and leaving no remainder, is certainly attractive, but, we are compelled to admit, is meaningless. For addition essentially implies that the numbers to be added must be finite in number; it is possible to add up any finite number of the terms 1, x, x^2,..., but to add up the whole infinite series of terms is certainly not possible in any sense of the term "addition" hitherto considered. There is moreover a very reasonable doubt as to the propriety of omitting the remainder,—lost in the haze of an infinite process.

Nevertheless we feel that in some natural way the infinite series considered, $1 + x + x^2 + \ldots$, does arise in the process of division described, and, even if it were only for the sake of curiosity, the entire banishment of such infinite series seems undesirable.

47. The sum of a convergent series. The particular series considered arises out of an indefinitely continued process of division, at successive stages of which the quotients are 1, $1 + x$, $1 + x + x^2$,.... We have therefore an infinite sequence of quotients. This sequence, which has already been discussed in the preceding

section, is convergent and has $1/(1-x)$ for its unique limit if
(and only if) x lies between -1 and 1. Thus, though it would be
quite meaningless (at present) to say that the sum of the infinite
series $1 + x + x^2 + \dots$ is $1/(1-x)$ and to write

$$1 + x + x^2 + \dots = \frac{1}{1-x},$$

yet it is true to say that the *sequence* formed by taking $1, 2, 3, \dots$
terms of the series $1 + x + x^2 + \dots$ is *convergent* and has for its
unique *limit* $1/(1-x)$ (provided $|x| < 1$), and to write

$$\lim_{n \to \infty} (1 + x + x^2 + \dots + x^{n-1}) = \frac{1}{1-x}.$$

We may now, if, for the sake of brevity, we wish to introduce
terms more directly concerned with the series rather than with the
sequence, say that an infinite series is *convergent* if the corresponding sequence (whose terms are the sums of terms of the series
taken in order) is convergent; and we may give to the limit
of the sequence a special name—the *sum* of the series. If we do
this, we must realise that the words "convergent" and "sum" are
used in entirely new (though suggestive) senses and that, in
particular, to speak of the "sum" of an infinite series does not
suppose in any sense that the terms of an infinite series can be
added together; the "sum" of the infinite series means simply
and precisely the unique limit of the corresponding sequence
(if there is such a limit). We may go further in this simplification
of terminology and write even

$$1 + x + x^2 + \dots = \frac{1}{1-x} \ (\text{if } |x| < 1),$$

as a symbolic expression of the fact that the "sum" of the infinite
series $1 + x + x^2 + \dots$ is $\dfrac{1}{1-x}$, i.e., expressed fully, the statement
$1 + x + x^2 + \dots = \dfrac{1}{1-x}$ *means* precisely that the sequence $1, 1+x,$
$1 + x + x^2, \dots$ is convergent and has $\dfrac{1}{1-x}$ for its unique limit.

The student may at first find this endowing of words with
a meaning totally different from their original meaning somewhat
confusing; and it will probably be better for him not to use the
shortened forms of expression until thoroughly familiar with the

actual meaning of the facts expressed. But this process of using words in this way is so essential a feature of mathematics that the student is advised gradually to familiarise himself with it.

48. Definitions and necessary condition for convergence. Series may arise in many ways and be of a variety of types. The particular series just considered is a specially simple case. We now consider convergence of series more generally.

Let
$$u_1 + u_2 + u_3 + \ldots \ldots\ldots\ldots\ldots\ldots\ldots(1)$$
be any infinite series, of which the *terms* are u_1, u_2, u_3,.... Let s_n represent the sum (or *partial sum*) of the first n terms of this series, i.e.
$$s_n = u_1 + u_2 + u_3 + \ldots + u_n.$$
Then we say that the infinite series (1) is *convergent* if the sequence
$$s_1, s_2, s_3, \ldots \ldots\ldots\ldots\ldots\ldots\ldots(2)$$
is convergent, and, in that case, the unique limit of the sequence (2) is called the *sum* of the series (1).

It is clear that the behaviour of the sequence (2) will depend very closely on that of the terms of the series (1). For example, the sequence (2) will be steadily increasing if and only if the terms u_2, u_3, u_4, ... of the series (1) are all positive; the sequence will be steadily decreasing if the terms of the series (after the first) are all negative; the terms of the sequence will alternately increase and decrease if the terms of the series are alternately positive and negative; and so on. It will be our concern to endeavour, as far as is easily possible, to find rules and methods by which the convergence (or otherwise) of the series is deduced directly from a knowledge of the terms of the series.

One simple central fact is at once noticeable in this regard: *An infinite series $u_1 + u_2 + u_3 + \ldots$ cannot be convergent unless the sequence of its terms, u_1, u_2, u_3, ..., has the unique limit zero;* for the sequence whose nth term, s_n, differs from its $(n-1)$th term, s_{n-1}, by u_n, clearly can have a unique limit only if $u_n \to 0$.

This simple fact may be sufficient to shew at once that a series is not convergent, but it must be pointed out that a series in which the terms tend to zero is not *necessarily* convergent, and that some such series (e.g. (vi), p. 67 below) are in fact not convergent. This *necessary condition for convergence* cannot therefore be used to establish convergence.

Other simple conditions of convergence depend on the type of the series concerned. We proceed to discuss two simple and important types of series.

49. Series of positive terms. Necessary and sufficient condition. Let us consider any series,

$$u_1 + u_2 + u_3 + \dots \quad \dots\dots\dots\dots\dots\dots(3),$$

whose terms are all positive (or zero), and investigate conditions under which such a series may be proved to be convergent or not[*].

The sequence $\qquad s_1, s_2, s_3, \dots \quad \dots\dots\dots\dots\dots\dots\dots(4),$

formed of the partial sums of the series, is necessarily steadily increasing (or non-decreasing) for $s_{n+1} - s_n = u_n \geq 0$ for all integral values of n. The necessary and sufficient condition, therefore, that the sequence (4) should be convergent is that it should be bounded, i.e. that a fixed number K can be found such that all the terms s_n of the sequence are less than K (see p. 43 above).

Hence the series (3) will be convergent if and only if a fixed number K can be found such that, for all values of n (no matter how large),

$$s_n \equiv u_1 + u_2 + u_3 + \dots + u_n < K.$$

Or *the necessary and sufficient condition that a series of positive terms should be convergent is that a number (K) can be found to exceed the sum of any number of terms of the series* beginning at the first.

50. Examples of series of positive terms. Let us consider some examples of series of positive terms.

(i) Take first the geometrical progression

$$1 + x + x^2 + \dots$$

in which the common ratio x is positive.

The terms of this series are all positive and the sum of the first n terms (if $x \neq 1$)

$$= s_n = 1 + x + x^2 + \dots + x^{n-1}$$
$$= \frac{1 - x^n}{1 - x} = \frac{1}{1 - x} - \frac{x^n}{1 - x}.$$

---

[*] The slight modifications needed to cover the case of series whose terms are all *negative* are left to the student.

If x, being positive, is less than 1, the number $x^n/(1-x)$ is necessarily positive, and it will follow that

$$s_n < 1/(1-x)$$

for all values of n.

The condition of convergence is satisfied (with $K = 1/(1-x)$) and therefore the series is convergent for all positive values of x less than 1. If, however, $x > 1$, the number $x^n/(1-x)$ is negative, and it will not follow that $s_n < 1/(1-x)$. Moreover we can easily prove that no number, K, can be found in this case so that $s_n < K$ for all values of n; for, if $x > 1$,

$$s_n = 1 + x + x^2 + \ldots + x^{n-1}$$
$$> 1 + 1 + 1 + \ldots + 1$$
$$= n;$$

and, whatever number K be taken, n can be taken to exceed it; i.e. by adding together sufficient terms of the series we can obtain a sum exceeding any assigned number whatever, however great. The necessary and sufficient condition of convergence is not satisfied and therefore the series is not convergent for values of x exceeding 1. It is similarly evident that when $x = 1$ the series is again not convergent.

(ii) The series $1 + \dfrac{1}{2} + \dfrac{1}{2}\dfrac{1}{2^2} + \dfrac{1}{3}\dfrac{1}{2^3} + \dfrac{1}{4}\dfrac{1}{2^4} + \ldots$

has its terms all positive and the sum of the first n terms

$$= s_n = 1 + \frac{1}{2} + \frac{1}{2}\frac{1}{2^2} + \frac{1}{3}\frac{1}{2^3} + \ldots + \frac{1}{n-1}\frac{1}{2^{n-1}}$$
$$< 1 + \tfrac{1}{2} + (\tfrac{1}{2})^2 + (\tfrac{1}{2})^3 + \ldots + (\tfrac{1}{2})^{n-1}.$$

This last expression is the sum of the first n terms of a geometrical progression with common ratio $\tfrac{1}{2}$ and therefore equals

$$\frac{1 - (\tfrac{1}{2})^n}{1 - \tfrac{1}{2}} = 2 - (\tfrac{1}{2})^{n-1},$$

which is necessarily less than 2, no matter how great n may be. Hence s_n is bounded, and the series is convergent.

(iii) The series $1 + 1 + \dfrac{1}{1\,.\,2} + \dfrac{1}{1\,.\,2\,.\,3} + \dfrac{1}{1\,.\,2\,.\,3\,.\,4} + \ldots$

or $\qquad\qquad 1 + \dfrac{1}{1!} + \dfrac{1}{2!} + \dfrac{1}{3!} + \dfrac{1}{4!} + \ldots$

has its terms all positive.

$$s_n = 1 + \frac{1}{1!} + \frac{1}{2!} + \cdots + \frac{1}{(n-1)!}$$

$$= 1 + \frac{1}{1} + \frac{1}{2} + \frac{1}{2.3} + \cdots + \frac{1}{2.3 \ldots (n-1)}$$

$$< 1 + \frac{1}{1} + \frac{1}{2} + \frac{1}{2^2} + \cdots + \frac{1}{2^{n-2}},$$

in which the rth term is $(\frac{1}{2})^{r-2}$; because, if $r > 2$, the rth term of the original series

$$= 1/(r-1)! = 1/[2.3 \ldots (r-1)] < 1/(2.2 \ldots 2) = 1/(2^{r-2});$$

whence, by summing the geometrical progression,

$$s_n < 1 + 2 - (\tfrac{1}{2})^{n-2}$$
$$= 3 - (\tfrac{1}{2})^{n-2} < 3.$$

Hence s_n is bounded and the series is convergent.

(iv) The series $1 + \dfrac{1}{1.2} + \dfrac{1}{2.3} + \dfrac{1}{3.4} + \cdots$

also has its terms all positive.

$$s_n = 1 + \frac{1}{1.2} + \frac{1}{2.3} + \cdots + \frac{1}{(n-1).n}.$$

In this case a little trial will soon convince us that we cannot replace the terms of s_n by the corresponding terms of any geometrical progression, with common ratio less than 1, whose terms would exceed the terms of s_n; we could argue that

$$s_n = 1 + \frac{1}{1.2} + \frac{1}{2.3} + \frac{1}{3.4} + \cdots + \frac{1}{(n-1).n}$$
$$< 1 + (\tfrac{1}{2}) + (\tfrac{1}{2})^2 + (\tfrac{1}{2})^2 + \cdots + (\tfrac{1}{2})^2$$
$$= 1 + \tfrac{1}{2} + \frac{n-2}{2^2},$$

but this number increases indefinitely as n increases and therefore we cannot by these means find a number K such that, for all values of n, $s_n < K$.

However, it happens exceptionally in this case that we can actually find an algebraic expression for s_n, thus:

$$s_n = 1 + \frac{1}{1.2} + \frac{1}{2.3} + \cdots + \frac{1}{(n-1).n}$$
$$= 1 + \left(\frac{1}{1} - \frac{1}{2}\right) + \left(\frac{1}{2} - \frac{1}{3}\right) + \cdots + \left(\frac{1}{n-1} - \frac{1}{n}\right) = 2 - \frac{1}{n}.$$

Hence, for all values of n, $s_n < 2$, and therefore the series is convergent.

(v) The series $\qquad 1 + \dfrac{1}{2^2} + \dfrac{1}{3^2} + \dfrac{1}{4^2} + \ldots$

has all its terms positive.

$$s_n = 1 + \frac{1}{2^2} + \ldots + \frac{1}{n^2}.$$

Attempts either to find an algebraic expression for the sum of the n terms of s_n, or to find a geometrical progression with a common ratio less than 1 whose terms exceed those of s_n, will, in this case, fail; but we notice that the terms of this series, after the first, are all less than the corresponding terms of the series (iv) just considered, i.e.

$$1/2^2 < 1/1.2, \ 1/3^2 < 1/2.3, \ldots 1/n^2 < 1/(n-1).n;$$

and therefore

$$s_n = 1 + \frac{1}{2^2} + \frac{1}{3^2} + \ldots + \frac{1}{n^2}$$

$$< 1 + \frac{1}{1.2} + \frac{1}{2.3} + \ldots + \frac{1}{(n-1).n}$$

$$< 2,$$

by the preceding work; and therefore the series is convergent.

(vi) The *harmonic series* $\quad 1 + \dfrac{1}{2} + \dfrac{1}{3} + \dfrac{1}{4} + \ldots$

has all its terms positive and

$$s_n = 1 + \frac{1}{2} + \frac{1}{3} + \ldots + \frac{1}{n}.$$

As with the series (v), the methods which have succeeded with series (i)—(iv) will not apply. Also, here we cannot argue that the terms of this series are less than the corresponding terms of one of the series dealt with and so prove the convergence as with the series (v). The fact is that this series is not convergent, for it is possible, by taking n sufficiently great, to make s_n (i.e. the sum of n terms of the series) exceed any number whatever, however great.

In fact: the sum of the first two terms $= 1 + \dfrac{1}{2} = 1\frac{1}{2}$, the sum of the first 2^2 (i.e. 4) terms $= 1 + \dfrac{1}{2} + \dfrac{1}{3} + \dfrac{1}{4}$

$$> 1 + \dfrac{1}{2} + \left(\dfrac{1}{4} + \dfrac{1}{4}\right)$$

$$= 1 + \dfrac{1}{2} + \dfrac{2}{4} = 2,$$

the sum of the first 2^3 (i.e. 8) terms

$$= 1 + \dfrac{1}{2} + \dfrac{1}{3} + \dfrac{1}{4} + \dfrac{1}{5} + \dfrac{1}{6} + \dfrac{1}{7} + \dfrac{1}{8}$$

$$> 1 + \dfrac{1}{2} + \left(\dfrac{1}{4} + \dfrac{1}{4}\right) + \left(\dfrac{1}{8} + \dfrac{1}{8} + \dfrac{1}{8} + \dfrac{1}{8}\right)$$

$$= 1 + \dfrac{1}{2} + \dfrac{2}{4} + \dfrac{4}{8} = 1 + \tfrac{1}{2} + \tfrac{1}{2} + \tfrac{1}{2} = 2\tfrac{1}{2},$$

$$\dotsb,$$

the sum of the first 2^m terms

$$= 1 + \dfrac{1}{2} + \left(\dfrac{1}{3} + \dfrac{1}{4}\right) + \dots + \left(\dfrac{1}{2^{m-1}-1} + \dots + \dfrac{1}{2^m}\right)$$

$$> 1 + \dfrac{1}{2} + \dfrac{2}{4} + \dots + \dfrac{2^{m-1}}{2^m}$$

$$= 1 + \tfrac{1}{2} + \tfrac{1}{2} + \dots + \tfrac{1}{2},$$

there being m terms $\tfrac{1}{2}$,

$$= 1 + \dfrac{m}{2}.$$

Thus, by adding the first 2^m terms of the series, we get a sum greater than $1 + \dfrac{m}{2}$;—whatever integer m may be, however great.

Hence we shall get a sum exceeding any number, K, if we add the first 2^{2M} terms of the series, where M is any integer not less than the number K, for the sum of 2^{2M} terms

$$> 1 + \dfrac{2M}{2} > M;$$

e.g. $s_n > 5$ if $n \geqslant 2^{10} = 1024$, i.e. the sum of the first 1024 terms of the series exceeds 5; or $s_n > 100$ if $n \geqslant 2^{200}$, or $s_n > 1000$ if $n \geqslant 2^{2000}$. The necessary and sufficient condition for convergence is not satisfied and therefore the series is not convergent.

51. Evaluation of the sum. The above examples will suffice to shew the nature of convergent (and non-convergent) series of positive terms, and to shew how, in the most usual cases, the

question as to whether or not such a series is convergent may be
settled. We will now consider the question of the actual evaluation
of the sums of such series as are convergent, partly for its own
sake, and partly for the light it throws on the question of convergence itself. In cases (e.g. series (i) and (v)) where a simple precise
formula for s_n can be obtained, the sum,—being the limit of the
sequence whose nth term is s_n,—can generally be obtained precisely.

Thus, in (i), if $x < 1$,

$$s_n = \frac{1}{1-x} - \frac{x^n}{1-x},$$

and, as we saw above ((iii), p. 50), $\frac{x^n}{1-x} \to 0$ as n increases indefinitely, and therefore $s_n \to \frac{1}{1-x}$, i.e. the sum of the infinite
series is precisely $1/(1-x)$.

Or, in (iv), $s_n = 2 - 1/n$, and we know that $1/n \to 0$, and therefore
$s_n \to 2$, and the sum of the series is 2.

In cases, however, where a formula for s_n cannot be found, the
precise evaluation of the limit of the sequence $s_1, s_2, \ldots$ will not, in
general, be possible, and we have therefore, in general, to be content
with approximations to the actual sum. If, as is most often the
case, the sum of the series is an irrational number, this approximate evaluation is necessarily the most that is possible if (as is
usual) we wish to express the result as a decimal. What is essential
in such approximate evaluation of the sum of a series (as with the
limit of a sequence) is that we should know with certainty the
degree of accuracy of our approximation, and that we should be
able to obtain approximations of any desired degree of accuracy.

52. Estimate of error. Let us consider series (ii) above, viz.

$$1 + \frac{1}{2} + \frac{1}{2}\frac{1}{2^2} + \frac{1}{3}\frac{1}{2^3} + \ldots.$$

Call the sum of this infinite series s and let s_n denote, as before,
the sum of the first n terms.

We will call the difference between the sum of the series (s)
and the sum of the first n terms (s_n) (i.e. $s - s_n$) *the error after
n terms*[*] of the series, and we will denote it by E_n; so that

$$s = s_n + E_n.$$

[*] Often called the *remainder after n terms*.

We will try to obtain an estimate to the value of E_n for any value of n.

We have evidently that E_n is the sum of the infinite series beginning with the $(n+1)$th term of the original series, viz.

$$\frac{1}{n}\frac{1}{2^n} + \frac{1}{n+1}\frac{1}{2^{n+1}} + \cdots,$$

which is clearly the unique limit and upper bound of the sequence whose $(m+1)$th term is

$$\frac{1}{n}\frac{1}{2^n} + \frac{1}{n+1}\frac{1}{2^{n+1}} + \cdots + \frac{1}{n+m}\frac{1}{2^{n+m}},$$

m being any positive integer.

The sum of these $m+1$ terms is clearly less than

$$(\tfrac{1}{2})^n + (\tfrac{1}{2})^{n+1} + \cdots + (\tfrac{1}{2})^{n+m},$$

i.e. $(\tfrac{1}{2})^n \left[\dfrac{1-(\tfrac{1}{2})^{m+1}}{1-\tfrac{1}{2}} \right]$, i.e. $(\tfrac{1}{2})^{n-1} - (\tfrac{1}{2})^{n+m}$,

which is less than $(\tfrac{1}{2})^{n-1}$, no matter what integer m may be (i.e. no matter how great).

Hence $\qquad\qquad E_n < (\tfrac{1}{2})^{n-1}$.

If now we add up say the first five terms of the given series:

$$
\begin{array}{r}
1 \cdot 00 \\
\cdot 50 \\
\cdot 12 \\
\cdot 04 \\
\cdot 02 \\
\hline
1 \cdot 68
\end{array}
$$

we know that this number is less than the sum of the infinite series, but that the error, E_5, is less than $(\tfrac{1}{2})^{5-1}$, i.e. $\cdot 0625$, which $< \cdot 07$*. Therefore the true sum of the infinite series lies between $1 \cdot 68$ and $1 \cdot 75$, and the sum of the series correct to the first decimal place is certainly known to be $1 \cdot 7$, the five terms of the series considered sufficing for this degree of accuracy.

Moreover, if we desired to find the sum of the series to within $\cdot 0001$ say, it is easily seen, by using the estimate for the error $E_n < (\tfrac{1}{2})^{n-1}$, that the first 15 terms will suffice; for

$$(\tfrac{1}{2})^{n-1} < \cdot 0001 \text{ if } 2^{n-1} > 10,000,$$

* Strictly s_5 exceeds $1 \cdot 68$ by something less than $\cdot 003$, but the sum of this error and E_5 is still less than $\cdot 07$.

which is the case if $n-1 \geqslant 14$, i.e. if $n \geqslant 15$; and therefore E_{15}, which is less than $(\frac{1}{2})^{14}$, is less than ·0001, and the sum of the infinite series is given by the sum of the first fifteen terms with an error less than ·0001.

Evidently the sum of the series can be found correct to *any* desired degree of accuracy simply by ascertaining what value of n will suffice to make E_n less than the assigned degree of error and adding up the first n terms of the series.

The student may have noticed that in passing from the expression

$$\frac{1}{n}\frac{1}{2^n} + \frac{1}{n+1}\frac{1}{2^{n+1}} + \dots + \frac{1}{n+m}\frac{1}{2^{n+m}}$$

to the expression $\quad \dfrac{1}{2^n} + \dfrac{1}{2^{n+1}} + \dots + \dfrac{1}{2^{n+m}}$

we have made a very generous allowance;—not only is the first of these expressions less than the second, but it is very considerably less, being in fact, almost as evidently, less than one nth of it. We can indeed assert that

$$\frac{1}{n}\frac{1}{2^n} + \frac{1}{n+1}\frac{1}{2^{n+1}} + \dots + \frac{1}{n+m}\frac{1}{2^{n+m}}$$
$$< \frac{1}{n}[(\tfrac{1}{2})^n + (\tfrac{1}{2})^{n+1} + \dots + (\tfrac{1}{2})^{n+m}],$$

and from this we deduce that

$$E_n < \frac{1}{n}\frac{1}{2^{n-1}}.$$

This estimate for E_n is clearly less than the former estimate, $(\frac{1}{2})^{n-1}$; by using this estimate we can assure ourselves that the error after five terms, E_5, is not only less than $(\frac{1}{2})^4$ (i.e. ·0625), but is actually less than $\frac{1}{5}(\frac{1}{2})^4$ (i.e. ·0125); so that, by summing the first five terms, we know that the sum of the infinite series lies between 1·68 and 1·70.

Further, in order to obtain the sum of the infinite series to within ·0001, we need only have

$$\frac{1}{n}(\tfrac{1}{2})^{n-1} \leqslant \text{·0001, i.e. } n\,.\,2^{n-1} \geqslant 10,000,$$

which is so if $n \geqslant 11$; so that, by using this new, closer, estimate for the error, we know that 11 terms will suffice, instead of the 15 terms which appeared to be necessary when we used the less accurate estimate for the error.

We could go further and get even closer estimates to the error, but in this case, to obtain any considerably closer estimate would entail an amount of labour not commensurate with the advantages to be gained. We have always, in estimating the error after n terms of a series, to find an estimate as close as possible without involving an inordinate amount of labour.

Consider now series (iii) above,

$$1 + \frac{1}{1!} + \frac{1}{2!} + \frac{1}{3!} + \frac{1}{4!} + \dots$$

We have
$$E_n = \frac{1}{n!} + \frac{1}{(n+1)!} + \dots$$
$$< (\tfrac{1}{2})^{n-1} + (\tfrac{1}{2})^n + (\tfrac{1}{2})^{n+1} + \dots$$
$$= (\tfrac{1}{2})^{n-1} \frac{1}{1 - \tfrac{1}{2}} = (\tfrac{1}{2})^{n-2} \quad \dots\dots\dots\dots\dots(a),$$

or, to obtain a closer estimate,

$$E_n = \frac{1}{n!} + \frac{1}{(n+1)!} + \dots$$
$$= \frac{1}{n!}\left[1 + \frac{1}{n+1} + \frac{1}{(n+1)(n+2)} + \dots\right]$$
$$< \frac{1}{n!}\left[1 + \frac{1}{n+1} + \frac{1}{(n+1)^2} + \dots\right]$$
$$= \frac{1}{n!} \frac{1}{1 - \dfrac{1}{n+1}} = \frac{n+1}{n \cdot n!} \quad \dots\dots\dots\dots\dots(b).$$

From the first estimate, (a), we see that, to calculate the sum of the series (iii) to within ·0001 say, 16 terms will suffice, for $(\tfrac{1}{2})^{n-2} \leqslant ·0001$ if $2^{n-2} \geqslant 10{,}000$, which is so if $n - 2 \geqslant 14$.

From the second estimate, (b), we see that, for the same degree of accuracy, 8 terms will in fact suffice, for

$$\frac{n+1}{n \cdot n!} \leqslant ·0001 \quad \text{if} \quad n! \frac{n}{n+1} \geqslant 10{,}000,$$

which is so if $n \geqslant 8$.

Actual calculation gives

$$s_8 = 1\cdot000000$$
$$+ 1\cdot000000$$
$$+ \cdot500000$$
$$+ \cdot166667$$
$$+ \cdot041667$$
$$+ \cdot008333$$
$$+ \cdot001389$$
$$+ \cdot000198$$
$$= 2\cdot718254$$

which therefore gives the sum of the series (iii) with an error less than ·0001.

The sum of this series is e. (See pp. 51—56 above.)

For the series (v) above,

$$1 + \frac{1}{2^2} + \frac{1}{3^2} + \dots,$$

we have

$$E_n = \frac{1}{(n+1)^2} + \frac{1}{(n+2)^2} + \dots$$
$$< \frac{1}{n(n+1)} + \frac{1}{(n+1)(n+2)} + \dots$$
$$= \left(\frac{1}{n} - \frac{1}{n+1}\right) + \left(\frac{1}{n+1} - \frac{1}{n+2}\right) + \dots$$
$$= \frac{1}{n};$$

whence, in order to obtain the sum of the series to within ·0001, 10,000 terms will suffice.

This series is, as a matter of fact, only very slowly convergent, and it is easily seen that no substantially closer approximation to the error can be found, for

$$E_n = \frac{1}{(n+1)^2} + \frac{1}{(n+2)^2} + \dots$$
$$> \frac{1}{(n+1)(n+2)} + \frac{1}{(n+2)(n+3)} + \dots = \frac{1}{n+1},$$

so that the sum of the first 9,999 terms will differ from the sum of the series by *more* than ·0001.

We can argue, if we wish, that, since $\frac{1}{n+1} < E_n < \frac{1}{n}$, the sum of the series

lies between $s_n + \frac{1}{n+1}$ and $s_n + \frac{1}{n}$, so that the sum, though differing from s_n by

more than $\frac{1}{n+1}$, is yet known from s_n to within $\frac{1}{n} - \frac{1}{n+1}$, i.e. $\frac{1}{n(n+1)}$, which

is less than ·0001 if $n \geqslant 100$; so that the sum of the series may be found to within ·0001 by adding only the first 100 terms.

53. Diminishing series of alternating signs. Consider the series

$$u_1 - u_2 + u_3 - \ldots \quad \ldots\ldots\ldots\ldots\ldots\ldots\ldots(5),$$

in which the terms are any real numbers of alternating signs, subject to the condition that the moduli of the terms form a monotonely decreasing, or non-increasing, sequence, or, more precisely, that

$$u_1 \geqslant u_2 \geqslant u_3 \geqslant \ldots \geqslant 0 \quad \ldots\ldots\ldots\ldots\ldots\ldots(6).$$

The sequence of partial sums

$$s_1, s_2, s_3, \ldots \quad \ldots\ldots\ldots\ldots\ldots\ldots\ldots(7),$$

where $\qquad s_n = u_1 - u_2 + u_3 - \ldots \pm u_n,$

is not steadily increasing, but is alternately increasing and decreasing.

By resort to the necessary and sufficient condition of convergence of general sequences (p. 48 above) we can see at once that our *series* (5) *will converge if* (*and only if*) *the sequence* $u_1, u_2, u_3, \ldots$ *has the unique limit zero.*

For the sequence (7) is convergent if $|s_{n+p} - s_n| <$ any arbitrary positive number ϵ, if n is sufficiently great, for all positive integral values of p, i.e. if

$$|u_{n+1} - u_{n+2} + u_{n+3} - \ldots \pm u_{n+p}| < \epsilon,$$

which is so if $\qquad |u_{n+1}| < \epsilon$

[because, if p is odd,

$$u_{n+1} - u_{n+2} + \ldots \pm u_{n+p} = u_{n+1} - (u_{n+2} - u_{n+3}) - \ldots - (u_{n+p-1} - u_{n+p})$$

and, if p is even,

$$u_{n+1} - u_{n+2} + \ldots \pm u_{n+p} = u_{n+1} - (u_{n+2} - u_{n+3}) - \ldots - (u_{n+p-2} - u_{n+p-1}) - u_{n+p},$$

and therefore, all the bracketed and unbracketed numbers here in either case being positive or zero,

$$|u_{n+1} - u_{n+2} + \ldots \pm u_{n+p}| \leqslant |u_{n+1}|];$$

i.e. the sequence (7) is convergent if $u_{n+1} \to 0$ as n increases indefinitely; i.e. if $u_n \to 0$ as n increases indefinitely.

Or we may argue directly thus *:

Let us divide the sequence (7) into two sequences:

the sequence of *odd* partial sums $s_1, s_3, s_5, \ldots \ldots$ (7 a)

and   the sequence of *even* partial sums $s_2, s_4, s_6, \ldots \ldots$ (7 b).

The sequence (7 a) is steadily decreasing and bounded below, because the typical term

$$s_{2n+1} = u_1 - u_2 + u_3 - \ldots + u_{2n+1}$$
$$= u_1 - u_2 + u_3 - \ldots + u_{2n-1} - (u_{2n} - u_{2n+1}) \leqslant s_{2n-1}$$

because $u_{2n} - u_{2n+1} \geqslant 0,$

and $s_{2n+1} = (u_1 - u_2) + (u_3 - u_4) + \ldots + (u_{2n-1} - u_{2n}) + u_{2n+1} \geqslant 0.$

Therefore (p. 43) the sequence (7 a) has a lower bound and unique limit, S say; or $s_1, s_3, s_5, \ldots \searrow S$.

Similarly the sequence (7 b) has an upper bound and unique limit, S' say; or $s_2, s_4, s_6, \ldots \nearrow S'$.

Furthermore, every number s_{2n+1} of the sequence (7 a) $\geqslant$ every number s_{2m} of the sequence (7 b), for

$$s_{2n+1} - s_{2m} = (u_{2m+1} - u_{2m+2}) + \ldots + (u_{2n-1} - u_{2n}) + u_{2n+1} \text{ if } n \geqslant m,$$

or $= (u_{2n+2} - u_{2n+3}) + \ldots + (u_{2m-2} - u_{2m-1}) + u_{2m} \text{ if } n < m.$

Therefore the lower bound S of the first sequence (7 a), which exceeds (or equals) every number less than all the numbers $s_1, s_3, \ldots$ of the sequence (7 a), must exceed (or equal) every number of the second sequence (7 b); and therefore S exceeds, or is equal to, the upper bound S' of this sequence. Thus $S \geqslant S'$.

But, on the other hand, we have also $S \leqslant S'$. For $s_{2n+1} \geqslant S$ and $s_{2n} \leqslant S'$, and therefore

$$u_{2n+1} = s_{2n+1} - s_{2n} \geqslant S - S'$$

for all values of n; and this would be impossible if $S > S'$ because, by hypothesis, u_n (and therefore u_{2n+1}) tends to zero.

Hence the two limits S and S' are identical.

Finally this common unique limit (S or S') of the two sequences (7 a) and (7 b) must evidently be the unique limit of the original sequence (7); and the proof that the series (5) is convergent under the conditions stated is complete. That the series is convergent *only* if $u_n \to 0$ follows from the simple general necessary condition for convergence (p. 63 above).

* The student is advised to sketch rough graphs, as suggested in the footnote to p. 57 above, to represent the sequences (7), (7 a) and (7 b).

We may state this theorem shortly thus:

The series $\qquad u_1 - u_2 + u_3 - \ldots$

converges if $u_n \searrow 0$.

In dealing with series of this type it is again necessary to discuss the *error* after n terms. In this case the discussion is particularly simple.

We have in fact

$$E_n = \pm (u_{n+1} - u_{n+2} + u_{n+3} - \ldots),$$

which is clearly less in absolute value than the term u_{n+1}, and has the same sign as that term*.

Or we may argue that the sum of the series lies between the sum of any odd number of terms and of any even number, with the same conclusion, viz.:

The error after n terms of a convergent series of the type (5) of alternating signs is less than the next (the $(n+1)th$) term.

The remarks of pp. 71—72 above, concerning the scope for choice in the degree of closeness of the estimate for the error, are applicable here also. In view of the simplicity of this estimate, however, it is not usually advisable to attempt to obtain any closer estimate for series of this kind.

54. Examples of diminishing series of alternating signs. As an example, consider the geometrical progression

(*a*) $\qquad\qquad 1 + x + x^2 + \ldots$

in which the common ratio, x, is negative.

The terms alternate in sign. The sequence whose nth term is $|x|^{n-1} \searrow 0$ if and only if $|x| < 1$.

Therefore the condition for convergence is fulfilled if x is negative and greater than -1.

(*b*) The series $\qquad 1 - \dfrac{1}{3!} + \dfrac{1}{5!} - \ldots$

has its terms alternating in sign and steadily decreasing to the limit zero. It is therefore convergent and the error after n terms is less than the $(n+1)$th term, i.e. $E_n < \dfrac{1}{(2n+1)!}$, in absolute value.

---

* Excluding the case when the terms from the $(n+1)$th all $= 0$.

To calculate the sum of this series correct to within ·0001 it will suffice to take the first four terms, for

$$E_n < ·0001 \quad \text{if} \quad (2n+1)! \geqslant 10{,}000,$$

i.e. if $n \geqslant 4$.

We have
$$
\begin{aligned}
s_4 &= 1·00000 - ·16666 \\
&+ ·00833 - ·00020 \\
&= 1·00833 - ·16686 \\
&= ·84147;
\end{aligned}
$$

giving the sum of the series (b) with an error of less than ·0001.

(c) The series
$$1 - \frac{1}{2} + \frac{1}{3} - \frac{1}{4} + \dots$$

has its terms alternately positive and negative and steadily decreasing to zero in absolute value, and is therefore convergent. $E_n < \dfrac{1}{n+1}$ in absolute value; so that, to obtain the sum correct to within ·0001, 9,999 terms will suffice.

This series is only very slowly convergent; it is easily proved, in fact, that $E_n > \frac{1}{2} \cdot \dfrac{1}{n+1}$ in absolute value, so that, in order to obtain the sum to within ·0001, *at least* 4,999 terms will be needed.

55. Absolutely convergent series. Series which do not belong to either of the above two types may give more difficulty. There is, however, one important type of series which is in fact substantially as simple as that of series with only positive terms.

If a series $\qquad u_1 + u_2 + u_3 + \dots$ ....(8)
is convergent, and the series would remain convergent if all the negative terms it may have were replaced by their moduli, then the series (8) *is said to be absolutely convergent.*

It is seen that not all convergent series are absolutely convergent. For example, the series (c) above is convergent, whilst if its negative terms are replaced by their positive absolute values the series becomes

$$1 + \frac{1}{2} + \frac{1}{3} + \frac{1}{4} + \dots,$$

which is the harmonic series (v), p. 67 above, and is not convergent.

A theorem of fundamental importance in connection with absolutely convergent series is:

If a series of positive terms $u_1 + u_2 + u_3 + \ldots$ is convergent, then, if the signs of any of the terms are altered, the series is still convergent.

This theorem can be proved easily by an appeal to the necessary and sufficient condition of convergence (p. 48) or thus:

For convenience, let us suppose that the modified series is

$$u_1 - u_2 - u_3 + u_4 - u_5 + u_6 + \ldots.$$

The positive terms of this series, viz.

$$u_1 + u_4 + u_6 + \ldots,$$

if not finite in number, will form a convergent series of positive terms, because the sum of any number of them is clearly bounded, being less than the sum of the original series. Call the sum of this series (or of the finite number of terms) P. Similarly the negative terms $-u_2 - u_3 - u_5 - \ldots$ form a convergent series (or a finite number of terms) having a sum $-N$ say. If, now, enough terms of the series are taken, i.e. for a certain value of n and all larger values, if s_n' is the sum of the first n terms of the modified series,

$$s_n' > (P - \epsilon) - N \text{ and } s_n' < P - (N - \epsilon)$$

for all positive values (however small) of the arbitrary number ϵ; i.e. s_n' lies between $(P - N) - \epsilon$ and $(P - N) + \epsilon$.

It follows that $s_n' \to P - N$ and the modified series is convergent, having $P - N$ for its sum.

The argument is plainly general.

From this theorem we are able to reduce the consideration of absolute convergence of series to the consideration of series of positive terms. We have clearly that *if the series of moduli* $|u_1| + |u_2| + |u_3| + \ldots$ *is convergent then the series* $u_1 + u_2 + u_3 + \ldots$ *is also convergent (and absolutely convergent)*. Thus we could argue that the series (*a*), p. 76, is convergent (and absolutely convergent) if x is negative and greater than -1, because, when the terms of the series are replaced by their moduli, the series is the series of positive terms (i), p. 64, which is known to be convergent; or again, the series (*b*), p. 76, is convergent (and absolutely convergent) because the series of positive terms formed by replacing all its terms by their moduli, viz. $1 + \dfrac{1}{3!} + \dfrac{1}{5!} + \ldots$, is convergent (because the sum of any number of terms of this series is clearly less than the sum of the same number of terms of the series (iii), p. 65, and is therefore known to be bounded).

56. Comparison theorems and tests for convergence. Series of positive terms are therefore seen to be of added importance. In view of this we add here two theorems by which the convergence (or otherwise) of a series of positive terms (or, therefore, the absolute convergence or otherwise of any series) may be established in some cases of frequent occurrence.

I. *If $c_1 + c_2 + c_3 + \ldots$ is a convergent series of positive terms and if every term, e.g. u_n, of another series of positive terms $u_1 + u_2 + u_3 + \ldots$ is less than the corresponding term, c_n, of the first series, then the second series, $u_1 + u_2 + u_3 + \ldots$, is also convergent.*

II. *If $c_1 + c_2 + c_3 + \ldots$ is a convergent series of positive terms, and if, for all values of n, the ratio $\dfrac{u_{n+1}}{u_n}$ of two consecutive terms of another series of positive terms, $u_1 + u_2 + u_3 + \ldots$, is less than the corresponding ratio $\dfrac{c_{n+1}}{c_n}$ for the first series, then the second series, $u_1 + u_2 + u_3 + \ldots$, is also convergent.*

The proof of I is almost intuitive and is left to the student. For II we see that the sum of the first n terms of the second series

$$= u_1 + u_2 + u_3 + \ldots + u_n$$

$$= u_1 \left(1 + \frac{u_2}{u_1} + \frac{u_3}{u_2}\frac{u_2}{u_1} + \frac{u_4}{u_3}\frac{u_3}{u_2}\frac{u_2}{u_1} + \ldots + \frac{u_n}{u_{n-1}}\frac{u_{n-1}}{u_{n-2}}\ldots\frac{u_3}{u_2}\frac{u_2}{u_1} \right)$$

$$< \frac{u_1 c_1}{c_1} \left[1 + \frac{c_2}{c_1} + \frac{c_3}{c_2}\frac{c_2}{c_1} + \frac{c_4}{c_3}\frac{c_3}{c_2}\frac{c_2}{c_1} + \ldots + \frac{c_n}{c_{n-1}}\frac{c_{n-1}}{c_{n-2}}\ldots\frac{c_3}{c_2}\frac{c_2}{c_1} \right]$$

$$= \frac{u_1}{c_1} (c_1 + c_2 + c_3 + \ldots + c_n),$$

which is less than a fixed number, K, however great n may be; whence the theorem follows from the necessary and sufficient condition for the convergence of series of positive terms (p. 64).

Knowing that a geometrical progression of common ratio k is convergent if $0 < k < 1$, we have, as direct corollaries of these two theorems, the following practical tests for convergence of series of positive terms or for absolute convergence:

Cauchy's test: *The series $u_1 + u_2 + u_3 + \ldots$ is absolutely convergent if, for all values of n, $\sqrt[n]{|u_n|} \leqslant$ some fixed number k, which is less than 1.*

d'Alembert's (ratio) test: *The series $u_1 + u_2 + u_3 + \ldots$ is absolutely convergent if, for all values of n, the ratio $|u_{n+1}|/|u_n| \leqslant$ some fixed number k, which is less than* 1.

For the series may be compared with the convergent series $1 + k + k^2 + \ldots$.

In using these tests a possible value of the number k should always be determined; it is *not true* that if the ratio $|u_{n+1}|/|u_n| < 1$ the series is absolutely convergent. (See Ex. 11 below.)

57. Operations with series. It is sometimes desired to effect arithmetical operations on series; e.g. to add or multiply two series together. The following two theorems are fundamental:

A. *If the series $u_1 + u_2 + u_3 + \ldots$ is convergent and has the sum U, and the series $v_1 + v_2 + v_3 + \ldots$ is convergent and has the sum V, then the series $(u_1 + v_1) + (u_2 + v_2) + (u_3 + v_3) + \ldots$ is also convergent and has the sum $U + V$.*

B. *If the series $u_1 + u_2 + u_3 + \ldots$ is absolutely convergent and has the sum U, and the series $v_1 + v_2 + v_3 + \ldots$ is absolutely convergent and has the sum V, then the series*

$$u_1 v_1 + (u_1 v_2 + u_2 v_1) + (u_1 v_3 + u_2 v_2 + u_3 v_1) + \ldots$$

is absolutely convergent and has the sum UV.

Theorem A is simple and the proof is left to the student.

In theorem B the "product series"

$$u_1 v_1 + (u_1 v_2 + u_2 v_1) + (u_1 v_3 + u_2 v_2 + u_3 v_1) + \ldots$$

is the series, as we should normally write it, if we wrote down the two series $u_1 + u_2 + u_3 + \ldots$ and $v_1 + v_2 + v_3 + \ldots$ and proceeded to multiply them systematically, as a "long multiplication" sum. The nth term of this series is therefore

$$u_1 v_n + u_2 v_{n-1} + \ldots + u_{n-1} v_2 + u_n v_1.$$

To prove theorem B, we suppose first that all the terms of the series are positive.

We then have that the sum of the first n terms of the product series

$$= u_1 v_1 + (u_1 v_2 + u_2 v_1) + \ldots + (u_1 v_n + \ldots + u_n v_1)$$
$$< (u_1 + u_2 + \ldots + u_n)(v_1 + v_2 + \ldots + v_n)$$
$$< UV,$$

whence it follows that the product series is convergent and its sum $\leqslant UV$.

But the sum of n terms of the product series

$$= u_1 v_1 + (u_1 v_2 + u_2 v_1) + \ldots + (u_1 v_n + \ldots + u_n v_1)$$
$$> (u_1 + \ldots + u_{\frac{n}{2}})(v_1 + \ldots + v_{\frac{n}{2}})$$

if n is even, or

$$\geqslant (u_1 + \ldots + u_{\frac{n+1}{2}})(v_1 + \ldots + v_{\frac{n+1}{2}})$$

if n is odd, and, by taking n sufficiently great,

$$u_1 + \ldots + u_{\frac{n}{2}} \text{ (or } u_1 + \ldots + u_{\frac{n+1}{2}})$$

and
$$v_1 + \ldots + v_{\frac{n}{2}} \text{ (or } v_1 + \ldots + v_{\frac{n+1}{2}})$$

differ from U and V respectively by arbitrarily little, and therefore the product differs from UV by arbitrarily little; whence it follows that the sum of the product series $\geqslant UV$. Therefore the sum of the product series $= UV$. Q.E.D.

If now the terms are not all positive, let us denote their positive absolute values by $u_1', u_2', \ldots; v_1', v_2', \ldots$.

The difference between n terms of the product series, viz.

$$u_1 v_1 + (u_1 v_2 + u_2 v_1) + \ldots + (u_1 v_n + \ldots + u_n v_1),$$

and the product $\quad (u_1 + u_2 + \ldots + u_n)(v_1 + v_2 + \ldots + v_n)$

consists of a certain set of terms, which, if replaced by their positive absolute values, would be precisely the difference between

$$u_1' v_1' + (u_1' v_2' + u_2' v_1') + \ldots + (u_1' v_n' + \ldots + u_n' v_1')$$

and
$$(u_1' + u_2' + \ldots + u_n')(v_1' + v_2' + \ldots + v_n'),$$

which is known to tend to zero as n increases indefinitely, by the part of the theorem just proved.

Therefore the difference between

$$u_1 v_1 + (u_1 v_2 + u_2 v_1) + \ldots + (u_1 v_n + \ldots + u_n v_1)$$

and the product $\quad (u_1 + u_2 + \ldots + u_n)(v_1 + v_2 + \ldots + v_n)$

must, *a fortiori*, tend to zero; whence it follows at once that the product series is convergent and has the sum UV. Q.E.D.

We do not here lay stress on these theorems (nor on the comparison theorems and tests for convergence proved above) but they are nevertheless of great importance and utility. We have only touched the fringe of the subject of convergence, but the results obtained and the methods discussed will suffice for our purposes*.

* For further information on convergence of series and sequences the student is referred to Hardy's *Pure Mathematics* and to Bromwich's *Infinite Series*.

58. We have now laid the foundations of analysis. That is to say, we have placed on a sound basis the notion of *real number*. In §§ 1—3 we observed the nature and fundamental properties of the different types of real number, and in §§ 6 and 7 we have considered a systematic method of effectively representing all real numbers (rational and irrational). We shall find later that this notion of real number, and the considerations involved in its discussion, lead to other notions of far-reaching importance—such as that of a function of a real variable, continuity, differentiation, etc.—and that many other similar notions, acquired from other sources, are satisfactorily explicable by means of our knowledge of the real number. In the next chapter we shall consider the somewhat special problem of logarithms—in particular the problem of their evaluation—in the course of which we shall be led to consider certain aspects of the notion of function and the behaviour of functions, which it will be our concern in the third and fourth chapters to consider more fully.

EXAMPLES VII.

1. Discuss the convergence of the following series, giving in each case an estimate to the error after n terms and finding the number of terms sufficient to obtain the sum of the series to within ·0001 ; calculate to this degree of accuracy the sums of the series (a), (c), (e), (f), (i), and (j):

$$(a) \quad 1 + \frac{1}{2}\frac{1}{2} + \frac{1}{3}\frac{1}{2^2} + \frac{1}{4}\frac{1}{2^3} + \dots$$

$$(b) \quad 1 + \frac{10}{1} + \frac{10^2}{2!} + \frac{10^3}{3!} + \dots$$

$$(c) \quad \frac{1}{10} + \frac{1}{2^2}\frac{1}{10^2} + \frac{1}{3^2}\frac{1}{10^3} + \frac{1}{4^2}\frac{1}{10^4} + \dots$$

$$(d) \quad 1 + \frac{1}{2^3} + \frac{1}{3^3} + \frac{1}{4^3} + \dots$$

$$(e) \quad 1 - \frac{2}{2} + \frac{3}{2^2} - \frac{4}{2^3} + \dots$$

$$(f) \quad \frac{1}{10} - \frac{1}{2^2}\frac{1}{10^2} + \frac{1}{3^2}\frac{1}{10^3} - \frac{1}{4^2}\frac{1}{10^4} + \dots$$

$$(g) \quad 1 - \frac{1}{2^2} + \frac{1}{3^2} - \frac{1}{4^2} + \dots$$

$$(h) \quad 1 - \frac{10}{1} + \frac{10^2}{2!} - \frac{10^3}{3!} + \dots$$

$$(i) \quad \frac{1}{2} - \frac{1}{2}\frac{1}{2^2} + \frac{1}{3}\frac{1}{2^3} - \frac{1}{4}\frac{1}{2^4} + \dots$$

$$(j) \quad 1 - \frac{1}{2!} + \frac{1}{4!} - \frac{1}{6!} + \dots$$

$$(k) \quad 1 - \frac{1}{\sqrt{2}} + \frac{1}{\sqrt{3}} - \frac{1}{\sqrt{4}} + \dots.$$

2. Shew that the series

$$(a) \quad 1 + 2 + 2^2 + 2^3 + \dots$$

$$(b) \quad 1 + \frac{2}{2} + \frac{2^2}{3} + \frac{2^3}{4} + \dots$$

$$(c) \quad 1 + \frac{1.3}{1.2} + \frac{1.3.5}{1.2.3} + \dots$$

$$(d) \quad 1 + \frac{1}{\sqrt{2}} + \frac{1}{\sqrt{3}} + \frac{1}{\sqrt{4}} + \dots$$

$$(e) \quad 1 - 1 + 1 - 1 + \dots$$

$$(f) \quad 1 - 2 + 3 - 4 + \dots$$

are not convergent.

3. Shew that the series

$$1 + 1 - \frac{1}{2!} - \frac{1}{3!} + \frac{1}{4!} + \frac{1}{5!} - \frac{1}{6!} - \frac{1}{7!} + \dots$$

(in which two positive terms are followed by two negative terms) is convergent, and find its sum correct to two decimal places.

4. Discuss the convergence (or otherwise) of the series:

$$(a) \quad \frac{\sin 1°}{1} + \frac{\sin 2°}{2^2} + \frac{\sin 3°}{3^2} + \dots,$$

$$(b) \quad \sin 1° + \sin 2° + \sin 3° + \dots,$$

$$(c) \quad \frac{\sin 1°}{1} + \frac{\sin 2°}{2} + \frac{\sin 3°}{3} + \dots.$$

[By Abel's equality lemma (p. 26, Ex. 3)

$$\frac{\sin 1°}{1} + \frac{\sin 2°}{2} + \dots + \frac{\sin n°}{n}$$

$$= s_1\left(\frac{1}{1} - \frac{1}{2}\right) + s_2\left(\frac{1}{2} - \frac{1}{3}\right) + \dots + s_{n-1}\left(\frac{1}{n-1} - \frac{1}{n}\right) + s_n\frac{1}{n},$$

where $s_1, s_2, \dots, s_n$ represent $\sin 1°$, $\sin 1° + \sin 2°$, ..., $\sin 1° + \dots + \sin n°$. All the sums $s_1, s_2, \dots s_n$ are positive and less than some number (< 180). Therefore $s_n \frac{1}{n} \to 0$, and the remaining terms

$$s_1\left(\frac{1}{1} - \frac{1}{2}\right) + \dots + s_{n-1}\left(\frac{1}{n-1} - \frac{1}{n}\right)$$

are the first $n-1$ terms of a positive convergent series. The series (c) is therefore convergent.]

5. Shew that the series $1 + x + x^2/2 + x^3/3 + \dots$ is convergent if $-1 \leqslant x < 1$.

6. Shew that the series

$$(a) \quad 1 + x + x^2/2! + x^3/3! + \dots$$

$$(b) \quad 1 - x^2/2! + x^4/4! - \dots$$

$$(c) \quad x - x^3/3! + x^5/5! - \dots$$

are convergent for all real values of x.

7. Shew that the series

$$1 + nx + \frac{n(n-1)}{2!}x + \frac{n(n-1)(n-2)}{3!}x^2 + \ldots,$$

where n is any real number other than 0 or a positive integer, is convergent for all values of x between -1 and 1.

Under what circumstances is the series convergent also for $x=1$ and for $x=-1$?

8. Shew that if $u_1 + u_2 + u_3 + \ldots$ is a series of positive terms, and if, for all values of n sufficiently great, the ratio $\frac{u_{n+1}}{u_n} \geq 1$, then the series is not convergent.

9. If $u_1 + u_2 + u_3 + \ldots$ is a series of positive terms, shew that if $\sqrt[n]{u_n} \geq 1$ for all values of n sufficiently great, then the series is not convergent.

10. Use d'Alembert's ratio test to establish the absolute convergence, for all values of the real number x, of the series of Ex. 6.

11. Shew that no number k, as required in the tests of pp. 79—80, can be found for the series

$$(a) \quad \frac{1}{2} + \frac{1}{3} + \frac{1}{4} + \ldots$$

and

$$(b) \quad \frac{1}{2^2} + \frac{1}{3^2} + \frac{1}{4^2} + \ldots,$$

though in each case $\sqrt[n]{u_n} < 1$ and $u_{n+1}/u_n < 1$ for all values of n.

[The tests will not apply. We know otherwise (p. 67) that (a) is not convergent but that (b) is convergent.]

12. If a series is absolutely convergent the terms may be deranged (without omissions) in any way without affecting the convergence or the sum. This is not true of a series which, though convergent, is not absolutely convergent.

[Non-absolutely convergent series are said to be *conditionally convergent*.]

13. Shew that the conditionally convergent series

$$(a) \quad 1 - \frac{1}{2} + \frac{1}{3} - \frac{1}{4} + \ldots,$$

when deranged into the series

$$(b) \quad 1 + \frac{1}{3} - \frac{1}{2} + \frac{1}{5} + \frac{1}{7} - \frac{1}{4} + \frac{1}{9} + \frac{1}{11} - \frac{1}{6} + \ldots$$

(whose law of formation is evident), remains convergent, but that the sum of series (b) is greater than 1 whilst that of series (a) is less than 1.

14. Shew that the absolutely convergent series

$$(a) \quad 1 - \frac{1}{2^2} + \frac{1}{3^2} - \frac{1}{4^2} + \ldots,$$

when deranged into the series

$$(b) \quad 1 + \frac{1}{3^2} - \frac{1}{2^2} + \frac{1}{5^2} + \frac{1}{7^2} - \frac{1}{4^2} + \ldots,$$

remains absolutely convergent, and that the sums of the two series are identical.

CHAPTER II

LOGARITHMS

§ 1. INDICES

59. Exponentiation with positive integral indices. Laws of indices. In the preceding chapter we have considered the four cardinal operations of arithmetic—addition, subtraction, multiplication and division—applied to real numbers. A fifth operation—that of *exponentiation*—will now be considered.

Just as the repetition of the operation of addition leads to multiplication (by a whole number) so the operation of exponentiation is arrived at by the repetition of multiplication. In symbols, a being any real number and m any positive integer, we write a^m to mean the continued product of m a's, and we call a^m the mth *power* of a, a being the *base*, and m the *index*. It is essential here (as it was in the first instance in the case of multiplication) that the index m should be a positive integer.

The laws governing calculations with powers are easily obtained. The addition and subtraction of two powers are operations which cannot in general be simplified to any extent, but multiplication and division lead to the laws expressed in symbols as

(I) $a^m \times a^n = a^{m+n}$, and

(II) $a^m \div a^n = a^{m-n}$ if $m > n$;

m and n being positive integers and a any real number whatever. To prove (II) for example we have, direct from the definition of the powers a^m, a^n, and a^{m-n}, that

$$a^m \div a^n = (a \times a \times a \times \ldots) \div (a \times a \times a \times \ldots),$$

there being m a's in the first bracket and n a's in the second,

$$= a \times a \times a \times \ldots,$$

there being $(m - n)$ a's,

$$= a^{m-n},$$

and the law (II) is established.

A third law arises from the application of the operation of exponentiation to powers; in symbols it is

(III) $(a^m)^n = a^{mn},$

a being, as before, any real number, and m and n any positive integers.

The proof of this is also simple, thus

$$(a^m)^n = a^m \times a^m \times a^m \times \dots,$$

there being n terms a^m in this product, which, therefore, by law (I),

$$= a^{m+m+m+\dots},$$

there being n terms m in the sum $m + m + m + \dots$ figuring as index, whence

$$(a^m)^n = a^{mn},$$

and the theorem is proved.

The law corresponding to the distributive law, viz.

$$(abc)^m = a^m b^m c^m,$$

a, b and c being any real numbers, is also easily established.

Observing that the operation of extracting the root is inverse to exponentiation, if we assume that roots necessarily exist and are unique *, we can deduce certain laws for this operation, such as, for example, that

$$\sqrt[m]{a^{mn}} = a^n, \text{ and } \sqrt[m]{(\sqrt[n]{a})} = \sqrt[n]{(\sqrt[m]{a})} = \sqrt[mn]{a};$$

but it is essential here to introduce restrictions on the numbers a, m, n; for if m or n were 2, for example, and a negative, $\sqrt[m]{a}$ or $\sqrt[n]{a}$ would be meaningless as real numbers.

60. Negative integral and zero indices. Our laws of indices deal with the ordinary simple operations, but slight consideration shews that all such operations are not covered by the laws. Our law (II) for division is valid only if $m > n$, or, in other words, such an operation as $a^2 \div a^3$ is not covered by the law. In the spirit of generalisation, however, there are two points which suggest themselves: firstly, such an operation is not in the least meaningless, it being evident that $a^2 \div a^3$ (or in fact $a^m \div a^n$ if $n = m + 1$) is equal to $1/a$ (provided $a \neq 0$); and secondly, the law (II) ceases to be applicable in this case only because of the meaninglessness of such a symbol as a^{-1}.

These two remarks suggest that we may "generalise" the notion of exponentiation to include "powers" with negative or

---

* That this is the case if all the numbers (and roots) occurring are restricted to be positive will be proved below (pp. 88—89).

zero indices, by ascribing to the meaningless symbol a^{m-n} (where $m \leqslant n$) the quite definite meaning a^m/a^n. That is, we use the equality

$$a^{m-n} = \frac{a^m}{a^n},$$

which is a statement of *fact* in cases for which $m > n$, as the *definition* of the left-hand side of the equality in other cases. Any self-consistent definition of a hitherto meaningless symbol is logical, but it remains to prove (if possible) that our definition will preserve the same laws of indices as hold in the ordinary cases, for then our extension will prove useful, in that we shall be enabled to operate with both the old and new symbols (the positive and negative integral powers) with equal facility and to use, if need be, the new powers to help in discussions concerning the old powers.

Our definition may evidently be expressed in the simpler form:

$$\left.\begin{array}{l} a^{-m} = 1/a^m \text{ if } m \text{ is a positive integer,} \\ a^0 = 1, \end{array}\right\}$$

where a may be any real number except zero.

That law (I) still holds is seen as follows:

If m and n are positive integers,

$$a^m \times a^{-n} = a^m \times \frac{1}{a^n}$$

by our new definition;

i.e. $= a^{m-n}$ if $m > n$,

by law (II) for positive integral indices, or

$$= \frac{1}{a^{n-m}} \text{ if } m < n$$
$$= a^{m-n}$$

by our new definition.

If $m = n$, $a^m \times a^{-n} = 1$
$$= a^{m-n}$$

by our definition of a^0.

Also $a^{-m} \times a^{-n} = \frac{1}{a^m} \times \frac{1}{a^n}$

by our definition;

i.e. $= \frac{1}{a^{m+n}}$
$$= a^{-m-n}$$

by our definition.

That laws (II) and (III) also still hold can be proved similarly.

Our extension of the notion of exponentiation to negative integral and zero indices is thus a useful extension with no disadvantages.

61. Fractional indices. In the same way, assuming tentatively that roots can be defined uniquely*, the question as to the possibility of a useful extension of the notion of exponentiation to include fractional indices would lead to the definition

$$a^{p/q} = \sqrt[q]{a^p}, \quad a^{-p/q} = 1/(\sqrt[q]{a^p}),$$

p and q being positive integers; for we know that $a^{p/q}$, whenever p/q is a positive integer, does equal $\sqrt[q]{a^p}$, and by arguments similar to the above for negative indices it is seen that the same laws of indices continue to hold. Thus, to prove the law (I),

$$a^{p/q} \times a^{r/s} = \sqrt[q]{a^p} \times \sqrt[s]{a^r}$$

by definition,

$$= \sqrt[qs]{a^{ps}} \times \sqrt[qs]{a^{rq}}$$

from the laws for positive integral indices,

$$= \sqrt[qs]{(a^{ps} \times a^{rq})}$$

$$= a^{\frac{ps+rq}{qs}}$$

$$= a^{\frac{p}{q} + \frac{r}{s}}.$$

We argue similarly for negative fractional indices, and for the other two laws (II) and (III) in all cases. Our suggested definition therefore leads to the desired results whenever properly applicable.

62. Existence theorem for root extraction and fractional powers. The definition will be properly applicable only if the operations involved are possible—a somewhat drastic restriction if we were to confine ourselves to rational numbers, since (as we have shewn in the case of $\sqrt{2}$) the majority of roots of the type $\sqrt[q]{a}$ cannot exist as rational numbers. As we saw in the case of $\sqrt{2}$, however, it is easy to see that *any positive real number necessarily has one and only one positive real root of any order whatever*, i.e. whatever positive real number a, and whatever positive integer q may be, there exists one and only one positive real number b whose qth power is a, i.e. such that $b^q = a$, or $b = \sqrt[q]{a}$.

* The proof follows immediately below.

To prove this we observe that if all real numbers are classified into those positive numbers y whose qth powers exceed a and those numbers x which are negative or have their qth powers less than or equal to a, a real number, denoted by $(x \mid y)$, is defined in accordance with Dedekind's definition, because it is clear that such classification includes all real numbers and every x is less than every y (because if $x \geqslant y$ and both x and y are positive, $x^q \geqslant y^q$, and therefore if $x^q \leqslant a$ and $y^q > a$ we must have $x < y$). From the definition of multiplication of real numbers (p. 18) we see at once that the qth power of this number $(x \mid y)$, which is $(x^q \mid y^q)$, is the real number a; any positive number other than this number $(x \mid y)$ evidently has a qth power differing from a. Our statement—which we may call *the existence theorem for root extraction*—is therefore proved.

It follows from this existence theorem that if a is any positive real number, there exists one and only one positive qth root of a^p, i.e. one and only one positive p/qth power of a, viz. $a^{\frac{p}{q}} = \sqrt[q]{a^p}$. Our definition of a non-integral rational power of a real number, a, is therefore properly applicable in all cases when that number, a, is positive.

The restriction we have introduced—that the base, a, should be positive—is essential in any theory of indices confined to deal only with real numbers. It is also very desirable in any such theory to consider only *positive* values of the powers, i.e. for example, to consider $4^{\frac{1}{2}}$ or $\sqrt{4}$ to mean only $+2$ and not -2. In this course we shall adhere to these restrictions. With these restrictions *the power a^m, where a is any positive real number and m any rational number, necessarily exists and is unique*.

63. Irrational indices. The question now arises: Can powers with *irrational* indices be satisfactorily defined?

Let a be any positive real number and m any real number defined by the Dedekindian classification $(x \mid y)$, where the numbers x are all rational numbers not exceeding m and the numbers y all greater rational numbers. We have defined all such powers as a^x and a^y, x and y being rational, and it therefore seems natural to

---

* With the altered definition appropriate when complex numbers are used, this statement must be modified. See Appendix.

lay down the definition of the power a^m as the real number defined by the classification $(a^x \mid a^y)$*. This definition is in fact sound, but in order to assure ourselves that this is so we must first prove that the classification $(a^x \mid a^y)$ is in fact a Dedekindian classification, i.e. (1) that all rational (or real) numbers are included in the classification, and (2) that every number of the a^y class exceeds every number of the a^x class (or conversely).

The proof of (2) is not difficult. If x and y are rational numbers, say p/q and r/s, then $a^y > a^x$, i.e. $a^{r/s} > a^{p/q}$, if $a^{rq} > a^{ps}$; i.e. if $rq > ps$ if $a > 1$, or $rq < ps$ if $a < 1$; i.e. if $y > x$ or $y < x$ according as $a > 1$ or $a < 1$; so that if $y > x$, $a^y > a^x$ or $a^y < a^x$ according as $a > 1$ or $a < 1$.

To prove (1) we remark that, supposing $a > 1$, if b is any real number exceeding the upper bound of the set of numbers a^x, there is a positive number ϵ such that $b - a^x > \epsilon$ for all numbers a^x of the a^x class; but there are numbers x and y of the two classes respectively for which $a^y - a^x < \epsilon$, because
$$a^y - a^x = a^x (a^{y-x} - 1) < a^x (y - x)(a - 1)$$
by the inequality of Ex. 2, opposite, for rational index (taking $y - x < 1$),
$$< \epsilon \text{ if } y - x < \epsilon/(a-1)A,$$
where A is any number exceeding all the powers a^x. But $y - x$ can be taken as small as we like; it therefore follows that any such number b must exceed some number a^y of the upper class.

On the other hand any number less than the upper bound of the numbers a^x must be exceeded by some numbers of the lower class.

We are therefore justified in asserting that *the classification $(a^x \mid a^y)$ defines a real number* and in calling this number *the power a^m.*

64. It is easy to see that this definition of a^m agrees with the other definitions in those cases where the index m happens to be rational (or integral). That the same laws of indices hold—and that therefore there is no inconvenience in this introduction of the idea of irrational powers—is intuitively evident; the formal proof of the first law is as follows:

Let m and n be the real numbers defined by the classifications $(x_1 \mid y_1)$ and $(x_2 \mid y_2)$ of the rational numbers; then $a^m \times a^n$ is the product of the two real numbers $(a^{x_1} \mid a^{y_1})$ and $(a^{x_2} \mid a^{y_2})$, i.e. is the real number $(a^{x_1 + x_2} \mid a^{y_1 + y_2})$, where x_1, x_2, y_1, y_2 are given all values admissible.

* We use the notation $(a^x \mid a^y)$ to denote the classification, not merely of the numbers expressible in the forms a^x, a^y, but of *all* numbers respectively less than or greater than the numbers so expressed. Thus, if $a > 1$, the lower class consists of all numbers which $\leqslant$ any number of the type a^x, and the upper class of all numbers which $\geqslant$ any number of the type a^y.

But the classification $(x_1 + x_2 \,|\, y_1 + y_2)$ defines the real number $m + n$, and the classification $(a^{x_1+x_2} \,|\, a^{y_1+y_2})$ defines therefore the real number a^{m+n}; whence it follows that $a^m a^n = a^{m+n}$. Q.E.D.

We are now in a position to use and develop the properties of powers of any positive real number to any real index, and in doing so to assume any algebraic developments of the original laws of indices for positive integral powers, since the laws are formally the same. Such irrational powers are of vital importance, not only from the theoretical standpoint of the theory of functions, but also in the practical matter of logarithms. In future we shall use such irrational powers as may be desired without distinction except where special discussion is desirable.

EXAMPLES VIII.

1. Prove that the inequality (ii) of p. 29 above, viz. $(1+x)^n > 1 + nx$, holds if n be a negative integer provided $x > -1$. (See also Ex. 3, p. 132 below.)

2. Prove that the inequality (iii) of p. 29 above, that $\dfrac{a^n - b^n}{a - b}$ lies between na^{n-1} and nb^{n-1}, a and b being unequal and either both positive or both negative, is true for all real values of n, positive or negative, rational or irrational (except when $n = 0$ or 1, in which cases all the three numbers are equal).

[For the rational case use the inequality (iv) of p. 29.]

3. Prove that if a is any real number greater than 1 then $a^x > a^y$ if $x > y$ in cases: (a) when x and y are positive integers, (b) when x and y are any rational numbers, and (c) when x and y are any real numbers; and that if a is positive and less than 1 then $a^x < a^y$ under the same circumstances.

[This property may be described by saying that *the exponential function a^x is monotone*. Case (a) is proved by direct application of the fundamental laws of inequalities to the definition of a^x etc.; case (b) can be deduced from case (a) by raising a^x and a^y to the qsth power where x and y are taken to be p/q and r/s respectively, with simple extension to the case of negative indices; case (c) follows from case (b) and the definition of an irrational power as a Dedekindian classification obtained from rational powers.]

4. Shew that if x_1, x_2, x_3, ... is (a) any monotone sequence, or (b) any sequence, monotone or not, tending to the unique limit m, then the sequence of corresponding powers of a, viz. a^{x_1}, a^{x_2}, a^{x_3}, ... is convergent and tends to the unique limit a^m, where a denotes any positive real number, and x_1, x_2, ... are supposed all positive.

[This property is the *continuity of the exponential function*. The proof of this property is essentially contained in the proof that if $(x \,|\, y)$ is a Dedekindian classification then the corresponding classification $(a^x \,|\, a^y)$ is also Dedekindian. The proof of (a) may be expressed differently thus:

Let X be the upper bound of the sequence x_1, x_2, ... (supposed increasing) and suppose $a > 1$; the sequence a^{x_1}, a^{x_2}, ... is increasing and has its upper bound $\leqslant a^x$. We prove that the upper bound of this sequence actually $= a^x$ by shewing that there are numbers of the sequence exceeding any number (A) less than a^x. If b is any number less than X, $a^x - a^b = a^b(a^{x-b} - 1) < a^b(a-1)(X-b)$ by Ex. 2 above (taking $X - b < 1$). Hence if A is any number less than a^x and if $b\,(< X)$ is chosen so that

$$X - b < (a^x - A)/[a^x(a-1)],$$

the number a^b will be such that

$$a^x - a^b < a^x - A,$$

i.e. such that $a^b > A$. But there are numbers x_n of the sequence $x_1, x_2, x_3, \ldots$ which exceed b, and therefore there are numbers a^{x_n} of the sequence $a^{x_1}, a^{x_2}, \ldots$ which exceed a^b and therefore, *a fortiori*, exceed A, any number less than a^x.

The case when $a < 1$ can be dealt with similarly, or can be made to depend on the case discussed.

For case (b) we notice that the sequence $y_1, y_2, y_3, \ldots$, where $y_1 =$ upper bound of $x_1, x_2, x_3, \ldots$, $y_2 =$ upper bound of $x_2, x_3, x_4, \ldots$, $y_3 =$ upper bound of $x_3, x_4, x_5, \ldots, \ldots$, is steadily decreasing and has the unique limit X and that the corresponding sequence of lower bounds is increasing and has the same unique limit X; whence case (a) shews that the sequence $a^{y_1}, a^{y_2}, a^{y_3}, \ldots$ steadily decreases (and the corresponding sequence of lower bounds steadily increases) to the unique limit a^x. By shewing now that the upper bound of $a^{x_1}, a^{x_2}, a^{x_3}, \ldots$ equals a^{y_1}, etc. the proof is completed. See also p. 47 above.]

5. Shew that if $a > b > 0$ then $a^n > b^n$ if n is (a) any positive rational number, (b) any positive real number. [*Monotony of x^n.*]

6. Shew that if $x_1, x_2, x_3, \ldots$ is any convergent sequence of positive numbers having the limit a, and if n is any real number, the sequence $x_1{}^n, x_2{}^n, x_3{}^n, \ldots$ is convergent and has the limit a^n. [*Continuity of x^n.*]

[See the proof when n is a positive integer, Ex. 14, p. 60 above.]

7. Shew that if the sequence $x_1, x_2, x_3, \ldots$ tends to the unique limit a, and n is any real number, then the sequence

$$\frac{x_1{}^n - a^n}{x_1 - a}, \quad \frac{x_2{}^n - a^n}{x_2 - a}, \quad \ldots$$

tends to the limit na^{n-1}. [*Differentiability of x^n.*]

8. Taking as an alternative definition of a fractional power that $a^{p/q} = (\sqrt[q]{a})^p$, prove from first principles the laws of indices.

9. Shew that if the base, a, were allowed to be negative in the definition of fractional powers, or if double values for square roots (and other even roots) were admitted, $a^{p/q}$ and $a^{2p/2q}$ might differ. Shew also that the definition of $a^{p/q}$ as $\sqrt[q]{a^p}$ might apply when the alternative definition as $(\sqrt[q]{a})^p$ would not apply.

10. Shew that if $a^{\frac{p}{q}}$ when a is negative were defined as $\sqrt[q]{a^p}$ (or $(\sqrt[q]{a})^p$), $\frac{p}{q}$ having been reduced to its lowest terms, then, independently of the fact that

the exponential function a^x would not exist for some rational values of the index, the function would not be "continuous" for any value of x; and that similar difficulties would arise if the definition were modified to $\sqrt[2q]{a^{2p}}$ when q is even.

§ 2. LOGARITHMS

65. Definition and existence theorem for logarithms. We have seen that the equation $a^m = b$, in which a is any positive and m any real number, can always be solved for b in terms of a and m (b being, in fact, always positive). The operation involved is that of exponentiation—the raising of a to the mth power—an operation always possible within the system of real numbers, whether m be integral or not. If a be regarded as the unknown, the equation can still be solved,—again by the operation of exponentiation,— for $a = b^{1/m}$. If however m be regarded as the unknown, we have as yet no regular means of solving the equation. The operation involved is a new operation. Such a solution is however possible for all positive values of a and b (except for $a = 1$), as can be proved as follows:

If z is any real number, and a any positive real number other than 1, a^z exists and necessarily either $> b$ or $\leqslant b$.

Let x denote any real number for which $a^x \leqslant b$, and y any real number for which $a^y > b$.

Then the classification $(x \mid y)$ divides the system of real numbers into two classes, such that there are numbers in each class.

To prove this: If $a > 1$, let $a = 1 + c$. Then, if n is any positive integer, $a^n = (1+c)^n > 1 + nc$ by inequality (ii) of p. 29; and therefore n can be found so that $a^n > b$. This number n belongs to the y class. Also, since $b > 0$, another integer m can be found so that $a^{-m} = \dfrac{1}{(1+c)^m} < \dfrac{1}{1+mc} < b$; and therefore the number $-m$ belongs to the x class. If $a < 1$, we write $a = \dfrac{1}{1+c}$ and the proof proceeds on the same lines. If $a = 1$ the statement is of course untrue, for then $a^z = 1$ for all values of z.

Also (1) all real numbers are included in one or other of the classes, and (2) every $x <$ every y; because, if any $x \geqslant$ any y, the corresponding a^x would be $\geqslant$ the corresponding a^y (Ex. 3, p. 91 above)*, which would contradict the supposition that $a^x \leqslant b < a^y$.

Therefore this classification $(x \mid y)$ is a Dedekindian classification

---

* Assuming $a > 1$. If $a < 1$, the signs of inequality must be reversed.

and defines a real number, m say. The number a^m is then the number defined by the classification $(a^x \mid a^y)$, i.e. is the number b, i.e. $a^m = b$; and it is evident that there is only one such number m having this property.

Hence the equation $a^m = b$, in which a and b are any positive real numbers, and a is not equal to 1, necessarily has a unique real solution for m. This real number m is called the *logarithm of b to the base a*, and is written $\log_a b$*.

66. It is interesting to notice that, except in rare cases, one at least of the three numbers a, b, m must be irrational; and in fact that *the logarithm to a rational base (a) of a rational number (b) must be irrational, unless the numbers (a and b) are positive or negative integral powers of a single rational number.*

To prove this in the special case when $a = 10$ and $b = 2$—i.e. to prove that the logarithm of 2 to the base 10 is irrational—we argue: If $\log_{10} 2$ were rational it could be expressed as a fraction p/q, p and q being (positive or negative) integers, and we should have the relation $10^{p/q} = 2$; whence $10^p = 2^q$, an impossible relation, since 10^p is an integer ending in 0 (or a decimal consisting of the digit 1 following a number of 0's following the decimal point) whilst 2^q must be an integer ending in 2, 4, 8, or 6 (or a decimal whose last figure is 5). Therefore $\log_{10} 2$ is irrational. The proof of the general case is left to the student. The student will notice, in consequence of this property, that the introduction of irrational numbers (in particular as indices) was essential if we are to use logarithms at all generally.

67. Properties of logarithms. In virtue of our knowledge of the power, or exponential function, of which the logarithm is the inverse, we at once see that logarithms have the following two important properties:

(1) If the base a is greater than 1, and x and y are two positive numbers, then $\log_a x > \log_a y$ if and only if $x > y$; and if $a < 1$, $\log_a x > \log_a y$ if and only if $y > x$. We say in either case that *the logarithmic function is monotone.*

(2) If $x_1, x_2, \ldots$ is any sequence of positive numbers tending to the unique positive limit b, the corresponding sequence of logarithms, viz. $\log_a x_1, \log_a x_2, \ldots$, tends to the unique limit $\log_a b$. We say *the logarithmic function is continuous.*

---

* With the introduction of complex numbers logarithms can be defined even if the number or the base is negative (or complex). Such logarithms are not unique. See Appendix.

The proofs of these properties are almost immediate.

The proof of (2) may be expressed in terms of monotone sequences (see pp. 47—48) thus:

Because   the sequence $x_1, x_2, \ldots \to b$,

we know that   the sequence $U_1, U_2, \ldots \searrow b$,

and   the sequence $L_1, L_2, \ldots \nearrow b$,

where U_1 represents the upper bound of $x_1, x_2, x_3, \ldots$, U_2 the upper bound of $x_2, x_3, x_4, \ldots$, etc., and L_1, L_2, etc. represent the corresponding lower bounds.

Hence   the sequence $\log_a U_1, \log_a U_2, \log_a U_3, \ldots \searrow \log_a b$,

and   the sequence $\log_a L_1, \log_a L_2, \log_a L_3, \ldots \nearrow \log_a b$,

supposing $a > 1$* and using property (1) to prove that if $y_1, y_2, \ldots \nearrow b$, then

$$\log_a y_1, \log_a y_2, \log_a y_3, \ldots \nearrow \log_a b\,\dagger.$$

But $\log_a U_1 =$ upper bound of $\log_a x_1, \log_a x_2, \log_a x_3, \ldots$; etc.

Therefore the sequence

(upper bound of $\log_a x_1, \log_a x_2, \ldots$), (upper bound of $\log_a x_2, \log_a x_3, \ldots$),

$$\ldots \searrow \log_a b,$$

and the sequence

(lower bound of $\log_a x_1, \log_a x_2, \ldots$), (lower bound of $\log_a x_2, \log_a x_3, \ldots$),

$$\ldots \nearrow \log_a b,$$

whence the sequence $\log_a x_1, \log_a x_2, \ldots \to \log_a b$. Q.E.D.

We notice further that, whatever the base (a), $\log_a 1 = 0$, because $a^0 = 1$.

Thus we know that *if the base (a) is greater than 1, every positive number b has a logarithm, which is positive or negative according as the number is greater than or less than 1, and that the logarithm increases steadily (and continuously) as the number increases.* If the base is less than 1, the logarithm is positive or negative according as the number is less or greater than 1 and the logarithm decreases as the number increases.

---

* The modifications when $a < 1$ are evident.

† That the sequence $\log_a y_1, \log_a y_2, \ldots$ is increasing and therefore has its upper bound as its unique limit and that this upper bound $\leqslant \log_a b$ is evident from (1). That this upper bound is $\log_a b$ is proved thus: if B is any number less than $\log_a b$, $a^B < a^{\log_a b} = b$, and therefore some of the numbers $y_1, y_2, \ldots$ exceed a^B and therefore some of the numbers $\log_a y_1, \log_a y_2, \ldots$ exceed B, so that B is less than the upper bound of the sequence $\log_a y_1, \log_a y_2, \ldots$ and the actual upper bound $= \log_a b$. Similar remarks apply to decreasing sequences and to lower bounds.

It should be noticed that negative numbers have no logarithms and that the base of the logarithms must in all cases be positive*.

We shall in this course, wherever for the sake of definiteness appears desirable, always assume that the base of the logarithms is greater than 1; the modifications necessary to cover the case of a base less than 1 will be found to be nowhere vital. In practice the base is invariably taken to be either e (2·718...) or 10.

68. Laws of logarithms. It is of natural interest (and of practical utility) to discover laws concerning the logarithms to a fixed base of various combinations of different numbers. We have in fact already found three laws of indices, which in view of the fact that logarithms are indices, can evidently be restated as laws of logarithms.

Thus we have firstly, from the first law that $a^m \times a^n = a^{m+n}$, that the logarithm to the same base a, of the product of two numbers, is the sum of the two logarithms of those numbers; or

$$\log_a (x \times y) = \log_a x + \log_a y,$$

x and y being any two positive numbers; it having been seen that, whatever numbers x and y may be, their logarithms, m and n, exist such that $a^m = x$ and $a^n = y$.

Secondly, $a^m \div a^n = a^{m-n}$ gives at once

$$\log_a \left(\frac{x}{y}\right) = \log_a x - \log_a y.$$

As in the case of indices this law is only a particular case of the first law, for we have, in virtue of the relation $1/a^n = a^{-n}$, that $\log_a (1/y) = - \log_a y$, and thence

$$\log_a \left(\frac{x}{y}\right) = \log_a \left(x \times \frac{1}{y}\right) = \log_a x + (- \log_a y) = \log_a x - \log_a y.$$

Thirdly, the law $(a^m)^n = a^{mn}$ gives

$$\log_a (x^n) = n \log_a x,$$

n being any real number.

To sum up these three laws : *logarithms of products and quotients are equivalent to sums and differences of logarithms, and the logarithm of a power is equivalent to the product of a number and a*

---

* With definitions in terms of complex numbers this statement is untrue. See Appendix.

logarithm. If we happen to know the logarithms to any base of all numbers we need, we can therefore replace the operation of multiplication by that of addition, division by subtraction, and exponentiation by multiplication. In numerical calculations involving such operations of multiplication, etc.—operations which are tedious if the numbers concerned are large (or have many significant figures)—we can therefore simplify the work by using logarithms, thereby reducing the operations of multiplication, etc. to the simpler operations of addition, etc. Since in practical work numbers with many figures are of frequent occurrence, the reduction in labour involved in such a use of logarithms is of considerable importance, provided only that tables of logarithms can be constructed and used with ease.

69. Logarithms to different bases. Common logarithms. There is so far in our logarithms an element of arbitrariness—the base a may be any positive number whatever. The choice of the most suitable base is not of much theoretical* but of considerable practical importance. If we take for example $a = 2$, we have evidently (straight from the definition) the following logarithms to the base 2 :

$$\log_2 1 = 0, \quad \log_2 2 = 1, \quad \log_2 4 = 2, \quad \log_2 8 = 3,$$

$\log_2 10 =$ an irrational number greater than 3 and less than 4,

$$\log_2 20 = \log_2 2 + \log_2 10 = 1 + \log_2 10, \quad \log_2 80 = 3 + \log_2 10,$$

and generally

$$\log_2 (2^n x) = n + \log_2 x, \text{ and } \log_2 (10^n x) = n \log_2 10 + \log_2 x.$$

The last two relations shew that if a number is multiplied by a power (positive or negative) of 2 its logarithm is modified only by the addition of the index of that power, whereas if multiplied by a power of 10 the logarithm is modified by the addition of a multiple of the irrational logarithm of 10. Or, if the logarithm of a number to the base 2 is known, the logarithms of all numbers got by multiplying (or dividing) the number by powers of 2 are also known immediately, whereas the logarithms of the numbers got by multiplying by the powers of 10 are not so known. Since, however, the normal system of notation is the decimal notation,

* But see § 4. below.

it is easier to see the similarity between numbers whose ratio is a power of 10 than that between numbers whose ratio is a power of 2. If we chose 2 for the base of our logarithms we should therefore lose the particular advantages of our customary notation. It is therefore preferable to discard 2 as our standard base of logarithms and to choose 10 for base. We then have the relation $\log_{10}(10^n x) = n + \log_{10} x$, so that if we find (from tables) the single logarithm $\log_{10} x$ (e.g. $\log_{10} 1\cdot2345$) we have immediately all the logarithms of the numbers got by shifting the decimal point in x (e.g. $\log_{10} 1234\cdot5$ or $\log_{10} \cdot0012345$), and it becomes thereby sufficient for the tables of logarithms to take no cognizance of the position of the decimal point but to tabulate only for values of x (with a sufficient number of digits) within a certain range (e.g. between 1 and 10). It is for this reason that 10 is nearly always chosen as the base of tables of logarithms designed for numerical calculation. Such logarithms are called *common logarithms**. If ever the decimal system of notation were displaced by the duo-decimal system it would be necessary to replace all tables of common logarithms by tables of logarithms to the base 12.

It is nevertheless desirable in some kinds of work to use a base other than 10. The investigation of the relation between logarithms of the same number to different bases is, therefore, of practical importance.

An appeal to the definition of logarithms shews at once that:
If a and b are the bases (positive) of two systems of logarithms, and x any given positive number, then $\log_b x = \log_b a \times \log_a x$, *or*
$$\log_b x = \frac{1}{\log_a b} \log_a x.$$

For, if $\log_a x = \alpha$ and $\log_b x = \beta$, we have $a^\alpha = x = b^\beta$, whence
$$\alpha \log_a a = \log_a x = \beta \log_a b \text{ and } \alpha \log_b a = \log_b x = \beta \log_b b,$$
giving　　$\log_b x = \log_b a \times \log_a x$ and $\log_b x = \dfrac{1}{\log_a b} \log_a x.$

We can now at once obtain the logarithm of any number to any base if only we know the logarithm of both the number and the base to any particular base.

* In elementary and practical work the suffix denoting the base (10) of common logarithms is usually omitted. We shall follow this custom in the next section, but not elsewhere in this course. In theoretical work the suffix is usually omitted when the base is the irrational number e, but we shall avoid this practice.

EXAMPLES IX.

1. Given that $\log_{10} 2 = \cdot 30103$, find the common logarithms of
$$32,\ 8,\ 1/2,\ 1/8,\ \sqrt{2},\ 1/(\sqrt[7]{2}).$$

2. Given that $\log_{10} 3 = \cdot 47712$ and $\log_{10} 2 = \cdot 30103$, find the common logarithm of
$$\sqrt[6]{\frac{3^7 2^{10}}{6^{2/3}}}.$$

3. Given that $\log_{10} e = \cdot 4343$ ("e" being the irrational number, defined above (p. 51), which is the base of the Napierian logarithms), and given the common logarithms of the preceding questions, calculate $\log_e 2$ and $\log_e 3$.

4. Shew that $\log_a b$ lies between
$$\frac{a\,(b-1)}{b\,(a-1)} \text{ and } \frac{b-1}{a-1}.$$

Hence prove directly that if $h_1, h_2, h_3, \dots$ is a sequence of positive numbers tending to the unique limit zero, then the sequence whose nth term is $\log_{10}(1+h_n)$ has zero as unique limit; deduce the continuity of the logarithmic function for all values of x.

5. Shew that the logarithm of a fixed number is a continuous function of the base (i.e. if $a_1, a_2, a_3, \dots \to a$ then $\log_{a_1} b, \log_{a_2} b, \log_{a_3} b, \dots \to \log_a b$) for all positive values of the base except unity; and monotone for all values of the base exceeding unity (i.e. $\log_{a_1} b > \log_{a_2} b$ if and only if $a_1 > a_2$ or else if and only if $a_1 < a_2$).

6. If $a > 1$ and n is any real number and x positive, prove:
   (i) $a^x > x^n$ and $a^{-x} < x^n$ for all values of x greater than a certain value;
   (ii) $x^n > \log_a x$ for all values of x sufficiently great;
   (iii) $x^n > |\log_a x|$ for all (positive) values of x sufficiently small.
[For (i) put $x^n = X$. $a^x = (a^{x/x})^x > X$ by Ex. 10, p. 36.]

§ 3. LOGARITHMIC TABLES

70. By the use of logarithms the operations of multiplication, division, and exponentiation can be replaced by the operations of addition, subtraction, and multiplication. In practical calculations these latter operations, as explained above (p. 97), are simpler than the former. If the logarithms of *all* numbers could be tabulated once for all, such calculations would be made easy, for all that would be necessary in any particular calculation would be to look up directly in the tables the logarithms required and the numbers corresponding to given logarithms. Such a complete tabulation is, however, manifestly impossible. But it might be possible to tabulate all the integers up to, say, 10,000, and their logarithms correct to, say, four decimal places. Before entering on the question as to whether (and how) such a tabulation can actually be carried out, it will be well to inquire whether such limited tables can be of sufficient utility to make the labour of compilation worth while.

71. The first remark which suggests itself in this connection is that any tables of logarithms, however accurate, cannot be used for arithmetical calculations where *absolute* accuracy is required (except in certain trivial cases),—for, as we have seen, logarithms are in general irrational numbers and therefore cannot be obtained accurately as decimals, but only to some degree of approximation (though this degree of approximation is variable at our choice). This lack of absolute accuracy is not however a vital objection to the use of logarithmic tables, for, generally speaking, numerical calculations are met with only in practical work (in the physical sciences and elsewhere) where the accuracy of the actual data of the calculations is by no means absolute;—in some surveying problems the direct measurements may be in error to the extent of one per cent., and in very few measurements are the errors known to be as small as one in a million. It would be absurd to desire absolute accuracy in calculations on such approximate data.

Thus, if only the tables can be so constructed and used as to preserve any desired degree of accuracy, the lack of absolute accuracy will be of no consequence. Our object will be attained if the successive numbers tabulated are so chosen as to differ only by the range of error to be allowed for in the data of the calculations, so that there will be a separate entry giving the logarithm of every distinguishable number, and, if, at the same time, the logarithms tabulated are obtained to such a degree of approximation as to render successive entries of logarithms distinguishable; that is to say, if the numbers for successive entries are taken sufficiently near together and the corresponding logarithms are tabulated to a sufficient number of decimal places. If, in using such a table, we encounter a logarithm which does not occur in the table, it will lie between two of the entries of logarithms and therefore, in view of the monotoneness of the logarithmic function, the corresponding number will lie between the two corresponding number-entries; and this will suffice for our needs. One proviso only is needed,—that the number of decimal places to which the tables are carried must be the same throughout and sufficient for the accuracy of the least favourable part of the tables. For work of different degrees of accuracy different tables will be needed,—e.g. four-figure tables for most practical work, or seven-figure tables for more accurate work. As has been noticed before, there is no need in a table of common logarithms to tabulate numbers beyond a certain stage, since, e.g., in four-figure tables 16013 would be indistinguishable from 16010, whose logarithm is $1 + \log 1601$; so that in four-figure tables we need only tabulate logarithms of integers from 1000 to 9999.

It should be observed that it does not follow that the sum (say) of a number of logarithms will have the same degree of accuracy as the several logarithms. It is true that in some cases the various errors may compensate each other, but on the other hand it is possible that they may be all of defect (or of excess) and thus add up, or accumulate, to a considerable total error. In performing calculations we must of course allow for this possible accumulation of errors, and it is best always to write down the limits between which the numbers and logarithms occurring are known to lie. Thus, for example, if we know that a measurement x, given as $2 \cdot 000$, is liable to an error (of excess or defect) of $\cdot 001$, we read from the tables that $\log 1 \cdot 999 = \cdot 3008$ and $\log 2 \cdot 001 = \cdot 3012$, to four places of decimals (i.e. with a possible error of $\cdot 00005$). All we can guarantee about $\log x$ from these data is that $\cdot 30075 \leqslant \log x \leqslant \cdot 30125$. If we work with inequalities of this kind we shall know with certainty the outside limits of error of the result. This method can be simplified in practice.

72. Tables of common logarithms. In tables of common logarithms in actual use certain conventions are used, which it is desirable to mention here. The logarithm of any number will in general consist of an integral part and a decimal part; the integral part is called the *characteristic* and the decimal part the *mantissa* of the logarithm; in the case of a negative logarithm the characteristic is always taken to be the negative integer next less than the logarithm itself, and the mantissa is the *positive* decimal by which the logarithm exceeds the characteristic. Thus for $\log 20$, which

$= 1\cdot30103$, the characteristic and mantissa are respectively 1 and $\cdot30103$; for $\log\cdot5$, which $= -\cdot30103 = -1 + \cdot69897$, the characteristic and mantissa are respectively -1 and $\cdot69897$. Such a logarithm is written $\bar{1}\cdot69897$.

Tables of logarithms tabulate only the mantissae, and the decimal point is usually omitted; the characteristics and decimal points have to be added. The determination of the characteristic of a common logarithm is simple, for we have from the laws of logarithms that if, e.g., $\log 2 = \cdot3010$, then

$$\log 20 = \log (10 \times 2) = \log 10 + \log 2 = 1\cdot3010,$$
$$\log 2000 = \log (10^3 \times 2) = 3 + \log 2 = 3\cdot3010,$$
$$\log\cdot2 = \log (2/10) = \log 2 - \log 10 = \log 2 - 1 = \bar{1}\cdot3010,$$
$$\log\cdot00002 = \log (2/10^5) = \log 2 - 5 = \bar{5}\cdot3010;$$

or if $\log 1\cdot302 = \cdot1146$, then

$$\log 130\cdot2 = 2\cdot1146, \quad \log\cdot001302 = \bar{3}\cdot1146.$$

The characteristic of the logarithm of a number greater than 1 is always one less than the number of digits preceding the decimal point; the characteristic of the logarithm of a number less than 1 is always negative, and is numerically one more than the number of zeros immediately following the decimal point.

In most published tables space is saved by the use of a "column of differences," the precise form of which varies with the publication, and the use of which will give no difficulty to the student.

The "principle of proportional parts" is of considerable use in mathematical tables in order (e.g.) to obtain the logarithm of a number intermediate to two numbers tabulated. It is, briefly, that if the difference between the intermediate number and the smaller of the two tabulated numbers is e.g. 1/10 or 2/10 of the difference between the two tabulated numbers, then the corresponding difference between the intermediate logarithm and the smaller of the two tabulated logarithms is 1/10 or 2/10 of the difference between the tabulated logarithms. The investigation of the degree of accuracy of this principle and its general justification depend on the notions of differential calculus and will not be given here. Generally speaking, the principle will be reliable only in parts of the table where the "differences" do not vary rapidly.

The student will probably already have had sufficient exercise in the performance of calculations with the help of logarithms, but in any case he will be able to find, in the course of practical work

in natural science or elsewhere, practical examples for practice*.
In all cases the student should use tables appropriate to the
degree of accuracy of the work in question, and should so use
the tables that he will be able to assign with certainty outside
limits of error.

73. Construction of tables. Having satisfied ourselves of the
utility of such tables as those contemplated, it remains for us to
investigate how such tables can be compiled. That is to say, we
wish to calculate the logarithms to any desired degree of accuracy
(e.g. within ·00005 or ·00000005) of all numbers of a certain number
(e.g. 4 or 7) of significant figures. *Our problem is that of calculating
the logarithm of any given number to any desired degree of accuracy.*

For definiteness we will suppose that the logarithms required
are common logarithms, i.e. logarithms to the base 10.

The logarithms of certain numbers are known at once; e.g.

$$\log 10 = 1, \ \log 100 = 2, \ \log 1000 = 3, \ \log 1 = 0,$$
$$\log \cdot 1 = -1, \ \log \cdot 01 = -2, \text{ etc.}$$

To find the logarithm of another number, e.g. 2, we have firstly
that $\log 1 < \log 2 < \log 10$, i.e. that $\log 2$ is positive and less than 1;
but we notice further that $2^3 = 8 < 10$ and therefore $2 < 10^{\frac{1}{3}}$ and
therefore $\log 2 < 1/3$, i.e. $\log 2 < \cdot 3$. Also $2^4 = 16 > 10$ and therefore
$2 > 10^{\frac{1}{4}}$, whence $\log 2 > 1/4 = \cdot 25$. We have thus already obtained
$\log 2$ correct to one decimal place, viz. $\log 2 = \cdot 3$.

For further accuracy we may try repeating this process of raising
2 to some integral power and comparing with a power of 10; we
have

$$2^5 = 32, \ 2^6 = 64, \ 2^7 = 128, \ 2^8 = 256, \ 2^9 = 512, \ 2^{10} = 1024;$$

whence we have, selecting the one actually useful result,

$$2^{10} = 1024 > 1000 \text{ and therefore } \log 2 > \frac{1}{10} \log 1000 = \cdot 3.$$

From these results we have found that $\log 2$ lies between ·3 and ·3.
We can find other logarithms similarly.

74. Two-figure tables. Evidently this process has limitations,
which must be removed if it is to be useful for our purpose.

* Worked examples and detailed instructions on the use of logarithmic tables will
be found in any good modern book on elementary algebra.

We could, as a last resort, calculate the values of $2^{\frac{1}{2}}$, $2^{\frac{1}{4}}$, $2^{\frac{1}{8}}$, ... by repeated application of the operation of extracting the square root,—or we could calculate the values of $2^{\frac{1}{10}}$, $2^{\frac{2}{10}}$, $2^{\frac{3}{10}}$, etc. and then $2^{\frac{1}{100}}$, $2^{\frac{2}{100}}$, ... by the fundamental method of "trial" (e.g. to find $2^{\frac{1}{10}}$ we argue $1^{10} = 1 < 2$, $1 \cdot 1^{10} = 2 \cdot 6 \ldots > 2$; therefore $2^{\frac{1}{10}}$ lies between $1 \cdot 0$ and $1 \cdot 1$, etc.). But clearly, no matter what simplifications be made in this method, the work involved is prohibitive.

We have used the fact (e.g.) that $8 < 10$ to shew that $\log 8$ (i.e. $3 \log 2$) $< \log 10$; but we have no information as to *how much* $\log 8$ is less than $\log 10$. Clearly what is needed is some estimate of the magnitude of the difference between two neighbouring logarithms, such as $\log 10 - \log 8$, or, what comes to the same thing, an estimate of $\log (1 + x)$, where x is fairly small, for

$$\log 10 - \log 8 = \log (10/8) = \log (1 + \tfrac{1}{4}).$$

Let us consider this question in the simplified form when x is the reciprocal of a positive integer; i.e. let us try to find an estimate of $\log \left(1 + \dfrac{1}{n}\right)$, where n is a positive integer.

Taking first the case mentioned, that of $\log (1 + \tfrac{1}{4})$, we notice that the successive powers of $(1 + \tfrac{1}{4})$ are $5/4$, $25/16$, $125/64$, etc., and of these, the third, viz. $(1 + \tfrac{1}{4})^3 = 125/64$, differs only slightly from 2 and that $3 \log (1 + \tfrac{1}{4}) < \log 2$, i.e. $3 - 9 \log 2 < \log 2$, whence $\log 2 > \cdot 3$,—a result which we found before by proceeding to the tenth power of 2. Similarly for other logarithms.

However, we know in general (see pp. 52—53 above) that if m is any positive integer not exceeding n,

$$(1 + 1/n)^m \leqslant (1 + 1/n)^n < 3,$$

and therefore in general

$$\log (1 + 1/n)^m < \log 3,$$

or indeed

$$\log (1 + 1/n)^n < \log 3,$$

i.e.

$$n \log (1 + 1/n) < \log 3, \text{ or } \log (1 + 1/n) < \frac{\log 3}{n},$$

so that if we can obtain a close upper estimate to $\log 3$, we shall have here a close estimate for $\log (1 + 1/n)$ for all integral values of n, and there will be no need to carry out the process of raising the particular value of $(1 + 1/n)$ to the various powers each time.

Moreover, we see that the greater the number n, the more accurate will be the result. Using the evident fact that $\log 3 < \cdot 5$, we may state our result as

$$\log\left(1 + 1/n\right) < \frac{\cdot 5}{n} \text{ or } 1/2n.$$

We can also improve our results if we obtain an *under* estimate for $\log\left(1 + 1/n\right)$ as well as the over estimate found. A sufficiently good estimate is obtained at once if we notice that if n is any positive integer whatever,

$$(1 + 1/n)^n \geqslant 2,$$

and therefore $n \log\left(1 + 1/n\right) \geqslant \log 2$, which we know $> \cdot 3$; whence

$$\log\left(1 + 1/n\right) > \frac{\cdot 3}{n}.$$

Combining our two results, we have proved that $\log\left(1 + 1/n\right)$ lies between $\cdot 3/n$ and $\cdot 5/n$; whence we have $\log\left(1 + \dfrac{1}{n}\right)$ correct to within an error of less than $\cdot 2/n$.

Certain logarithms can now be found to a considerable degree of accuracy; for example $\log 101 = \log 100 + \log\left(1 + 1/100\right)$ and therefore lies between $2 \cdot 003$ and $2 \cdot 005$.

The logarithms of the integers from 1 to 10 can now be determined to a substantial degree of accuracy. Thus $\log 2$ can be found correct to three decimal places thus:

$2^{10} = 1024$, which lies between 1020 and 1025, and therefore $10 \log 2$ lies between $\log 1020$ and $\log 1025$, i.e. between

$$3 + \log\left(1 + 20/1000\right) \text{ and } 3 + \log\left(1 + 25/1000\right),$$

i.e. between $3 + \log\left(1 + 1/50\right)$ and $3 + \log\left(1 + 1/40\right)$, whence $10 \log 2$ lies between $3 + \cdot 3/50$ and $3 + \cdot 5/40$. Therefore $\log 2$ lies between $\cdot 3 + \cdot 0006$ and $\cdot 3 + \cdot 00125$, i.e. $\log 2$ lies between $\cdot 3006$ and $\cdot 30125$, whence, correct to three decimal places, $\log 2 = \cdot 301$.

There is now no real difficulty in calculating the logarithms of all integers, say up to 100, to a degree of accuracy corresponding roughly to "two-figure tables." The student is advised, as an exercise, to calculate by this method the common logarithms of all the integers up to 20; there is scope for much ingenuity in choice of the particular numerical relations used, but the student will not have much difficulty in obtaining the results correct to within $\cdot 005$.

75. Limitations of method. There is still a fundamental draw-back to this method of calculating logarithms. We cannot obtain *unlimited* accuracy.

The possible error in our approximation to $\log(1+1/n)$, on which the calculations rest, is $\cdot 2/n$, which, even in the favourable case of

$$\log 101 = 2 + \log(1 + 1/100),$$

allows an error of ·002. We can improve the method by realising that the result used—that $\log(1+1/n)$ lies between $\cdot 3/n$ and $\cdot 5/n$—is not the best possible result of the kind. For we can easily shew that log 3 is not only less than ·5 but less than ·48. We know also that the inequality $(1+1/n)^n < 3$ is not the best result of its kind;—it is in fact not hard to prove directly that $(1+1/n)^n < 2\frac{3}{4}$ for all values of the positive integer n, or we know already (see pp. 51—55) that the upper bound of the increasing sequence whose nth term is $(1+1/n)^n$ is e, which, correct to six places of decimals $= 2\cdot718282$, and therefore $(1+1/n)^n < e < 2\cdot72$.

We can therefore replace $\cdot 5/n$ in our upper estimate for $\log(1+1/n)$ by $\cdot 48/n$ or $\log 2\cdot72/n$, or even $\log e/n$.

At the same time, the lower estimate ($\cdot 3/n$) for $\log(1+1/n)$ can be increased. Using the result of Ex. 7 of p. 59 above, viz. that the increasing sequence whose nth term is $\left(1 - \dfrac{1}{n+1}\right)^{n+1}$ has $1/e$ for its upper bound, we have

$$\left(1 - \frac{1}{n+1}\right)^{n+1} < 1/e, \text{ i.e. } \left(\frac{n}{n+1}\right)^{n+1} < 1/e,$$

and therefore

$$\left(\frac{n+1}{n}\right)^{n+1} > e, \text{ or } (1+1/n)^{n+1} > e;$$

whence $\log(1+1/n) > \dfrac{\log e}{n+1}$ for all positive integral values of n. Our result can now be modified to

$$\frac{\log e}{n+1} < \log\left(1 + \frac{1}{n}\right) < \frac{\log e}{n}.$$

If we had the precise value of $\log e$, we should now have an estimate for $\log(1+1/n)$ with an error of less than $\log e/n - \log e/(n+1)$, i.e. $\dfrac{\log e}{n(n+1)}$. Since $\log e$ is certainly less than $\frac{1}{2}$, our possible error is seen to be less than

$$\frac{1}{2n(n+1)}.$$

If n is large, this error is very small, and indeed is much less than our previously obtained possible error; e.g. if $n = 100$, we obtain $\log(1+1/100)$ with an error of less than $\dfrac{1}{2\,.\,100\,.\,101}$, i.e. less than ·00005, as compared with ·002.

The degree of accuracy of this method is however still limited.

Evidently further modifications of a vital character are necessary before we achieve the unlimited accuracy we need. In the next section we shall see how the result that $\log(1 + 1/n)$ lies between $\log e/(n+1)$ and $\log e/n$ can be extended and developed to gain this end.

EXAMPLES X.

1. Given that the square, fourth, eighth, 16th, 32nd, 64th, 128th, 256th, and 512th roots of 2 are (to three places of decimals) $1\cdot414$, $1\cdot189$, $1\cdot090$, $1\cdot044$, $1\cdot022$, $1\cdot011$, $1\cdot006$, $1\cdot003$, and $1\cdot001$ respectively, calculate the logarithms of all the integers from 1 to 20 to the base 2, correct to two decimal places.

2. Shew, by the methods of the text, that $\log 3$ lies between $\cdot476$ and $\cdot478$ and $\log 7$ between $\cdot844$ and $\cdot847$.

3. Construct that portion of a "two-figure table" of common logarithms which gives the logarithms of integers between 30 and 40.

4. Use tables to calculate the sum of 20 terms of the geometrical progression having 13 for its first term and $1\cdot045$ for common ratio. State the degree of accuracy of the result.

5. Given the four-figure values of the common logarithms of 8 and 21, obtained from tables, and that $\log e$ lies between $\cdot4$ and $\cdot5$, use the relation $13^2 = 169$ to obtain, by the method of the text, $\log 13$ correct to four decimal places.

6. Calculate the common logarithms correct to three decimal places of $10\cdot1$, $10\cdot2$, $10\cdot3$, $10\cdot4$, $10\cdot5$, $10\cdot6$, $10\cdot7$, $10\cdot8$, $10\cdot9$, and 11.

7. Calculate the logarithms of the integers from 1 to 12, to the base 12, correct to one decimal place.

8. Calculate to one place of decimals:

$(\log 20 - \log 10)/10$, $(\log 15 - \log 10)/5$, $(\log 12 - \log 10)/2$,
$(\log 11 - \log 10)/1$, $(\log 10 - \log 9)/1$, $(\log 10 - \log 8)/2$;

and to two places of decimals:

$(\log 10\cdot5 - \log 10)/\cdot5$, $(\log 10\cdot4 - \log 10)/\cdot4$, $(\log 10\cdot3 - \log 10)/\cdot3$,
$(\log 10\cdot2 - \log 10)/\cdot2$, $(\log 10\cdot1 - \log 10)/\cdot1$, $(\log 10 - \log 9\cdot9)/\cdot1$,
and $(\log 10 - \log 9\cdot8)/\cdot2$.

§ 4. THE LOGARITHMIC SERIES

76. Proof of inequality. In view of the rôle which the relation proved in the last section, viz.

$$\frac{\log e}{n + 1} < \log(1 + 1/n) < \frac{\log e}{n},$$

plays in the development of the logarithmic series—which is the means whereby we are to be enabled to calculate logarithms to an unlimited degree of accuracy—we here set out the proof in full.

Firstly, the two sequences, whose typical terms are $(1+1/n)^n$ and $(1-1/n)^{-n}$, are respectively steadily increasing and decreasing, are convergent and have the same unique limit e.

Combining the two proofs by using the alternative signs $\pm$, $\mp$ in the respective sense, we have, if n is any positive integer greater than 1,

$$\left(1 \pm \frac{1}{n+1}\right)^{n+1} = \left[1 \pm \frac{1}{n} \mp \frac{1}{n(n+1)}\right]^{n+1}$$

$$= \left(1 \pm \frac{1}{n}\right)^n \left(1 \pm \frac{1}{n}\right)\left[1 \mp \frac{n}{(n \pm 1)\,n\,(n+1)}\right]^{n+1}$$

$$> \left(1 \pm \frac{1}{n}\right)^n \frac{n \pm 1}{n}\left[1 \mp \frac{n(n+1)}{n(n+1)(n \pm 1)}\right],$$

by the inequality (ii) of p. 29, since $-\dfrac{1}{(n+1)(n \pm 1)} > -1$.

Hence $\left(1 \pm \dfrac{1}{n+1}\right)^{n+1} > \left(1 \pm \dfrac{1}{n}\right)^n \dfrac{(n \pm 1)n}{n(n \pm 1)} = \left(1 \pm \dfrac{1}{n}\right)^n.$

Hence the sequences whose typical terms are $\left(1 + \dfrac{1}{n}\right)^n$ and $\left(1 - \dfrac{1}{n}\right)^n$ are increasing.

These sequences are bounded above because, as proved on p. 52,

$$\left(1 + \frac{1}{n}\right)^n < 1 + 1 + \frac{1}{2!} + \frac{1}{3!} + \ldots + \frac{1}{n!} < 3,$$

and $$\left(1 - \frac{1}{n}\right)^n < 1.$$

The sequences, therefore, have unique limits. That of the first is denoted by e^*.

Finally $\left(1 - \dfrac{1}{n}\right)^n = \left(\dfrac{n-1}{n}\right)^{n-1} \dfrac{n-1}{n}$

$$= \frac{1}{\left(1 + \dfrac{1}{n-1}\right)^{n-1}} \frac{1}{1 + \dfrac{1}{n-1}}$$

$$\to \frac{1}{e}$$

because $\left(1 + \dfrac{1}{n-1}\right)^{n-1} \to e$ and $1 + \dfrac{1}{n-1} \to 1.$

* The irrational number e, thus defined, has been proved (p. 55) to be approximately 2·718282.

Therefore $\left(1 - \dfrac{1}{n}\right)^{n} \nearrow \dfrac{1}{e}$, and we have established the two results

$$\left(1 + \frac{1}{n}\right)^{n} \nearrow e, \quad \left(1 - \frac{1}{n}\right)^{-n} \searrow e. \quad \text{Q.E.D.}$$

Secondly, from this double result, we deduce at once, since

$$(1 + 1/n)^{n} < e < \left(1 - \frac{1}{n+1}\right)^{-(n+1)},$$

that $\quad n \log_a \left(1 + \dfrac{1}{n}\right) < \log_a e < - (n+1) \log_a \left(1 - \dfrac{1}{n+1}\right),$

where the base a is any number greater than 1;

i.e. $\qquad n \log_a \left(1 + \dfrac{1}{n}\right) < \log_a e < (n+1) \log_a \dfrac{n+1}{n},$

i.e. $\qquad n \log_a (1 + 1/n) < \log_a e < (n+1) \log_a (1 + 1/n),$

or $\log_a (1 + 1/n)$ lies between $\log_a e/(n+1)$ and $\log_a e/n$; and the relation at the head of the section is proved, where the logarithms may be taken to any base greater than 1.

77. Extension of inequality. Omitting from this point the suffix a, the relation we have proved is

$$\frac{\log e}{n+1} < \log (n+1) - \log n < \frac{\log e}{n}.$$

From it we deduce at once the relation

$$\frac{p}{q+p} \log e < \log (q+p) - \log q < \frac{p}{q} \log e$$

or $\qquad \dfrac{\log e}{q+p} < \dfrac{\log (q+p) - \log q}{p} < \dfrac{\log e}{q},$

where p and q are any two positive integers.

For $\quad \log (q+p) - \log q = \log (q+p) - \log (q+p-1)$

$$+ \log (q+p-1) - \log (q+p-2)$$
$$+ \cdots\cdots\cdots\cdots\cdots\cdots$$
$$+ \log (q+2) - \log (q+1)$$
$$+ \log (q+1) - \log q,$$

and to each of the differences of logarithms occurring, the proved relation applies; whence

$$\log (q+p) - \log q < \log e \left(\frac{1}{q+p-1} + \frac{1}{q+p-2} + \cdots + \frac{1}{q}\right)$$

$$< \frac{p}{q} \log e,$$

there being p terms in the bracket, each at most $1/q$; and

$$\log(q+p) - \log q > \log e \left(\frac{1}{q+p} + \frac{1}{q+p-1} + \cdots + \frac{1}{q+2} + \frac{1}{q+1}\right)$$
$$> \frac{p}{q+p} \log e.$$

The relation stated is proved.

78. Fundamental inequality. From this we deduce that *if h is any rational number greater than -1, $\log(1+h)$ lies between $h \log e$ and $h \log e/(1+h)$;* for the rational number h can be expressed as a fraction $\pm p/q$ where p and q are positive integers[*] and

$$\log(1+h) = \log\left(1 \pm \frac{p}{q}\right) = \log(q \pm p) - \log q, \text{ which lies between}$$

$\pm p \log e/(q \pm p)$ and $\pm p \log e/q$, i.e. $\log(1+h)$ lies between $h \log e/(1+h)$ and $h \log e$.

Finally *if h is any real number greater than -1, $\log(1+h)$ lies between $h \log e/(1+h)$ and $h \log e$.*

For, if h is defined as the Dedekindian classification of the rational numbers $(x\,|\,y)$, we have

$$h = (x\,|\,y), \quad h \log e = (x \log e\,|\,y \log e), \quad h \log e/(1+h) = \left(\frac{x \log e}{1+x}\,\Big|\,\frac{y \log e}{1+y}\right)$$

and

$$\log(1+h) = (\log[1+x]\,|\,\log[1+y]),$$

where, in the last two classifications, x is restricted to exceed -1, and the classes determining $h \log e/(1+h)$ are completed by the addition of all numbers less than and greater than those expressed.

But we know, by the result just proved for rational numbers, that

$$\frac{x \log e}{1+x} < \log(1+x) < x \log e,$$

and

$$\frac{y \log e}{1+y} < \log(1+y) < y \log e;$$

whence, by direct reference to the geometrical[†] or arithmetical idea of Dedekindian classifications,

$$\left(\frac{x \log e}{1+x}\,\Big|\,\frac{y \log e}{1+y}\right) < (\log[1+x]\,|\,\log[1+y]) < (x \log e\,|\,y \log e),$$

i.e.

$$\frac{h \log e}{1+h} < \log(1+h) < h \log e. \quad \text{Q.E.D.}$$

By an evident slight modification of this result we obtain at once *the fundamental inequality*

$$\frac{\log e}{x+h} < \frac{\log(x+h) - \log x}{h} < \frac{\log e}{x},$$

---

[*] If $h = \dfrac{-p}{q}$, $q > p$.   [†] i.e. by reference to the straight line of Fig. 1.

if x and h are any positive real numbers, or

$$\frac{\log e}{x} < \frac{\log (x + h) - \log x}{h} < \frac{\log e}{x + h},$$

if x is any positive real number and h any negative real number greater than $-x$.

This fundamental inequality gives upper and lower estimates for the ratio

$$\frac{\log (x + h) - \log x}{h},$$

and determines not only the difference $\log (x + h) - \log x$, but also this ratio $[\log (x + h) - \log x]/h$, with an error which is small if the number h is small.

79. Limit of incrementary ratio of log x. The ratio

$$\frac{\log (x + h) - \log x}{h}$$

is a kind of *relative difference* between the logarithms of the numbers x and $x + h$, compared, that is, with the difference between the numbers themselves. If the difference h between two numbers ($x + h$ and x) is small, we knew before that the difference between the corresponding logarithms is also small (see p. 94 above), but we had previously no knowledge as to how this difference between the logarithms compares with the difference between the numbers. By a study of this "relative difference" we shall be likely to obtain more precise knowledge of the logarithmic relation and of the values of the logarithms themselves. In conformity with the custom of the differential calculus we shall call this ratio the *incrementary ratio of* $\log x$ *from* x *to* $x + h$, or *over the range* $(x,\ x + h)$.

Let us consider this incrementary ratio

$$\frac{\log (x + h) - \log x}{h}$$

for a sequence of values of h (other than zero) having the unique limit zero. The corresponding sequence of values of $\log e/(x + h)$ clearly is also convergent and has the unique limit $\log e/x$*. But, by our fundamental inequality, the incrementary ratio

$$\frac{\log (x + h) - \log x}{h}$$

---

* The proof of this is left to the student.

lies between $\log e/x$ and $\log e/(x + h)$; whence it follows at once that:

If $h_1, h_2, \ldots$ be any sequence of real numbers tending to the unique limit zero (and such that $x + h_n > 0$ for all values of n), then the corresponding sequence of incrementary ratios,

$$\frac{\log(x + h_1) - \log x}{h_1}, \quad \frac{\log(x + h_2) - \log x}{h_2}, \quad \ldots,$$

is convergent and tends to the limit $\dfrac{\log e}{x}$, *x being any positive real number.*

80. The student will have noticed that as the sequence for h tends to zero, the numerators of the incrementary ratios (viz. $\log(x+h) - \log x$) themselves tend to zero, and it might be tempting to argue that therefore the ratios tend to $\dfrac{0}{0}$. But this expression, $0/0$, as we know, is quite meaningless, whereas the limit to which the incrementary ratios have been proved to tend (viz. $\log e/x$) is a perfectly definite number. The student should find no difficulty in this and will realise that the *limit* of a sequence such as $\dfrac{\log(x+h_n) - \log x}{h_n}$, corresponding to a sequence h_n, which tends to zero, means something totally different from the *value* of the expression $\dfrac{\log(x+h_n) - \log x}{h_n}$ when h_n is put equal to zero, which, as we have said, is meaningless.

81. Graph of log x. Inclination of tangent. Derivative. A recourse here to the ideas of *graphs* will be suggestive and useful. From the knowledge of logarithms we have so far acquired, we can sketch a rough graph of $\log x$. It is as drawn in Fig. 2.

Let P, Q be the points on the graph corresponding to the values $x, x + h$ of the variable x; PM, QN the respective ordinates; so that

$$MP = \log x, \quad NQ = \log(x + h)^*.$$

The incrementary ratio of $\log x$ from x to $x + h$ is

$$\frac{\log(x + h) - \log x}{h} = \frac{NQ - MP}{MN} = \frac{RQ}{PR} = \tan R\hat{P}Q,$$

where $R\hat{P}Q$ is the angle of inclination of the chord PQ to the x axis; i.e. *the incrementary ratio from x to $x + h$ is the trigonometrical tangent of the inclination to the x axis of the chord joining the points on the graph corresponding to the values x and $x + h$ of the variable x.*

---

* In the figure, h is taken to be positive. This is unnecessary, provided x and $x + h$ are both positive and the usual conventions as to sign are followed.

If we take a sequence of values of h, say $h_1, h_2, h_3, \ldots$, tending to the unique limit zero, we obtain a sequence of chords $PQ_1, PQ_2, PQ_3, \ldots$, angles $R\hat{P}Q_1, R\hat{P}Q_2, R\hat{P}Q_3, \ldots$, and incrementary ratios $[\log (x + h_1) - \log x]/h_1$, $[\log (x + h_2) - \log x]/h_2$,

$$[\log (x + h_3) - \log x]/h_3, \ldots,$$

such that

$$\frac{\log (x + h_1) - \log x}{h_1} = \tan R\hat{P}Q_1,$$

$$\frac{\log (x + h_2) - \log x}{h_2} = \tan R\hat{P}Q_2,$$

$$\frac{\log (x + h_3) - \log x}{h_3} = \tan R\hat{P}Q_3,$$

etc.

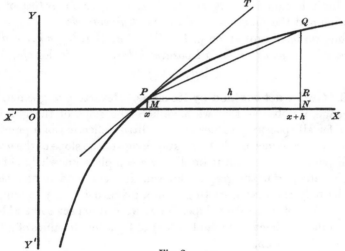

Fig. 2.

The points $Q_1, Q_2, \ldots$ visibly tend to approach P indefinitely closely along the graph, and the chords $PQ_1, PQ_2, \ldots$ tend to coincide with the tangent PT to the graph at P; so that the angles $R\hat{P}Q_1, R\hat{P}Q_2, \ldots$ tend to coincide with the angle $R\hat{P}T$, which the tangent at P makes with the x axis.

Therefore the incrementary ratios, which are the same as $\tan R\hat{P}Q_1$, $\tan R\hat{P}Q_2, \ldots$, tend to $\tan R\hat{P}T^*$; or *the limit of the sequence of*

* This is, of course, not meant to be a *proof*.

*incrementary ratios is the trigonometrical tangent of the inclination
to the x axis of the tangent PT to the graph at the point P (x, log x)
on the graph corresponding to the value x of the variable.*

It is customary to call this trigonometrical tangent simply the
slope or *gradient* of the graph at the point P.

Thus we have seen—intuitively and without any kind of proof—
that the limit of the incrementary ratios represents this important
property of the graph, its slope.

In view of its importance we give this limit of the sequence of incrementary ratios a name,—the *differential coefficient* or the *derivative*
with respect to x of the function log x,—and we write it D log x.

We have in fact, for any function*, that if the sequence of incrementary ratios (defined as above for log x) has a unique limit,
this limit is called the *differential coefficient*, or *derivative*, of the
function and the function is said to be *differentiable*.

Thus we have proved, on p. 112 (Par. **79**), that *log x has a derivative with respect to x for all positive values of x, viz.* log e/x,

or $$D \log x = \log e/x.$$

82. Graph determined by its slope. Reverting to geometrical
language, we observe that we have here the slope of the graph of
log x, for all (positive) values of x. A little reflection of a practical
nature will convince us that this knowledge of the slope of the graph
at all points carries with it implicitly a complete knowledge of the
graph if any part of the graph is known. It seems evident that there
can be only *one* graph starting from a given point and having, for
every value of x, a given slope. Thus we can expect to be able to
determine the behaviour (and value) of log x for all values of x by a
use of this relation,

$$D \log x = \log e/x,$$

giving the slope.

Arithmetically this seems at least slightly hopeful as a way of
calculating log x for all values of x, because the expression giving
the slope (viz. log e/x) is in a very simple form in terms of the
calculable expression $1/x$ (assuming that we know the single logarithm log e). *We will therefore consider this relation and how it may
be possible to deduce from it the nature of* log x.

* y is a function of x if y is given in terms of x. A function may be represented
by a graph.

83. Modification of problem. We will first modify the problem in two ways:

For the sake of brevity we will consider *logarithms to the base e*; for then all the preceding work is valid and we have $\log_e e = 1$, so that our relation giving the derivative of $\log_e x$ is

$$D \log_e x = 1/x.$$

We shall find also (because $\log_e 1 = 0$) that it will be more convenient to consider $\log_e (1 + x)$,—with $x > -1$,—rather than $\log_e x$. It is clear that

$$D \log_e (1 + x) = \frac{1}{1 + x} \quad \ldots\ldots\ldots\ldots\ldots(1),$$

because

$$\frac{\log_e (1 + x + h) - \log_e (1 + x)}{h}$$

lies between $1/(1 + x)$ and $1/(1 + x + h)$, whence the limit of the sequence of incrementary ratios is $1/(1 + x)$.

Our problem is to find, if possible, expressed in calculable form, a function of x which has $1/(1 + x)$ for its derivative for all values of x (exceeding -1). It would then be necessary for us to *prove* that the function obtained actually is identical with $\log_e (1 + x)$.

84. Polynomial approximations to derivative. We have

$$D \log_e (1 + x) = \frac{1}{1 + x} = 1 - x + x^2 - \ldots \pm x^n \mp \frac{x^{n+1}}{1 + x},$$

where n is any positive integer and the upper or lower signs are to be taken according as n is even or odd.

Since we are assuming $x > -1$, let $-\theta$ be a negative number between -1 and x, so that $0 < \theta < 1$; we have

$$\left| \frac{x^{n+1}}{1 + x} \right| < \left| \frac{x^{n+1}}{1 - \theta} \right| \quad \text{(if } x \neq 0\text{)},$$

and therefore $D \log_e (1 + x)$ lies between

$$1 - x + x^2 - \ldots \pm x^n$$

and

$$\left. 1 - x + x^2 - \ldots \pm x^n \mp \frac{x^{n+1}}{1 - \theta} \right\} \quad \ldots\ldots\ldots\ldots(2),$$

for all values of x exceeding $-\theta$, and of these expressions the second is the greater, except for positive values of x and even values of n taken together*.

---

* If $x = 0$, $D \log_e (1 + x)$ is equal to both of the expressions (2).

85. Differentiation of simple functions involved. Now*, the incrementary ratio of the function x from x to $x + h$ is $[(x + h) - x]/h$, i.e. 1; and therefore the sequence of incrementary ratios for this function x, corresponding to the sequence $h_1, h_2, h_3, \ldots$, is $1, 1, 1, \ldots$, which is convergent and has the unique limit 1: or *the function x is differentiable and its derivative is 1 for all values of x.*

Again, the incrementary ratio of the function x^2 from x to $x + h$ is $[(x + h)^2 - x^2]/h$, i.e. $2x + h$; and therefore the sequence of incrementary ratios is $2x + h_1, 2x + h_2, \ldots$, which is evidently convergent and has the unique limit $2x$ if the sequence $h_1, h_2, \ldots$ has the unique limit 0; or *the function x^2 is differentiable and its derivative is $2x$ for all values of x.*

Also the incrementary ratio for x^3 is $[(x + h)^3 - x^3]/h$, i.e.
$$3x^2 + 3hx + h^2;$$
and therefore the sequence of incrementary ratios, viz.
$$3x^2 + 3h_1 x + h_1^2, \; 3x^2 + 3h_2 x + h_2^2, \ldots,$$
is convergent and has the unique limit $3x^2$; or *the function x^3 is differentiable and its derivative is $3x^2$ for all values of x.* And so on.

In general, if n is any positive integer, the incrementary ratio of the function x^n from x to $x + h$ is $[(x + h)^n - x^n]/h$, which, by the inequality (iii) of p. 29, lies between $n(x + h)^{n-1}$ and nx^{n-1}. But the sequence
$$n(x + h_1)^{n-1}, \; n(x + h_2)^{n-1}, \ldots$$
is convergent and tends to nx^{n-1} if the sequence $h_1, h_2, \ldots \to 0$; for $[n(x + h)^{n-1} - nx^{n-1}]/h$ lies between $n(n-1)(x + h)^{n-2}$ and $n(n-1)x^{n-2}$, whence the numerator, $n(x + h)^{n-1} - nx^{n-1}$, is less numerically than the numerically greater of $n(n-1)(x + h)^{n-2}h$ and $n(n-1)x^{n-2}h$, and therefore the sequence
$$n(x + h_1)^{n-1} - nx^{n-1}, \; n(x + h_2)^{n-1} - nx^{n-1}, \ldots \to 0,$$
whence the sequence
$$n(x + h_1)^{n-1}, \; n(x + h_2)^{n-1}, \ldots \to nx^{n-1}.$$
Therefore the sequence of incrementary ratios
$$\frac{(x + h_1)^n - x^n}{h_1}, \; \frac{(x + h_2)^n - x^n}{h_2}, \ldots$$

---

* The student familiar with the facts of elementary differential calculus will realise the truth of the equalities (3) and (4) below without the discussion of Par. **85.**

itself is convergent and tends to nx^{n-1}; i.e. *the function x^n is differ-entiable and its derivative is nx^{n-1} for all values of x.*

We notice further that the incrementary ratios of $2x$, $2x^2$, ... $2x^n$ are twice the incrementary ratios of x, x^2, ... x^n; and therefore

$$D2x = 2Dx = 2, \; D2x^2 = 2Dx^2 = 4x, \; ... \; D2x^n = 2Dx^n = 2nx^{n-1};$$

and in general, if k is any fixed (positive or negative) number (i.e. independent of x),

$$Dkx = kDx = k, \; Dkx^2 = kDx^2 = 2kx, \; ... \; Dkx^n = kDx^n = knx^{n-1}.$$

Hence, in particular, we have

$$Dx = 1, \; D\left(\tfrac{1}{2}x^2\right) = \tfrac{1}{2} . 2x = x, \; D\left(\tfrac{1}{3}x^3\right) = \tfrac{1}{3} . 3x^2 = x^2, \; ...$$

$$D\left(\frac{1}{n+1} x^{n+1}\right) = \frac{1}{n+1} (n+1) x^n = x^n.$$

Finally if we have a function, such as $3x + 4x^2 - 6x^3$, composed of several of the functions just considered combined by signs of addition and subtraction, the incrementary ratio of the compound function is just the same combination of the incrementary ratios of the separate functions as the combined function is of the separate functions. Such a compound function is therefore differentiable and its derivative is the same combination of the derivatives of the separate functions as the compound function is of the separate functions. In the case instanced,

$$\begin{aligned}
D\left(3x + 4x^2 - 6x^3\right) &= D\left(3x\right) + D\left(4x^2\right) - D\left(6x^3\right) \\
&= 3Dx + 4Dx^2 - 6Dx^3 \\
&= 3 . 1 + 4 . 2x - 6 . 3x^2 \\
&= 3 + 8x - 18x^2.
\end{aligned}$$

86. Applying these results to the compound function

$$x - x^2/2 + x^3/3 - ... \pm x^{n+1}/(n+1),$$

n being any positive integer, we have

$$\begin{aligned}
D\left[x - x^2/2 + ... \pm x^{n+1}/(n+1)\right] &= Dx - D\left(x^2/2\right) + D\left(x^3/3\right) \\
&\quad - ... \pm D\left[x^{n+1}/(n+1)\right] \\
&= Dx - \tfrac{1}{2}Dx^2 + \tfrac{1}{3}Dx^3 - ... \pm \frac{1}{n+1} Dx^{n+1} \\
&= 1 - \tfrac{1}{2} . 2x + \tfrac{1}{3} . 3x^2 - ... \pm \frac{1}{n+1} (n+1) x^n \\
&= 1 - x + x^2 - ... \pm x^n \quad\quad\quad ....(3),
\end{aligned}$$

which is the first of the two expressions (2) of p. 115; and

$$D\left[x - x^2/2 + x^3/3 - \ldots \pm x^{n+1}/(n+1) \mp \frac{1}{1-\theta}\frac{x^{n+2}}{n+2}\right]$$

$$= Dx - \tfrac{1}{2}Dx^2 + \tfrac{1}{3}Dx^3 - \ldots \pm \frac{1}{n+1}Dx^{n+1} \mp \frac{1}{(n+2)(1-\theta)}Dx^{n+2}$$

$$= 1 - x + x^2 - \ldots \pm x^n \mp \frac{1}{1-\theta}x^{n+1} \quad\ldots\ldots\ldots\ldots\ldots\ldots\ldots\ldots(4),$$

which is the second of the two expressions (2).

87. Graphical illustration. Reverting once more temporarily to the geometrical ideas of graphs, let us interpret the results (2), (3) and (4) and see what we may expect to deduce from them. Let us consider first the simple case when $n = 1$.

Our results may now be stated: $D\log_e(1+x)$ lies between $D(x - x^2/2)$ and $D[x - x^2/2 + x^3/3(1-\theta)]$, where θ is some fixed number between 0 and 1,—except for $x = 0$ when all three expressions are equal.

Sketch the graphs of $y = x - x^2/2$ and $y = x - x^2/2 + x^3/3(1-\theta)$. The graphs are drawn roughly in Fig. 3 with $\theta = 2/3$.

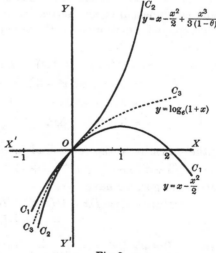

Fig. 3.

Compare with these two graphs,—which we will call C_1 and C_2,—that of $\log_e(1+x)$,—which we will call C_3.

When $x = 0$, $\log_e(1+x) = 0$, i.e. the graph C_3 of $\log_e(1+x)$ passes through the origin O, through which both the graphs C_1 and C_2 pass.

Also the slope of the graph of $\log_e(1+x)$ for all values of x is intermediate between the slopes of the two graphs C_1 and C_2.

It is therefore geometrically intuitively evident that *the graph C_3 lies entirely between the two graphs C_1 and C_2.*

Granting this intuition, we have obtained the result that $\log_e(1+x)$ *lies between* $x - x^2/2$ *and* $x - x^2/2 + x^3/3\,(1-\theta)$ *for all values of x exceeding* $-\theta$, θ *being any positive number less than* 1 *.

If now, for example, x is positive and less than ·1 say, θ can be taken to be 0 and $x^3 < ·001$, and therefore $x^3/3\,(1-\theta) < ·0003$, so that we have obtained the value of $\log_e(1+x)$ expressed in the simple rational calculable form $x - x^2/2$, correct to within ·0003.

This result is itself an improvement on the results of the last section. It is moreover evident that the argument used is general, and will shew similarly that $\log_e(1+x)$ lies between

$x - x^2/2 + \ldots \pm x^n/n$ and $x - x^2/2 + \ldots \pm x^n/n \mp x^{n+1}/(n+1)(1-\theta)$,

for all values of the integer n, however great*; and therefore, by taking n sufficiently great, *we can calculate the value of* $\log_e(1+x)$ *to any desired degree of accuracy* if x is small and positive, or in fact, if x is any number between -1 and $+1$ (i.e. $-1 < x < 1$).

88. General theorem on derivatives. We prefer however not to rely on "geometrical intuition,"—a misleading, if suggestive, process. It therefore behoves us now to substantiate, by a strict analytical proof, the deductions we have thus made. We must *prove* that if y_1, y_2, and y_3 are any three functions of x which are differentiable for all values of x in a certain range†, and whose derivatives throughout that range satisfy the relation

$$Dy_1 \leqslant Dy_3 \leqslant Dy_2,$$

and if the three functions all have the same value for the lower bound of the range, then throughout the range,

$$y_1 \leqslant y_3 \leqslant y_2.$$

This will follow at once if we can prove the following theorem.

If y is a function of x which, for all values of x in a certain range (a, b) and for the upper and lower bounds, b, a, of the range,

---

* Or equal to both expressions if $x=0$.

† A *range* is any connected set of values of x. The range (a, b) means all the real numbers between a and b, with or without the addition of either or both of the *upper* and *lower bounds*, b and a. If a range (a, b) includes the upper and lower bounds (as in the above theorem) it is called an *interval*, and the bounds a and b are called the *end-points*. If $a < x < b$, x is said to be *interior* to the range or interval (a, b).

is differentiable and has its derivative positive (or zero) (i.e. $Dy \geqslant 0$) whilst the function y has the value 0 when $x =$ the lower bound a, then $y \geqslant 0$ throughout the range; or, if $y = 0$ when $x =$ the upper bound b, then $y \leqslant 0$ throughout the range.

Because, assuming this theorem, if $Dy_1 \leqslant Dy_3$, we have $D(y_3 - y_1) \geqslant 0$, and if $y_1 = y_3$ for x equal to the lower bound of the range, $y_3 - y_1 = 0$ for that value of x, and therefore $y_3 - y_1 \geqslant 0$ throughout the range, i.e. $y_1 \leqslant y_3$; and similarly $y_3 \leqslant y_2$.

89. Proof. To prove the theorem we have:

For definiteness take the first part of the theorem and suppose that the range of values for x is $(0, K)$, where $K > 0$. If the theorem is untrue there will be at least one number (k say) for which the value of the function, $y(k)$ say, is negative.

The incrementary ratio of y from 0 to k, which

$$= [y(k) - y(0)]/k = y(k)/k,$$

would then be negative.

From this it would follow that at least one of the two incrementary ratios:

(1) that from 0 to $k/2$, viz. $y(k/2)/(k/2)$,

(2) that from $k/2$ to k, viz. $[y(k) - y(k/2)]/(k/2)$,

is also negative, and in fact $\leqslant y(k)/k$, for the sum of these two incrementary ratios is twice the incrementary ratio between 0 and k.

If (e.g.) the first of these ratios is negative and $\leqslant y(k)/k$, it will follow similarly that at least one of the two incrementary ratios:

(3) that from 0 to $k/4$, (4) that from $k/4$ to $k/2$,

is negative and $\leqslant y(k)/k$.

If (e.g.) the second of these ratios is negative and $\leqslant y(k)/k$, it will follow again that at least one of the two incrementary ratios:

(5) that from $k/4$ to $3k/8$, (6) that from $3k/8$ to $k/2$,

is negative and $\leqslant y(k)/k$.

And so on, repeating this halving process indefinitely.

We thus obtain an unending succession of negative incrementary ratios,—for the ranges $(0, k)$, $\left(0, \dfrac{k}{2}\right)$, $\left(\dfrac{k}{4}, \dfrac{k}{2}\right)$, ...,—which are such that the sequence of lower bounds $0, 0, k/4, \ldots$, is non-decreasing, is bounded above, and therefore $\nearrow$ a unique limit which $\geqslant 0$ and $\leqslant k$, and the sequence of upper bounds, $k, k/2, k/2, \ldots$, is non-increasing, is bounded below, and therefore ↘ a unique limit which ⩾ 0 and ⩽ k; and these two limits are the same, since the sequence of differences between corresponding terms of the two sequences, viz. k, $k/2$, $k/4$, …, → 0.

Call this common limit α.

This number α will belong to all the above ranges, and it will be possible to find a sequence, h_1, h_2, h_3, …, tending to the limit 0, such that the corresponding incrementary ratios,

$$\frac{y\,(\alpha + h_1) - y\,(\alpha)}{h_1}, \quad \frac{y\,(\alpha + h_2) - y\,(\alpha)}{h_2}, \quad …,$$

are all negative, and in fact, ⩽ the negative number $y\,(k)/k$. If this sequence of incrementary ratios has a limit, the limit must ⩽ this negative number $y\,(k)/k$.

But, by hypothesis, the function $y\,(x)$ is differentiable for the value α, and the derivative ⩾ 0, i.e. any such sequence of incrementary ratios as this has the unique limit $Dy\,(\alpha) \geqslant 0$.

Our supposition that there was a value of x, viz. k, for which $y\,(k) < 0$, involving the existence of such a sequence of incrementary ratios less than or equal to the negative number $y\,(k)/k$, is untenable; and the theorem is proved*.

90. Corollaries. The slightly modified theorem where the signs "⩾" and "⩽" are replaced by ">" and "<" respectively is proved similarly. The corresponding theorems when the derivative is negative (or not positive) are also proved similarly, or deduced by considering the function $-y$.

An important theorem which is needed below (Chapter IV, § 6, and elsewhere) is:

If the derivative Dy of a function y is zero throughout a range then the function y is constant throughout the range, i.e. y has the same value for all values of x belonging to the range.

---

* The student is unlikely on a first reading to appreciate fully the above arithmetical proof. The idea of the proof is more easily grasped if recourse be had once more to the graph. The above proof may then be expressed in the form:

If there is a point P of the graph (which passes through the origin O) *below* the x axis, then the chord OP will slope *downwards*, and therefore if Q be the point on the graph midway (horizontally) between O and P, at least one of the chords OQ, QP, must also slope downwards (and at least as steeply as OP); whence again if R be the mid-point of the graph between the ends of this downward-sloping chord, one at least of the chords so formed, having one extremity at R, will also slope downwards as steeply as OP. Proceeding in this way we see that there must then be a point, A say, at which the tangent to the graph slopes downwards,—contradictory to the supposition that the graph everywhere slopes upwards or is horizontal (because the derivative, i.e. the slope, is positive or zero).

For, if c is the value of y for x equal to the lower bound of the range, a say, and if y' denotes the function $y - c$, we have

$$Dy' \geqslant 0 \text{ and } Dy' \leqslant 0$$

throughout the range and

$$y' = 0 \text{ for } x = a;$$

whence

$$y' \geqslant 0 \text{ and } y' \leqslant 0,$$

i.e.

$$y' = 0, \text{ or } y = c$$

throughout the range.

The above theorems are true even if nothing is known as to the existence or values of Dy at the bounds a, b, provided the function y is known to be "continuous on the right" at a and "continuous on the left" at b, in the terminology of the next Chapter (p. 139). The above proof (modified) will apply.

91. Upper and lower approximations to $\log_e (1 + x)$. From (2), (3) and (4) above, we know that, for $x > -1$ (and $x \neq 0$), $D \log_e (1 + x)$ lies between $D \{x - x^2/2 + \dots \pm x^{n+1}/(n+1)\}$ and

$$D \left\{ x - x^2/2 + \dots \pm x^{n+1}/(n+1) \mp \frac{1}{1-\theta} \frac{x^{n+2}}{n+2} \right\}.$$

Also, of the two derivatives,

$$D \left[\log_e (1 + x) - \{x - x^2/2 + \dots \pm x^{n+1}/(n+1)\} \right],$$

and

$$D \left[\log_e (1 + x) - \left\{ x - x^2/2 + \dots \pm x^{n+1}/(n+1) \mp \frac{1}{1-\theta} \frac{x^{n+2}}{n+2} \right\} \right],$$

the first is positive and the second negative for all (non-zero) values of x greater than $-\theta$ (where θ is any number between 0 and 1)[*], whilst both $= 0$ for $x = 0$. Further the functions in square brackets, viz.

$$\log_e (1 + x) - \{x - x^2/2 + \dots \pm x^{n+1}/(n+1)\}$$

and

$$\log_e (1 + x) - \left\{ x - x^2/2 + \dots \pm x^{n+1}/(n+1) \mp \frac{1}{1-\theta} \frac{x^{n+2}}{n+2} \right\},$$

are zero for the value $x = 0$.

Therefore, by the theorem of p. 119, of these two functions last written, the first must be positive and the other negative for all

---

[*] Except when n is even; in which case the first and second of the two derivatives mentioned are respectively positive and negative only for negative values of x and are respectively negative and positive for positive values of x.

positive values of x, and the first negative and the other positive for all negative values of x exceeding $-\theta^*$.

Hence, whatever real number x may be greater than -1, $\log_e(1+x)$ lies between $x - x^2/2 + \ldots \pm x^{n+1}/(n+1)$ and

$$x - x^2/2 + \ldots \pm x^{n+1}/(n+1) \mp \frac{x^{n+2}}{(1-\theta)(n+2)} \quad \ldots\ldots(5),$$

where θ may be taken to be 0 if x is positive, or any positive number between $-x$ and 1 if x is negative; if $x = 0$, $\log_e(1+x)$ is equal to both the expressions (5). This is true for all values of the positive integer n.

92. Logarithmic series. If x lies between -1 and $+1$, or, more precisely, if $-1 < x \leqslant 1$, the difference between the two expressions (5), viz. $x^{n+2}/(1-\theta)(n+2)$, diminishes indefinitely as n increases indefinitely, and in fact,

$$\frac{x^{n+2}}{(1-\theta)(n+2)} \to 0 \text{ as } n \to \infty.$$

Hence the sequences whose nth terms are, respectively, the expressions (5) are, in these circumstances, convergent, and have $\log_e(1+x)$ for unique limit.

Or *the infinite series*

$$x - x^2/2 + x^3/3 - \ldots \quad \ldots\ldots\ldots\ldots\ldots\ldots(6)$$

is convergent for all values of x between -1 and 1 (1 included but -1 excluded) and its sum is then $\log_e(1+x)$.

The series (6) is called the *logarithmic series*.

By its help we are able to calculate logarithms to the unlimited degree of accuracy we need for the complete solution of the problem of the tabulation of logarithms, stated in the last section.

93. Calculation of Napierian logarithms. Certain Napierian logarithms,—i.e. logarithms to the base e,—can be calculated direct from the logarithmic series.

Thus, e.g. $\log_e 1{\cdot}1 = {\cdot}1 - ({\cdot}1)^2/2 + ({\cdot}1)^3/3 - \ldots$.

The error after n terms of this series (see p. 76 above) is less than the $(n+1)$th term, viz. $({\cdot}1)^{n+1}/(n+1)$, which $< {\cdot}00005$ if $n + 1 \geqslant 4$, i.e. if $n \geqslant 3$; so that $\log_e 1{\cdot}1$ is obtained correct to within ${\cdot}00005$ from the first three terms of this series; giving $\log_e 1{\cdot}1$ to four decimal places as ${\cdot}0953$.

---

* Except when n is even; in which case the first of the two functions is negative and the second positive for all values of x greater than $-\theta$.

Or again $\log_e 2 = 1 - 1/2 + 1/3 - \ldots$; but this series (see p. 77 above), though convergent, is only very slowly convergent, and in order to guarantee an error less than ·00005 it would be necessary to sum the first 20,000 terms; or even to obtain the result correct to within ·1, we need to sum the first nine terms.

In such cases, however, the problem may be tackled in modified ways. One simple way in which the logarithmic series may be modified for this purpose is as follows:

$$\log_e 2 = \log_e (4/2) = \log_e \frac{3+1}{3-1} = \log_e \left(\frac{1+1/3}{1-1/3}\right)$$

$$= \log_e (1+1/3) - \log_e (1-1/3)$$

$$= \left(\frac{1}{3} - \frac{1}{2}\frac{1}{3^2} + \frac{1}{3}\frac{1}{3^3} - \ldots\right) - \left(-\frac{1}{3} - \frac{1}{2}\frac{1}{3^2} - \frac{1}{3}\frac{1}{3^3} - \ldots\right)$$

$$= 2\left(\frac{1}{3} + \frac{1}{3}\frac{1}{3^3} + \frac{1}{5}\frac{1}{3^5} + \ldots\right) \quad \ldots\ldots\ldots\ldots\ldots\ldots\ldots\ldots(7),$$

by using theorem A of p. 80.

The error after n terms of this series

$$= 2\left(\frac{1}{2n+1}\frac{1}{3^{2n+1}} + \frac{1}{2n+3}\frac{1}{3^{2n+3}} + \ldots\right)$$

$$< 2\frac{1}{2n+1}\frac{1}{3^{2n+1}}(1 + 1/3^2 + 1/3^4 + \ldots)$$

$$< \frac{2 \cdot 1}{(2n+1)\, 3^{2n+1}\left(1 - \dfrac{1}{3^2}\right)} = \frac{2}{8(2n+1)\, 3^{2n-1}}$$

$$< \frac{1}{8n \cdot 3^{2n-1}};$$

and therefore, in order to obtain $\log_e 2$ correct to within ·1, one term of this series (7) will suffice, or, to obtain $\log_e 2$ correct to within ·000005, five terms of the series will suffice, for $8 \cdot 5 \cdot 3^9 > 20{,}000$ and therefore $1/8 \cdot 5 \cdot 3^9 < ·000005$.

The result is:

$$
\begin{aligned}
s_5 = \quad &·666667 \\
+ &·024691 \\
+ &·001646 \\
+ &·000131 \\
+ &·000011 \\
\hline
= &·693146
\end{aligned}
$$

and $E_5 < 1/8 \cdot 5 \cdot 3^9 < ·000002$; whence $\log_e 2$, correct to five decimal places, is ·69315.

In general, if x is any positive number greater than 1, integers m and n can always be found such that $m > n$ and $x = \dfrac{m+n}{m-n}$, $\left(\text{or, if } x < 1,\ x = \dfrac{m-n}{m+n}\right)$. We have then

$$\log_e x = \log_e \frac{m+n}{m-n} = \log_e \left(1 + \frac{n}{m}\right) - \log_e \left(1 - \frac{n}{m}\right)$$

$$= \left[\frac{n}{m} - \frac{1}{2}\frac{n^2}{m^2} + \frac{1}{3}\frac{n^3}{m^3} - \cdots\right] - \left[-\frac{n}{m} - \frac{1}{2}\frac{n^2}{m^2} - \frac{1}{3}\frac{n^3}{m^3} - \cdots\right]$$

$$= 2\left[\frac{n}{m} + \frac{1}{3}\frac{n^3}{m^3} + \frac{1}{5}\frac{n^5}{m^5} + \cdots\right].$$

The sum of this convergent series of positive terms can always be evaluated as above.

94. Calculation of common logarithms and $\log_{10} e$. In order to calculate *common logarithms* we need to calculate first the Napierian logarithms and transform by means of the formula (p. 98)

$$\log_{10} x = \log_{10} e \cdot \log_e x.$$

To calculate $\log_{10} e$ correct to four decimal places we have

$$\log_e 10 = \log_e (8 + 2) = \log_e 8 + \log_e (1 + \tfrac{1}{4})$$
$$= 3 \log_e 2 + \log_e (1 + \tfrac{1}{4})$$
$$= 2{\cdot}079435 + \left(\frac{1}{4} - \frac{1}{2}\frac{1}{4^2} + \cdots\right)$$

with an error of defect of less than ·000006 (p. 124).

Five terms of the series in brackets will suffice to ensure an error of less than ·00005, for $E_5 < 1/6 \cdot 4^6 = 1/24576 < {\cdot}00005$.

The sum of the first five terms of the series

$$= s_5 = \quad {\cdot}250000 - {\cdot}031250$$
$$+ {\cdot}005208 - {\cdot}000977$$
$$+ {\cdot}000195$$
$$= \quad {\cdot}255403 - {\cdot}032227 = {\cdot}223176.$$

Hence, with an error of defect of less than ·000006 or an error of excess of less than ·00005, we have

$$\log_e 10 = 2{\cdot}079435 + {\cdot}223176 = 2{\cdot}302611.$$

Finally $\log_{10} e = 1/\log_e 10 = 1/2{\cdot}3026 = {\cdot}43429$, with an error of less than ·00002; whence, to four decimal places, $\log_{10} e = {\cdot}4343$.

Common logarithms, such as $\log_{10} 2$, can now be found thus:
$$\log_{10} 2 = \log_{10} e \cdot \log_e 2$$
$$= {\cdot}4343 \times {\cdot}6931, \text{ with an error of less than } {\cdot}00004 \text{ (p. 124)},$$
$$= {\cdot}30101, \text{ with an error less than } {\cdot}00004,$$
$$= {\cdot}3010 \text{ to four places of decimals.}$$

The student will have no difficulty in calculating, correct to four places of decimals, the common logarithms of all the integers from

1 to 10 and those of the interpolated fractional numbers,—and thus to construct a "four-figure table" of common logarithms. In using series for such calculations it is essential (see Chapter I, § 7) always to estimate the error involved, so as to be able to guarantee the degree of accuracy obtained and to save the extra labour which would otherwise often be involved in evaluating terms of the series which are not in fact relevant.

EXAMPLES XI.

1. Use the result that, if x is positive, $D\left[\log_e(1+x)-x\right]$ is negative and $>-x$ to shew that, if x is positive, $\log_e(1+x)<x$ and $>x-x^2/2$, sketching the graphs of all the functions concerned.

2. Given that $D\left[\log_e(1+x)-x+x^2/2\right]$ lies between 0 and $\cdot01$ for all values of x between 0 and $\frac{1}{10}$, shew that $\log_e 1\cdot1$ lies between $\cdot095$ and $\cdot096$.

3. Obtain with the help of trigonometrical tables the angles of inclination to the x axis of the tangents to the graph of $\log_e x$ at the points where $x=1/100, 1/2, 1, 2, 100$; and of the tangent to the graph of $\log_{10} x$ at the point where $x=10$. Compare the trigonometrical tangent of this last angle with the numerical results of Ex. 8, p. 107.

4. Express in precise analytical language the theorems described as:

   (a) If the slope of a graph is at all points zero, then the graph is a straight line parallel to the x axis.

   (b) If the slope of a graph which passes through the origin lies between $\pm 1/10$ at all points, then, for all values of x between $\pm k$, the ordinate y of the point on the graph lies between $\pm k/10$.

Give strict analytical proofs of these theorems.

5. Express the theorem of p. 119, and the corollaries of pp. 121—122, in general geometrical language (as in Ex. 4). Give proofs of these theorems (a) in geometrical language, as in the footnote to p. 121, and (b) in precise analytical language.

6. Calculate to four places of decimals the common logarithms of the integers between 2000 and 2010.

7. Shew that if $y>1$ whilst $\sqrt[n]{y}$ differs from 1 by less than 10^{-s}, then, correct to within 10^{-2s+1}, $\log_e y=n(\sqrt[n]{y}-1)$, where n and s are positive integers and $n<20$.

Under what conditions is the formula $\log_{10} y=\dfrac{\sqrt[n]{y}-1}{\sqrt[n]{10}-1}$ correct to the same degree of approximation?

8. Use the formulae of Ex. 7 to calculate $\log_{10} e$ and $\log_{10} 3$ correct to four decimal places.

9. Prove the formula
$$\log_{10} y=\frac{(\sqrt[n]{y}-1)-\frac{1}{2}(\sqrt[n]{y}-1)^2+\ldots}{(\sqrt[n]{10}-1)-\frac{1}{2}(\sqrt[n]{10}-1)^2+\ldots},$$
where n is such an integer that $\sqrt[n]{y}$ and $\sqrt[n]{10}$ lie between 1 and 2, and use it to calculate $\log_{10} 2$ correct to four decimal places.

10. Shew that if the positive real number a is less than 1, the fundamental inequality of p. 110 remains true when a is the base of the logarithms, provided the signs of inequality are reversed.

11. Shew that if the sequence of real numbers (positive or negative)
$$x_1, \; x_2, \; x_3, \; \ldots \to 0,$$
then the sequence
$$(1+x_1)^{\frac{1}{x_1}}, \; (1+x_2)^{\frac{1}{x_2}}, \; \ldots \to e.$$

12. Shew that the logarithmic series is not convergent if $x>1$ or $x \leqslant -1$.

[If $|x|>1$ the terms of the series do not tend to zero. If $x=-1$ the series is the known non-convergent harmonic series (vi) of p. 67, with signs changed.]

13. Prove by induction that, if n is a positive integer, x^n is differentiable for all values of x, and $Dx^n = nx^{n-1}$. Also x^{-n} is differentiable (except for $x=0$), and $Dx^{-n} = -nx^{-n-1}$.

[The incrementary ratio for x^{n+1}, viz. $[(x+h)^{n+1} - x^{n+1}]/h$, can be rewritten
$$\frac{(x+h)^n - x^n}{h}(x+h) + x^n,$$
which tends to $(n+1)x^n$ if x^n is differentiable and $Dx^n = nx^{n-1}$. Similarly for x^{-n}.]

14. By considering the derivatives of the functions, $y_1 = 1 + nx$ and $y_2 = (1+x)^n$, and using the methods of the text (pp. 118—122), obtain an independent proof of the inequality (ii) of p. 29.

15. Obtain similar proofs of the inequalities (i) and (iii) of p. 29.

[For (i) consider the functions $y_1 = (x+b)/2$, $y_2 = \sqrt{(xb)}$. For (iii) consider $y_1 = nb^{n-1}(x-b)$, $y_2 = x^n - b^n$, $y_3 = nx^{n-1}(x-b)$.]

§ 5. THE EXPONENTIAL SERIES

95. Evaluation of powers. The problem inverse to that dealt with in the last two sections, viz. that of the evaluation of the powers of a given base (or anti-logarithms of numbers to a given base), leads to a similar solution. The power a^x can be expressed as the sum of an infinite series similar in type to the logarithmic series for $\log_e(1+x)$. A tabulation of anti-logarithms is not, of course, a practical necessity, once tables of logarithms are constructed, for it is easy (it is in fact customary) to use a table of logarithms for the dual purpose of finding logarithms of given numbers and of finding the numbers corresponding to given logarithms; but theoretically the solution of this problem is, like the solution of the corresponding problem for logarithms, of considerable interest and importance. The importance of the exponential function in the physical sciences is moreover sufficient justification for the study of this problem. The problem can be tackled on the same lines as those which have proved successful in the last section in the establishment of the logarithmic series.

96. Lemma on differentiation of a product. We need a simple lemma concerning the derivative of the product of two functions:

If u and v are any two functions of x which are differentiable, then the product uv is also differentiable and
$$D(uv) = u\,Dv + v\,Du.$$

To prove this we argue:

The incrementary ratio of uv from x to $x+h$ is
$$[u(x+h)\,v(x+h) - u(x)\,v(x)]/h$$
$$= u(x+h)\frac{v(x+h) - v(x)}{h} + v(x)\frac{u(x+h) - u(x)}{h},$$

where $u(x+h)$, etc. are written for the value of u corresponding to the value $x+h$ of x, etc.

But, if h takes on the values $h_1, h_2, \ldots$ of a sequence which tends to zero, the corresponding sequences for
$$[v(x+h) - v(x)]/h \text{ and } [u(x+h) - u(x)]/h$$
respectively tend to $Dv(x)$ and $Du(x)$, whilst the sequence for $u(x+h)$ must also $\to u(x)$, because the sequence for
$$u(x+h) - u(x) \text{ clearly} \to 0.$$

Hence the incrementary ratio for uv tends to
$$u(x)\,Dv(x) + v(x)\,Du(x). \qquad \text{Q.E.D.}$$

97. Derivative of e^x. Let x, h be any real numbers; put
$$y = e^x, \quad y + k = e^{x+h},$$
so that $x = \log_e y, \quad x + h = \log_e(y + k).$

The incrementary ratio for e^x with respect to x from x to $x+h$
$$= (e^{x+h} - e^x)/h = k/[\log_e(y + k) - \log_e y].$$

Now, if h has the values $h_1, h_2, \ldots$ of any sequence which tends to zero, k has the corresponding values
$$k_1 = e^{x+h_1} - e^x, \qquad k_2 = e^{x+h_2} - e^x,$$
$$= e^x(e^{h_1} - 1), \qquad = e^x(e^{h_2} - 1), \ldots,$$
and the sequence $k_1, k_2, \ldots$ is convergent and $\to 0$, for
$$(e^h - 1)/(e - 1)$$
lies between he^{h-1} and h, by the inequality of Ex. 2, p. 91.

But we have proved in the last section (p. 112), that if a sequence $k_1, k_2, \ldots \to 0$, the sequence

$$\{\log_e (y + k_1) - \log_e y\}/k_1, \quad \{\log_e (y + k_2) - \log_e y\}/k_2, \ldots \to 1/y.$$

Therefore the sequence of incrementary ratios for e^x from x to $x + h_1$, x to $x + h_2$, ..., which is

$$k_1/\{\log_e (y + k_1) - \log_e y\}, \quad k_2/\{\log_e (y + k_2) - \log_e y\},$$

tends to the unique limit y, i.e. e^x*.

The function e^x is differentiable and $De^x = e^x$.

Similarly we prove that e^{-x} is differentiable and $De^{-x} = -e^{-x}$.

A difficulty here arises in that the derivative obtained, e^x, is the function whose properties we are investigating and not a simpler function, as was the case with $\log_e (1 + x)$. It is clearly hopeless to expect to use the fact $De^x = e^x$ to *discover* a simple infinite series to represent e^x. We have, in fact, at this stage to resort to other methods to *suggest* what the result might be; and then we shall be able to use this method to *prove* the result.

98. Suggestion as to exponential series. We defined e as the limit of the sequence whose nth term is $(1+1/n)^n$. From our fundamental inequality of the last section it is easy to deduce that the sequence whose nth term is $(1+x/n)^{n/x}$ is also convergent and has the same limit e; whence it follows that the sequence whose nth term is

$$\left(1+\frac{x}{n}\right)^n \to e^x \quad \ldots\ldots\ldots\ldots\ldots\ldots\ldots\ldots\ldots (2).$$

But, by the binomial theorem (p. 22),

$$(1+x/n)^n = 1 + n\frac{x}{n} + \frac{n(n-1)}{2!}\frac{x^2}{n^2} + \ldots + \frac{x^n}{n^n}$$

$$= 1 + x + \frac{x^2}{2!}\left(1 - \frac{1}{n}\right) + \frac{x^3}{3!}\left(1 - \frac{1}{n}\right)\left(1 - \frac{2}{n}\right) + \ldots$$

$$+ \frac{x^n}{n!}\left(1 - \frac{1}{n}\right)\ldots\left(1 - \frac{n-1}{n}\right) \quad \ldots\ldots\ldots (3).$$

As $n \to \infty$ the first few terms of this expression (3) clearly $\to 1, x, x^2/2!, x^3/3!$, etc., and, though there appears to be some doubt as to how the terms near the end of the expression (3) behave, yet it at least seems plausible† that, as $n \to \infty$, the expression (3) tends to become the infinite series

$$1 + x + x^2/2! + x^3/3! + \ldots.$$

We will use this suggestion and proceed with the proof.

---

* The proof of the obvious theorem in sequences that if a sequence $s_1, s_2, \ldots \to$ a unique limit s (which $\neq 0$) then the sequence $1/s_1, 1/s_2, \ldots \to$ the unique limit $1/s$ is left to the student.

† See alternative proof below.

99. Proof of exponential expansion. For brevity write

$$s_n(x) \text{ for } 1 + x + x^2/2! + \ldots + x^n/n!,$$

n being any positive integer;

and y for $$e^x - s_n(x).$$

As in the last section (pp. 116—117), $s_n(x)$ is differentiable with respect to x and its derivative is

$$Ds_n(x) = D1 + Dx + D(x^2/2!) + D(x^3/3!) + \ldots + D(x^n/n!)$$

$$= D1 + Dx + D(x^2/2) + \frac{1}{2!}D(x^3/3) + \ldots + \frac{1}{(n-1)!}D(x^n/n)$$

$$= 0 + 1 + x + x^2/2! + \ldots + x^{n-1}/(n-1)!$$

$$= s_{n-1}(x).$$

Therefore

$$Dy = D\{e^x - s_n(x)\} = De^x - Ds_n(x) = e^x - s_{n-1}(x)$$

$$= e^x - s_n(x) + x^n/n! = y + x^n/n!.$$

Hence $$D(e^{-x}.y) = e^{-x}Dy + yDe^{-x},$$

by the lemma proved above,

$$= e^{-x}Dy - e^{-x}y = e^{-x}.x^n/n! \quad \ldots\ldots\ldots\ldots(4).$$

Therefore, if x is positive and less than some positive number, K, we have

$$0 < D(e^{-x}.y) < K^n/n! \ldots\ldots\ldots\ldots\ldots(5\,a),$$

or, if x is negative and greater than some negative number, $-K'$, we have

$$-e^{K'}K'^n/n! < D(e^{-x}.y) < 0$$

or $$\left.\begin{array}{c} 0 < D(e^{-x}.y) < e^{K'}K'^n/n! \end{array}\right\} \ldots\ldots\ldots\ldots(5\,b),$$

according as n is odd or even.

Now for the value $x = 0$, $y = e^x - s_n(x) = 1 - 1 = 0$ and $e^{-x}.y = 0$. Our theorem of the last section (p. 119) will therefore apply and we thus deduce that, for all positive values of x less than K, $e^{-x}.y$ is positive and less than $xK^n/n! < K^{n+1}/n!$; and for all negative values of x greater than $-K'$, $e^{-x}.y$ lies between 0 and $x(-K')^n e^{K'}/n!$, or, *a fortiori*, between 0 and $(-K')^{n+1}e^{K'}/n!$.

Now, whatever positive number K may be, $K^{n+1}/n! \to 0$ as $n \to \infty$; and similarly for $(-K')^{n+1}e^{K'}/n!$.*

---

* This may be proved thus: Let m be the integer next greater than K, so that $K/(m+1) < 1$. Then for all positive integral values of p,

$K^{m+p+1}/(m+p)! = (K^{m+1}/m!)\{K^p/(m+1)\ldots(m+p)\} < (K^{m+1}/m!)\{K/(m+1)\}^p;$

and therefore, as $n = m + p \to \infty$, $p \to \infty$

and $K^{n+1}/n' = K^{m+p+1}/(m+p)! < (K^{m+1}/m!)\{K/(m+1)\}^p \to 0.$

Hence as $n \to \infty$, $e^{-x} \cdot y \to 0$ and therefore $y \to 0$;

i.e. $\qquad\qquad e^x - s_n(x) \to 0$; whence $s_n(x) \to e^x$,

i.e. $\qquad\qquad 1 + x + x^2/2! + \ldots + x^n/n! \to e^x$,

or *the infinite series* $1 + x + x^2/2! + x^3/3! + \ldots$ *is convergent for all values of x and its sum is e^x.*

This series is called the *exponential series*.

100. Alternative proof. An alternative proof of this expansion can be obtained starting from the relation (3) above. We have only to prove that the sequence whose nth term is

$$1 + x + \frac{x^2}{2!}(1 - 1/n) + \frac{x^3}{3!}(1 - 1/n)(1 - 2/n) + \ldots + \frac{x^n}{n!}(1 - 1/n) \ldots \left(1 - \frac{n-1}{n}\right) \ldots (3)$$

tends to the sum of the infinite series $1 + x + x^2/2! + \ldots$, proved convergent.

If m is any number less than n we have that the expression (3)

$$\left.\begin{aligned} &= 1 + x + \frac{x^2}{2!}(1 - 1/n) + \ldots + \frac{x^m}{m!}(1 - 1/n) \ldots \left(1 - \frac{m-1}{n}\right) \\ &\quad + \frac{x^{m+1}}{(m+1)!}(1 - 1/n) \ldots (1 - m/n) + \ldots + \frac{x^n}{n!}(1 - 1/n) \ldots \left(1 - \frac{n-1}{n}\right) \end{aligned}\right\} \ldots \ldots (6).$$

Now, the line last written, if $x > 0$,

$$< \frac{x^{m+1}}{(m+1)!} + \frac{x^{m+2}}{(m+2)!} + \ldots + \frac{x^n}{n!}$$

$$< \frac{x^{m+1}}{(m+1)!}\left[1 + \frac{x}{m+1} + \left(\frac{x}{m+1}\right)^2 + \ldots + \left(\frac{x}{m+1}\right)^{n-m-1}\right]$$

$$< \frac{x^{m+1}}{(m+1)!}\frac{1}{1 - \dfrac{x}{m+1}} \text{ if } x < m+1$$

$$< \frac{x^{m+1}}{m!}\frac{1}{m+1-x};$$

or, if $x < 0$, the same terms $< x^{m+1}/(m+1)!$ numerically if $-x < m+2$.

Both the expressions $\dfrac{x^{m+1}}{m!}\dfrac{1}{m+1-x}$ and $\dfrac{x^{m+1}}{(m+1)!}$ are numerically less than any positive number ϵ, however small, if m is chosen sufficiently great,—as can be done (see footnote to p. 130).

The first line of expression (6) clearly tends to

$$1 + x + x^2/2! + \ldots + x^m/m! \text{ as } n \to \infty.$$

Hence, from (3),

$$(1 + x/n)^n = 1 + x + x^2/2! + \ldots + x^m/m! + A + B$$

(say), where $A \to 0$ as $n \to \infty$ and $B < \epsilon$ numerically.

But $\qquad\qquad\qquad (1 + x/n)^n \to e^x$.

Therefore $\qquad 1 + x + x^2/2! + \ldots + x^m/m! + B \to e^x$ as $n \to \infty$.

Therefore $1+x+x^2/2!+\dots+x^m/m!$ differs from e^x by less than ϵ; and it follows that the infinite series $1+x+x^2/2!+\dots$ is convergent and its sum $=e^x$. Q.E.D.

A third and simpler proof (using only the theorem of p. 119 and the relation $De^x=e^x$) is given in Ex. 5.

101. The exponential series differs from the logarithmic series in that it is convergent and its sum $= e^x$ for *all* real values of x, whereas the logarithmic series is convergent only for values of x such that $-1 < x \leqslant 1$ *. The rate of convergence of the exponential series will of course depend on the value of x. Thus if $x = \cdot 1$, the error after three terms will be less than $\cdot 0002$, whilst if $x = 10$, 33 terms will be needed to obtain the sum correct to within this same error.

The corresponding series for a^x and 10^x are:

$$a^x = e^{\log_e (a^x)} = e^{x \log_e a} = 1 + \log_e a + (x \log_e a)^2/2! + \dots$$

and $\quad 10^x = e^{x \log_e 10} = e^{(x/\mu)}$, where $\mu = \log_{10} e = \cdot 43429\dots$,

$$= 1 + \frac{x}{\mu} + \frac{1}{2!}\left(\frac{x}{\mu}\right)^2 + \dots.$$

EXAMPLES XII.

1. Calculate the common antilogarithms of $\cdot 5, 2\cdot 5, \bar{2}\cdot 5$ correct to within $\cdot 005$.

2. Evaluate to two significant figures e^x and e^{-x} for $x=0, \frac{1}{2}, 1, 2, 10, 100$,—using the exponential series or logarithmic tables.

3. Prove that if $c > -1$, then for all real values of x, $(1+c)^x > 1+xc$, except that, when $0 < x < 1$, the inequality is reversed, and when $x=0$ or 1, or $c=0$, $(1+c)^x = 1+xc$. (Compare inequality ii, p. 29 and Ex. 1, p. 91.)

[Consider the derivative $D[(1+c)^x - (1+xc)]$ and use the theorem of p. 119. Or apply method of Ex. 2, p. 36 to Ex. 2, p. 91. See also Ex. 14, p. 127.]

4. Shew that $Da^x = a^x . \log_e a$, where a is any positive number.

5. By argument similar to that of Ex. 3, prove successively that

$$e^x > 1, \ e^x > 1+x, \ e^x > 1+x+x^2/2!, \ e^x > 1+x+x^2/2!+x^3/3!, \dots$$

if x is positive, and

$$e^x < 1, \ e^x > 1+x, \ e^x < 1+x+x^2/2!, \dots$$

if x is negative. Deduce the exponential expansion.

[$De^x = e^x > 0$ for all values of x and $e^x = 1$ when $x=0$; therefore $e^x > 1$ for all positive values of x and $e^x < 1$ for all negative values of x. $D(e^x-1-x)=e^x-1$ and therefore >0 for $x>0$ and <0 for $x<0$, whilst $e^x-1-x=0$ when $x=0$; and therefore $e^x-1-x>0$ for all values of x. Again

$$D(e^x-1-x-x^2/2!)=e^x-1-x$$

* Ex. 12, p. 127 above.

and therefore >0 for all values of x, whilst $e^x - 1 - x - x^2/2! = 0$ when $x = 0$; therefore $e^x - 1 - x - x^2/2! > 0$ for $x > 0$ and <0 for $x < 0$. This process may be repeated indefinitely and gives $e^x > 1 + x + x^2/2! + \dots + x^n/n!$ for $x > 0$ and, if n is odd, also for $x < 0$; and $e^x < 1 + x + x^2/2! + \dots + x^n/n!$ for $x < 0$ if n is even. This suffices to establish the exponential series for negative values of x. To complete the proof for positive values of x we proceed: Also if $x < K$, $e^x < e^K = A$ say and we have as before $D(e^x - 1 - Ax) = e^x - A < 0$ for all values of x concerned; whence $e^x - 1 - Ax < 0$ for $x > 0$, and so on; whence $e^x < 1 + x + x^2/2! + \dots + x^n/n! + Ax^{n+1}/(n+1)!$. Since $Ax^{n+1}/(n+1)! \to 0$ as $n \to \infty$, it follows that the exponential series is convergent and its sum is e^x for all values of x. The student will find it instructive to sketch the graphs of the functions concerned, say for values of n up to 3 and x between -10 and 10.]

6. Prove that if $a > 1$, whatever number n may be, $a^x > x^n$ for all values of x exceeding a certain value.

[Let $y_n = a^x - x^n$, $y_{n-1} = Dy_n = a^x \log_e a - nx^{n-1}$, etc., $y_0 = a^x(\log_e a)^n - n!$. Then $Dy_1 = y_0 > 1$ if $x > [\log_e(1 + n!) - n \log_e \log_e a]/\log_e a = a_0$ say. For $x > a_0$, the slope of the graph of y_1 exceeds that of a straight line inclined at 45° to the x axis (because $\tan 45° = 1$) and therefore, however small (e.g. $-K$) y_1 may be when $x = a_0$, for some value of x (certainly for $x \geq a_0 + K$) y_1 is positive, and in fact $y_1 > 1$ for all values of x exceeding some value, say a_1,—by the theorem of p. 119. That is $Dy_2 > 1$ for $x > a_1$, and the argument can be repeated to prove $y_2 > 1$ for $x > a_2$; and so on. Finally $y_n > 1$ for $x > a_n$.]

7. If x is sufficiently great, $\log_e x < x^n < e^x < [x]! < x^x$, however great or small the *positive* number n may be ($n \neq 0$), or $\log_e x$ and e^x may be replaced by $\log_a x$ and a^x if a is any real number greater than 1. ($[x]$ denotes the integral part of x, i.e. the greatest integer not exceeding x.) In fact the sequences whose xth terms are the successive ratios $\log_e x/x^n$, etc. all have the unique limit 0. [See also Ex. 6, p. 99.]

8. The monotony of e^x and of $\log_e x$ are direct consequences (*via* the theorem of p. 119) of the facts $De^x = e^x > 0$ and $D\log_e x = 1/x > 0$ for all significant values of x, independently of any special knowledge of the functions e^x and $\log_e x$.

9. Shew that the series

$$\text{(i)} \quad 1 + \frac{x^2}{2!} + \frac{x^4}{4!} + \dots$$

and

$$\text{(ii)} \quad x + \frac{x^3}{3!} + \frac{x^5}{5!} + \dots$$

are convergent for all values of x and that the sums of the series are respectively $(e^x + e^{-x})/2$ and $(e^x - e^{-x})/2$.

CHAPTER III

FUNCTIONS

102. So far we have been concerned almost exclusively with *fixed* numbers: the letters employed have denoted numbers which, though most often indefinite, have been looked on as remaining the same throughout the operations applied to them. In many places, however, we have been implicitly concerned with the notion of a *variable*, i.e. of a letter capable of taking up various values. We wish now to direct our attention to the *variability* of the numbers.

The idea of a variable is itself of great practical utility in the sciences from the fact of the variability of almost all measured quantities. Thus the time, measured in solar or sidereal units, varies as the world's history progresses; or, while a train is travelling from one station to another, both the time and the distance travelled from the station vary; or, again, the temperature of a chemical mixture varies while the mixture chemically combines; or, the score in a cricket match varies as the game proceeds; and so on. While these practical variables are not capable of assuming all real values, as the completely general real variable is, yet they are capable of taking on all real values within certain ranges (except in the last example,—where the variable score is restricted to integral values). In addition, in such a case as that of the train, if we know the speed of the train we can express the variable distance from the station in terms of the variable time: or we may say that the distance is a *function* of the time.

In directing attention to the notion of variability we are led to the notion of a function. In this chapter we shall study these notions of variables and functions in various cases of simplicity and importance *.

§ 1. The Graph of x^2

103. Functions of a real variable. Let x be any real number, or,—as we may say in order to bring into prominence the possibilities of variability,—a real *variable*. We have already a graphical representation of this real variable in our straight line of Chapter I, § 3.

Suppose now y is a second number, or variable, depending on x; i.e. suppose we have some means whereby for every value of the

---

* We have to some extent already used the notions of variable and function in the last chapter. In this chapter we begin the study independently.

number x considered (which may include all real numbers), a definite corresponding number y can be found. Thus, for example, if we agreed that whatever x is y should be its double ($y = 2x$), we should know at once the number y corresponding to any given number x. When y can be determined in this way for every number x belonging to a certain class of numbers, we say that y *is a function of x*, defined for x belonging to that class. Thus if x is any real variable, $2x$, x^2, $1/x$ would be functions of x,—the first two defined for all real values of x, the last for all real values of x other than zero. We shall invariably use y to denote the *function* (or *dependent variable*), and x the original (or *independent*) *variable*. If the function is not specified, it is often denoted by $f(x)$. The value of y (or, simply, the number y) corresponding to a value, a, of x (or the number a) is then written $f(a)$*.

In analysis the functions with which we are mainly concerned are functions defined by means of algebraic and similar operations; that is to say such functions of x as

$$x^2, \; 2x^2 + 3x - 2, \; x/(1 - x), \; (1 + x)^3, \; (1 + x)^n,$$
$$10^x, \; a^x, \; \log_e x, \; \log_a x, \; (1 + 1/x)^x, \; (1 - 1/x)^x,$$

and sums of infinite series, such as

$$1 + x + x^2/2 + x^3/3 + \dots.$$

104. The function x^2. Let us begin with the function x^2.

For convenience we shall denote the function by y.

For different values of x, y will have different values. If we wish, we can draw up a *table* giving the values of y corresponding to any number of values of x, thus:

x	0	1	2	3	100	1/2	1/3	1/4	2/3	3/4	2/5	-1	-2	-3
y	0	1	4	9	10000	1/4	1/9	1/16	4/9	9/16	4/25	1	4	9

x	-100	$-1/2$	$-1/3$	$-1/4$	$-2/3$	$-3/4$	$-2/5$	......
y	10000	1/4	1/9	1/16	4/9	9/16	4/25	......

* In this definition of a function,—appropriate to the case of functions of a *real* variable,—it is implied that the value of y corresponding to a given value of x is *unique*; and the function is *one-valued*. For functions of a *complex* variable it is essential to consider *multiple-valued* functions,—where y may have more than one value corresponding to a single value of x. See Appendix. Multiple-valued functions of real variables are definable, but are not important in analysis.

Such a table, if arranged in orderly fashion (e.g. on the model of logarithmic and other tables), would give at a glance the value of y corresponding to any tabulated value of x. Such tables, however, do not give at a glance any kind of a *picture* of the function. To get such a picture a graphical representation is desirable. In Chapter I, § 3, we had occasion to represent the single real variable x by points along a straight line. We could represent our function by using a *second* such straight line for the values of y. But the extension best suited for the graphical representation of such a functional correspondence of one variable, y, with another, x, is that obtained by applying the elementary ideas of *Cartesian* (or *analytical* or *coordinate*) *geometry*.

105. The graph of x^2. The principle of Cartesian geometry is the correspondence of a point of a plane with a pair of numbers, in the same way as our representation of Chapter I, § 3 relied on the correspondence of a point in a line with a single number. The actual machinery effecting the correspondence is, to some extent, arbitrary; but the most useful method is to erect through a point O, called the *origin*, two mutually perpendicular straight lines, $X'OX$, $Y'OY$, called the *axes of reference*, and to assign the two numbers, or *coordinates*, x, y, to the point P, x (the *abscissa*) being the distance (measured in terms of some appropriate unit) from $Y'OY$ of the point P, and y (the *ordinate*) the distance from $X'OX$; with the convention that these distances are to be considered positive or negative according to their directions. It is customary to agree that, taking OX to be horizontal and OY vertical, x is positive if the point P represented by (x, y) lies to the right of $Y'OY$ and negative if to the left; and y is positive or negative according as P is above or below $X'OX$. See Fig. 4.

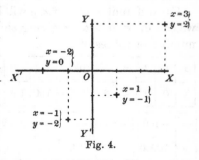

Fig. 4.

If now we take the pairs of numbers (x, y) which we have tabulated for the relation $y = x^2$, and mark off on the plane on which our axes OX, OY are drawn, the corresponding points, we obtain a figure like that of Fig. 5, consisting of an indefinite number of points.

It is seen, by taking more and more values of x,—e.g. by taking all x's differing by ·1, and then all x's differing by ·01, and then all x's differing by ·001, and so on,—that the points lie on some kind of *curve*. This curve is the *graph of the function* x^2. We cannot as yet say definitely that the graph is a *continuous* curve: it may be broken. But the graph in any case gives some kind of picture of the whole function simultaneously. Properties and questions at once suggest themselves which were obscured in the tabular representation.

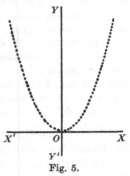

Fig. 5.

106. Monotony. In the first place it seems probable, from its appearance, that the graph slopes upwards everywhere to the right of the origin O,—or, in more precise language, if P and Q are two points on the graph (to the right of the origin) and Q is to the right of P, then Q is also above P. Is this surmise definitely provable?

That is, in analytical language:

Does it follow from the relation $y = x^2$ that if $x_1 > x_2$, x_1 and x_2 being any two positive real numbers, then $y_1 > y_2$, y_1 and y_2 being the y's corresponding to x_1 and x_2?

That is, can we prove that if

$$x_1 > x_2 > 0 \text{ then } x_1{}^2 > x_2{}^2?$$

The answer is evident. This inequality is indeed an immediate consequence of the law of inequalities that if $a > b$ and $c > d$, a, b, c, d being positive, then $ac > bd$; for we have $x_1 > x_2$ and $x_1 > x_2$ and therefore $x_1 x_1 > x_2 x_2$, i.e. $x_1{}^2 > x_2{}^2$.

At the same time we see that if $x_1 > x_2$ and x_1 and x_2 are both negative, then $x_1{}^2 < x_2{}^2$.

What we have proved may be stated: *the function x^2 steadily increases as x increases when x is positive and decreases as x increases when x is negative.* When a function of x either steadily increases (or never decreases) throughout a certain range of values of the variable x (e.g. for $a \leqslant x \leqslant b$) or steadily decreases (or never increases) throughout the range, we say that the function is *monotone* (or *monotonic*) throughout that range.

Monotone functions belong to a wider class of functions, known as *functions of bounded variation*. A function is said to be of bounded variation throughout a range (or over, or in, a range) if it can be expressed as the sum or difference of two functions each of which is bounded and monotone throughout the range. Thus the function x^2 is of bounded variation over any range because it can be expressed as the difference $y_1 - y_2$, where y_1 is the non-decreasing function which is zero for all negative values of x and equals x^2 for all positive (or zero) values of x, and y_2 is the non-decreasing function which equals $-x^2$ for all negative values of x and is zero for all positive (or zero) values of x.

Another definition is given in Ex. 14, p. 152 below.

107. Continuity. A second important question has already arisen: Is the curve *continuous*?

Let P (Fig. 6) be any point (x, y) on the graph to the right of the origin; Q_1 any other point (x_1, y_1) on the graph to the right of P.

We have

$$y = x^2, \ y_1 = x_1{}^2; \ x_1 > x, \ y_1 > y$$

by the property just proved.

If Q_2 be another point (x_2, y_2) on the graph horizontally between P and Q_1,— i.e. so that $x_1 > x_2 > x$,—we shall have also $y_1 > y_2 > y$.

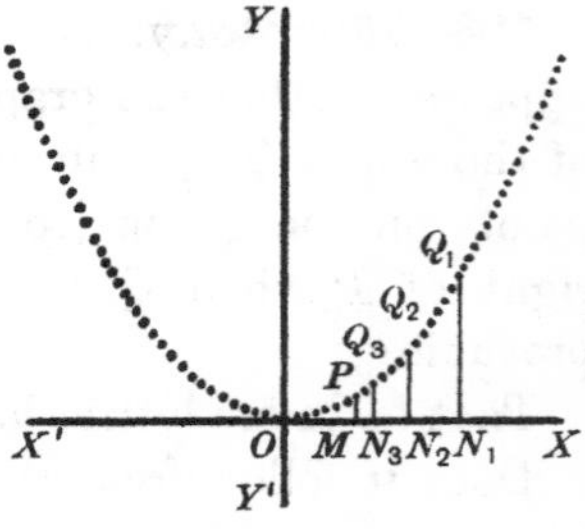

Take a *sequence* of points $Q_1, Q_2, Q_3, \ldots$

Fig. 6.

corresponding to the sequence of abscissae $x_1, x_2, x_3, \ldots$, which is such that $x_1 > x_2 > x_3 > \ldots > x$ and $x_1, x_2, x_3, \ldots \searrow x$. Then we have $y_1 > y_2 > y_3 > \ldots > y$ and $y_1, y_2, y_3, \ldots \searrow y$; because, if z is any number greater than y, $\sqrt{z} > \sqrt{y} = x$, and therefore $\sqrt{z}$ must exceed *some* number of the sequence $x_1, x_2, \ldots$, which $\searrow x$, so that z must exceed the corresponding number of the sequence $y_1, y_2, \ldots$; the lower bound and unique limit of the sequence $y_1, y_2, \ldots$ therefore $\leqslant y$; it evidently $\geqslant y$; therefore it $= y$.

Hence, if we take any sequence of abscissae, $x_1, x_2, \ldots$, such that the feet of the corresponding ordinates, viz. $N_1, N_2, \ldots$, tend to coincide from the right with M,—the foot of the ordinate from P,—the corresponding points $Q_1, Q_2, \ldots$ on the graph tend to coincide with P.

Whatever our ideas of continuity, or unbrokenness, may have been*, it seems clear that this property will satisfy them, at least

---

* E.g. if we look on a continuous curve as one which could be drawn without removing the pen from the paper.

as regards the continuity of the graph at the point P on the right-hand side. We have not so far[*] given any *definition* of continuity. Let us then lay down the definitions:

The graph of a function $f(x)$ is said to be continuous on the right at a point P (x, y) if, corresponding to any convergent decreasing sequence of abscissae $x_1, x_2, \ldots$ which $\searrow x$, the sequence of ordinates $y_1, y_2, \ldots$ (i.e. $f(x_1), f(x_2), \ldots$) is convergent and $\to y$, i.e. $f(x)$.

Similarly *the graph is continuous on the left at P if the sequence $y_1, y_2, \ldots \to y$ corresponding to any increasing sequence $x_1, x_2, \ldots$ which $\nearrow x$.*

The graph is continuous at P if it is both continuous on the right and continuous on the left.

108. Continuous functions. These definitions specify properties of the *function* defining the graph. We therefore express them directly in terms of the function:

The function $f(x)$ is continuous on the right for the value x (or at the point x) if, corresponding to any sequence $x_1, x_2, \ldots$ which $\searrow x$, the sequence $f(x_1), f(x_2), \ldots \to f(x)$; $f(x)$ *is continuous on the left* at x if $f(x_1), f(x_2), \ldots \to f(x)$ corresponding to any sequence $x_1, x_2, \ldots$ which $\nearrow x$; $f(x)$ *is continuous at x if it is continuous on the right and on the left at P*, or, expressed differently, $f(x)$ *is continuous at x if the sequence $f(x_1), f(x_2), \ldots \to f(x)$, corresponding to any sequence $x_1, x_2, \ldots$ which $\to x$*[†].

We have proved that our graph, $y = x^2$, or the function x^2, is continuous on the right at any point P to the right of the origin. It is proved similarly that it is also continuous on the left, and it is therefore continuous at P. It is easy to see that these properties hold equally well if P lies to the left of the origin, or if P actually is the origin O.

109. These two simple properties proved of the function x^2,—its partial monotony (decreasing for x negative and increasing for x positive) and its continuity,—are properties of considerable importance and are possessed (often only in a limited range) by all the most useful functions of elementary analysis. It would perhaps even seem somewhat unreasonable to expect to be able to use, for purposes of analysis, functions which do not have such properties.

[*] In this chapter.
[†] The proof that these last two statements are equivalent is left to the student.

Continuous functions possess the *fundamental property* :

If $f(x)$ is continuous at all points from a to b, and k is any number between $f(a)$ and $f(b)$, then there is at least one value of x between a and b for which $f(x)=k$.

This theorem,—which is easily verified for simple functions such as x^2,—is merely the precise analytical statement of the geometrically intuitive fact that a continuous graph cannot pass from one side of the straight line $y=k$ to the other without cutting it. We can avoid use of this theorem throughout the greater part of this course. A proof is outlined in Ex. 8, p. 151. See also Chapter IV, § 5.

110. Tangent and slope. There is a third question suggested by the graph. The graph is visibly steeper in some parts than others. Can we obtain an expression for this steepness or slope?

We must first settle, in more or less geometrical language, what we mean by the slope of the curve at a point P. To find practically the slope at the point P we should evidently draw the *tangent* and measure the angle it makes with OX. Agreeing to this we must now ask what is the tangent?

In elementary geometry (e.g. Euclid, Book III) the tangent to a circle is defined as the straight line perpendicular to the radius passing through the point concerned. Such a definition will clearly not do for such a curve as that under consideration. Instead the following more general definition is used:

A curve is said to have a tangent at a point P on it if the chords joining P to points Q_1, Q_2, etc. on the curve become closer to a fixed line PT through P as the points Q_1, Q_2, ... are taken nearer to P, so that the degree of closeness of approximation is greater than any assigned degree if only the points Q be taken sufficiently near P*; and in this the points Q may be on either side of P.

That this definition is the natural definition of a tangent to a curve the student will readily agree† Though it is still somewhat vague, let us try to apply it to our graph $y=x^2$.

Let P (Fig. 7) be the point (x, y) on the graph,—so that $y=x^2$. Let Q_1, Q_2, ... be the points (x_1, y_1), (x_2, y_2), ... where

$$x_1 = x + h_1, \ x_2 = x + h_2, \dots;$$
$$y_1 = x_1{}^2 = (x + h_1)^2 = y + k_1 \text{ say},$$
$$y_2 = x_2{}^2 = (x + h_2)^2 = y + k_2 \text{ say},$$

etc.

---

* I.e. for all points Q nearer to P than a certain distance depending on the assigned degree of closeness of approximation. It is not necessary that the chords should become *steadily* closer to PT.

† This definition is in fact given in many modern textbooks on Geometry.

Let the numbers h_1, h_2, ... be all positive, so that the points Q_1, Q_2, ... lie to the right of P; and let the sequence $h_1, h_2, \ldots \searrow 0$, so that $x_1, x_2, \ldots \searrow x$ and $y_1, y_2, \ldots \rightarrow y$ and $k_1, k_2, \ldots \rightarrow 0$, in virtue of the proved continuity of the graph on the right at P.

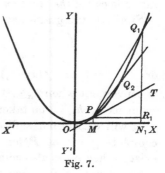

Fig. 7.

The angle $R_1\hat{P}Q_1$, which the chord PQ_1 makes with the x axis, OX, is determined by the ratio R_1Q_1/PR_1, which is in fact $\tan R_1\hat{P}Q_1$. Let us take this ratio as *the measure of the slope* of the chord PQ_1. The slope of the chord

$$PQ_1 = \frac{N_1Q_1 - N_1R_1}{PR_1} = \frac{y_1 - y}{x_1 - x} = \frac{k_1}{h_1}$$

$$= \frac{(x + h_1)^2 - x^2}{h_1} = 2x + h_1.$$

Corresponding to the sequence of numbers h_1, h_2, ... which tends to zero, i.e. corresponding to the sequence of points Q_1, Q_2, ... which tend to coincide with P, in the manner contemplated in the geometrical definition of the tangent, the chords PQ_1, PQ_2, ... have slopes $2x + h_1$, $2x + h_2$, ...

This sequence of slopes tends to a unique limit,—viz. $2x$. Hence the chords PQ_1, PQ_2, ... tend to coincide with a definite straight line through P,—viz. that line which has a slope $2x$. This line, PT, is independent of the particular sequence of points Q_1, Q_2, ... chosen, and clearly fulfils the conditions of the geometrical definition of the tangent, as regards the portion of the graph lying to the right of P.

We can now lay down the strict definitions for graphs in general.

The graph of a function $f(x)$ has a *tangent on the right* at a point $P(x, y)$ if, corresponding to any convergent decreasing sequence of positive numbers h_1, h_2, ... having the limit 0, the corresponding sequence of ratios (called *incrementary ratios*)

$$(y_1 - y)/h_1, \ (y_2 - y)/h_2, \ldots,$$

i.e. $\qquad [f(x + h_1) - f(x)]/h_1, \ [f(x + h_2) - f(x)]/h_2, \ldots,$

is convergent and has the same limit, whatever such sequence

$h_1, h_2, \ldots$ be taken. The limit of this sequence of incrementary ratios is called *the slope on the right at P*.

Similarly the graph has a *tangent on the left* at P if the sequence

$$\frac{f(x - h_1) - f(x)}{- h_1}, \quad \frac{f(x - h_2) - f(x)}{- h_2}, \ldots$$

or $\qquad [f(x) - f(x - h_1)]/h_1, \; [f(x) - f(x - h_2)]/h_2, \ldots$

is convergent when $h_n \searrow 0$, and has the same limit, whatever such sequence $h_1, h_2, \ldots$ be taken.

It will clearly not suffice for the existence of a tangent to the complete graph at P for the tangents on the right and left to exist;—they must also coincide.

The graph of $f(x)$ has a tangent at P if it has a tangent on the right and a tangent on the left and if these "semi-tangents" coincide. The slope of the tangent is the slope of either of these "semi-tangents."

Our graph, $y = x^2$, evidently has a tangent on the left, with the same slope $(2x)$ as that of its tangent on the right. It has therefore a (complete) tangent.

111. Differentiability on the right and on the left. Differential coefficients. These definitions apply primarily to properties of the function defining the graph. Hence we have the following definitions for functions:

A function $f(x)$ is differentiable on the right for the value x (or at the point x) if, corresponding to every positive decreasing sequence $h_1, h_2, \ldots$ which $\searrow 0$, the sequence of incrementary ratios

$$\frac{f(x + h_1) - f(x)}{h_1}, \quad \frac{f(x + h_2) - f(x)}{h_2}, \ldots$$

is convergent and has the same limit whatever such sequence $h_1, h_2, \ldots$ be taken. The limit of this sequence is then called *the differential coefficient of $f(x)$ on the right at x.*

A function $f(x)$ is differentiable on the left at x if, corresponding to any positive decreasing sequence $h_1, h_2, \ldots$ which $\searrow 0$, the sequence of incrementary ratios

$$\frac{f(x - h_1) - f(x)}{- h_1}, \quad \frac{f(x - h_2) - f(x)}{- h_2}, \ldots$$

is convergent and has the same limit whatever such sequence $h_1, h_2, \ldots$

be taken. The limit of this sequence is then called *the differential coefficient of $f(x)$ on the left at x*.

A function $f(x)$ *is differentiable at x (or for the value x) if it is differentiable on the right and differentiable on the left at x and if the differential coefficient on the right equals the differential coefficient on the left. The common value of the two "semi-differential coefficients" is then called the differential coefficient of $f(x)$ at x.*

The term *derivative* is often used for *differential coefficient*. Alternative notations for the differential coefficient of a function $f(x)$, or y, are:

$$Dy \text{ or } Df(x), \ D_x y \text{ or } D_x f(x), \ \frac{dy}{dx} \text{ or } \frac{df(x)}{dx}, \ f'(x).$$

The notation $\frac{dy}{dx}$ is most commonly used. We shall here use Dy or $Df(x)$.

112. Tangents parallel to y axis. There is one qualification needed. Geometrically a graph or a curve may have a tangent in any direction, and in particular parallel to either of the axes. If the tangent is parallel to the x axis, its slope is 0 and the function has a differential coefficient equal to zero. But if the tangent is parallel to the y axis, the slope is not represented by any of our real numbers. (It would be ∞ if this number had been introduced into our system.) The function concerned will not be differentiable, for the sequence of incrementary ratios would clearly be unbounded. We could get over this difficulty by introducing infinite differential coefficients,— under specified conditions. But the difficulty is avoided simply by interchanging the x and y axes.

113. Area bounded by graph of x^2 by simple process. A fourth question of interest arises. In elementary Euclidean geometry *areas* of plane figures are discussed. The areas there dealt with are bounded by straight lines. The consideration of other areas is however desirable. For many practical purposes the area enclosed between the x axis, a curve and two bounding ordinates (called the *area under the curve* between the ordinates) is of importance. Thus if a graph is drawn to represent the velocity of a moving particle

---

* It should be observed that the differential coefficient on the right at x may exist even if the function is not defined for values of the variable less than x; and similarly as regards the differential coefficient on the left.

at a given time, the area under the curve between two given ordinates will represent the distance covered in the time concerned; or, if x be taken to represent the distance, in a certain direction, moved through by a particle under the influence of a force, represented by y, the area under the curve will represent the work done by the force in the motion.

Let us consider the case of an area bounded above by a portion of our curved graph, $y = x^2$.

Let A be the point $(1, 1)$ on the graph $y = x^2$. (Fig. 8.)

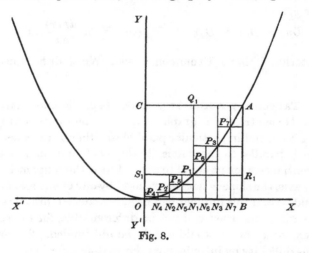

Fig. 8.

For definiteness we will consider the "curved triangle" OBA and investigate its "area."

As we did in building up a definition of the tangent, we will begin by assuming what appears geometrically to be evident and then prove strictly that a property which clearly ensures all that we need geometrically, actually is possessed by the graph. We shall then be able to lay down a general analytical definition of area to be applied to other graphs.

Complete the square $OBAC$. The area required of the "curved triangle" OBA clearly is less than the area of this square.

Calling the required area A, we have

$$0 < A < 1 \quad \ldots\ldots\ldots\ldots\ldots\ldots\ldots(1).$$

Draw the straight line Q_1N_1, bisecting OB at right angles, and let

$Q_1 N_1$ cut the graph at P_1. Through P_1 draw $S_1 P_1 R_1$ parallel to OX cutting BA, OY in R_1, S_1.

We have $A <$ the sum of the areas of the rectangles $N_1 B A Q_1$ and $O N_1 P_1 S_1$ and $>$ area of the rectangle $N_1 B R_1 P_1$, i.e.

$$\tfrac{1}{2}(\tfrac{1}{2})^2 = \tfrac{1}{2} N_1 P_1 < A < \tfrac{1}{2}(N_1 P_1 + BA) = \tfrac{1}{2}[(\tfrac{1}{2})^2 + 1] \dots\dots(2).$$

Bisect $O N_1$ and $N_1 B$ at right angles by straight lines $N_2 P_2$, $N_3 P_3$, cutting the graph at P_2, P_3.

Drawing through P_2, P_3 parallels to OX as before, we form altogether seven new rectangles,—of heights $N_2 P_2$, $N_1 P_1$, $N_3 P_3$, BA, and equal widths $OB/4 = \tfrac{1}{4}$. We have $A <$ sum of areas of rectangles

$$O N_2 . N_2 P_2 + N_2 N_1 . N_1 P_1 + N_1 N_3 . N_3 P_3 + N_3 B . BA$$

and $A >$ sum of areas of rectangles

$$O N_2 . 0 + N_2 N_1 . N_2 P_2 + N_1 N_3 . N_1 P_1 + N_3 B . N_3 P_3,$$

i.e. $\tfrac{1}{4}[0 + (\tfrac{1}{4})^2 + (\tfrac{1}{2})^2 + (\tfrac{3}{4})^2] < A < \tfrac{1}{4}[(\tfrac{1}{4})^2 + (\tfrac{1}{2})^2 + (\tfrac{3}{4})^2 + 1] \dots(3).$

Bisect again the bases of all these rectangles, obtaining four new points on the graph, say P_4, P_5, P_6, P_7, and corresponding new rectangles of width $OB/8 = \tfrac{1}{8}$. As before $A <$ sum of larger rectangles

$$O N_4 . N_4 P_4 + N_4 N_2 . N_2 P_2 + N_2 N_5 . N_5 P_5 + N_5 N_1 . N_1 P_1$$
$$+ N_1 N_6 . N_6 P_6 + N_6 N_3 . N_3 P_3 + N_3 N_7 . N_7 P_7 + N_7 B . BA$$
$$= \tfrac{1}{8}[(\tfrac{1}{8})^2 + (\tfrac{2}{8})^2 + \dots + (\tfrac{7}{8})^2 + 1] \dots\dots\dots\dots(4)$$

and $A >$ sum of smaller rectangles

$$O N_4 . 0 + N_4 N_2 . N_4 P_4 + N_2 N_5 . N_2 P_2 + N_5 N_1 . N_5 P_5$$
$$+ N_1 N_6 . N_1 P_1 + N_6 N_3 . N_6 P_6 + N_3 N_7 . N_3 P_3 + N_7 B . N_7 P_7$$
$$= \tfrac{1}{8}[0 + (\tfrac{1}{8})^2 + (\tfrac{2}{8})^2 + \dots + (\tfrac{7}{8})^2] \dots\dots\dots\dots(5).$$

Continuing this process we obtain a succession of results of this kind. Corresponding to the nth division we have that the area A lies between the sum of rectangles of total area

$$\frac{1}{2^n}\left[0 + \left(\frac{1}{2^n}\right)^2 + \left(\frac{2}{2^n}\right)^2 + \dots + \left(\frac{2^n - 1}{2^n}\right)^2\right] \dots\dots\dots(6),$$

and $\dfrac{1}{2^n}\left[\left(\dfrac{1}{2^n}\right)^2 + \left(\dfrac{2}{2^n}\right)^2 + \dots + 1\right] \dots\dots\dots\dots(7).$

The process may be continued indefinitely, and evidently provides closer and closer approximations to the desired area. In fact, the smaller rectangles have total areas

$$0, \tfrac{1}{2}, \tfrac{1}{2}(\tfrac{1}{2})^2, \tfrac{1}{4}[0 + (\tfrac{1}{4})^2 + (\tfrac{1}{2})^2 + (\tfrac{3}{4})^2], \tfrac{1}{8}[0 + (\tfrac{1}{8})^2 + (\tfrac{2}{8})^2 + \dots + (\tfrac{7}{8})^2], \dots$$

and, either by recourse to the figure, or directly from these arithmetical expressions, it is seen that this sequence last written is steadily increasing and bounded. It therefore tends increasingly to a unique limit,—say A_1.

Similarly the sequence of the sums of the areas of the larger rectangles, viz.

$$1, \; \tfrac{1}{2}[(\tfrac{1}{2})^2 + 1], \; \tfrac{1}{4}[(\tfrac{1}{4})^2 + (\tfrac{1}{2})^2 + (\tfrac{3}{4})^2 + 1], \; \dots,$$

tends decreasingly to a unique limit,—say A_2.

And these two limits, A_1, A_2, are identical, for the differences between the total areas of the larger and smaller rectangles form a sequence, viz. $1, \tfrac{1}{2}, \tfrac{1}{4}, \tfrac{1}{8}, \dots$, which $\to 0$.

This common limit, A_1 or A_2, evidently is the area sought, viz. A.

114. More general process. We can clearly generalise this process. We could have divided the interval OB* into any number (m) of parts,—say by the points $x_1, x_2, \dots x_m$ (where $x_m = OB = 1$),— erected perpendiculars through these points, and formed two sets of rectangles,—larger and smaller,—as before. If we denote by y_1 the ordinate through the point x_1, etc., so that $y_1 = x_1{}^2$, etc., the sum of the areas of the smaller set of rectangles formed equals

$$(x_1 - 0)\,0 + (x_2 - x_1)\,y_1 + (x_3 - x_2)\,y_2 + \dots + (x_m - x_{m-1})y_{m-1} \dots (8),$$

and the sum of the larger rectangles equals

$$(x_1 - 0)\,y_1 + (x_2 - x_1)\,y_2 + (x_3 - x_2)\,y_3 + \dots + (x_m - x_{m-1})\,y_m \dots (9).$$

The required area will lie between these two sums.

If we divide up further, by introducing additional points of division, we shall obtain new rectangles, which, as before, will form closer estimates to the required area,—from below and above respectively. Continuing the division indefinitely in such a way that the width of the widest rectangle decreases indefinitely, we shall obtain two sequences, whose typical terms are respectively

    (i) the sum of the areas of the smaller rectangles,—such as (8).

    (ii) the sum of the areas of the larger rectangles,—such as (9).

* An *interval* is any limited portion of the axis, such as OB, or N_1B, etc. See footnote, p. 119 above.

The sequence (i) steadily increases, is bounded above, and therefore ⌐ a limit, say B_1.

The sequence (ii) steadily decreases, is bounded below, and therefore ⌐ a limit, say B_2.

If Δ denotes the greatest of the widths of the rectangles, such as
$$x_1 - 0, \; x_2 - x_1, \; x_3 - x_2, \; \ldots \; x_m - x_{m-1},$$
at any stage, then the difference between the larger and smaller areas
$$= (x_1 - 0)(y_1 - 0) + (x_2 - x_1)(y_2 - y_1) + \ldots + (x_m - x_{m-1})(y_m - y_{m-1})$$
$$< \Delta . [(y_1 - 0) + (y_2 - y_1) + \ldots + (y_m - y_{m-1})],$$
because all the multiplying factors
$$y_1 - 0, \; y_2 - y_1, \; \ldots \; y_m - y_{m-1}$$
are positive,—the graph being known to be monotonely increasing. Therefore this difference $< \Delta y_m = \Delta$, because $y_m = OB^2 = 1$. The division being continued in such a way that the maximum width Δ ⌐ 0, it follows that this difference between the larger and smaller areas tends to 0, and therefore the two limits B_1, B_2, as before, are identical, $= B$ say.

It is indeed true that this limit B, whatever mode of division (of the type described) may have been adopted (provided the maximum width of the rectangles tends to zero), is identical with the limit A, obtained by the specialised mode of bisection first considered.

For it can be proved easily analytically*, or it is clear from Fig. 9, in which the letters P_1 etc., P_1' etc. refer to the two systems of division, that the total area of any set of "small rectangles" for the second system of division is less than the total area of any set of "large rectangles" for the first system, and it therefore follows that the limit $B_1 \leqslant$ the limit A_2; and similarly $B_2 \geqslant A_1$; i.e. $B \leqslant A$ and $B \geqslant A$, or $B = A$.

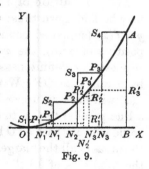

Fig. 9.

We have not quite proved that, if we divide up the interval anew at each stage (instead of subdividing the divisions already made), the sequences of total areas of small and large rectangles tend to this limit if the maximum width of the rectangles tends to zero. This is however true. For, if Δ is the maximum

---

* See Chapter IV, § 6 below.

width of any set of rectangles and S and s the total areas of the corresponding large and small rectangles, we have $S - A < S - s < \Delta$, as above, and therefore $S \to A$ (and similarly $s \to A$) as $\Delta \to 0$, no matter how the divisions are chosen. It is not however the case in general that $S \searrow A$ and $s \nearrow A$ as $\Delta \searrow 0$, as it is with the modes of subdivision adopted above.

115. We have now proved that the graph $y = x^2$ is such that if the region between $x = 0$ and $x = 1$ be divided up into strips by lines parallel to OY, then the total areas of the smaller and larger rectangles so formed, having one vertex on the graph, as above, form two sequences; and if the strips be taken successively narrower, so that the widths tend to zero, *these sequences are both convergent; and their limits are identical, and the same whatever the particular mode of division.*

This much has been proved analytically, without presupposing any idea of area except as applied to rectangles.

We have seen that this common limit will evidently represent the area of the region OBA, if any definition of area is given which agrees at all with our preconceived geometrical ideas. We may say then that we have proved that the area of this region (supposed, if possible, defined geometrically) is this common limit. Or, more logically,—in our desire for strict *analytical* definitions of all mathematical entities,—we may *define* the area of such a region as such a limit if (as is here the case) it exists.

116. Definition of area under a graph. The general definition may be laid down thus: Consider the region bounded by the graph $y = f(x)$ (supposed bounded and, in the first instance, everywhere positive or zero), the x axis, and the parallels $x = a$, $x = b$; and suppose, for definiteness, $a < b$. Divide the region up into strips by parallels to OY. Within any strip the values of the function $f(x)$ will have an upper bound *; with this value for ordinate draw a line parallel to OX to complete within that strip a "large" rectangle with its base on OY. Do this for every strip. Call the sum of the areas of all the "large" rectangles so formed S. Form similarly s, the total area of all the "small" rectangles of heights equal to the lower bounds of $f(x)$ in the various strips. If the strips be subdivided indefinitely, so that the width of the widest strip tends to zero, the corresponding numbers S and s will form two sequences. *If these*

---

* I.e. a least number not exceeded by any value of $f(x)$ in the strip.

two sequences have a common unique limit, which is the same for all such modes of division, the region is said to have an area. The common limit is called the *area* of the region. If $f(x)$ is anywhere negative, one or both of the upper and lower bounds of $f(x)$ in some of the strips will be negative. The "areas" of the corresponding rectangles in such strips are to be considered negative. The area of such a region will be positive if it lies entirely above the x axis and negative if entirely below the x axis. If it lies on both sides of the axis it may be positive or negative or zero.

117. Definitions of integrable function and definite integral. As in the case of the tangent to a graph and the differential co-efficient of the function concerned, so here we give a name to those functions whose graphs between two ordinates, $x = a$ and $x = b$ ($a < b$), bound a region which has an area in accordance with this definition. *Such a function $f(x)$ is said to be integrable between a and b. The limit which defines the area between the graph and the x axis, cut off between the ordinates a and b, is called the definite integral of $f(x)$ from a to b, or over the range (a, b), and is written* $\int_a^b f(x)\,dx$.

We have proved that the function x^2 is integrable between 0 and 1. The area required of the "curved triangle" $OBA = \int_0^1 x^2\,dx$.

118. Evaluation of area. The actual *evaluation* of the area of the region OBA,—now defined and proved to exist,—can in this special case be carried out directly from the above process:

The expression (6), p. 145, giving the total area of the smaller rectangles at the nth stage of bisection

$$= \frac{1}{N}\left[(1/N)^2 + (2/N)^2 + \dots + \left(\frac{N-1}{N}\right)^2\right],$$

where N is written for 2^n,

$$= \frac{1}{N^3}[1 + 2^2 + 3^2 + \dots + (N-1)^2].$$

The expression in the bracket can be proved,—by induction or otherwise,—to be equal to

$$\tfrac{1}{6}(N-1)N(2N-1) = N^3/3 - N^2/2 + N/6.$$

The expression (6) therefore equals $1/3 - 1/2N + 1/6N^2$.

As the process of division continues, as contemplated in the definition, $N \to \infty$ *, and the expression (6) $\to 1/3$. The required area of the region $OBA = 1/3$.

This direct method is, however, only possible in specially simple cases. In other cases the evaluation is carried out by a much more powerful indirect method resting on the property which may be described as the inverse character of integration and differentiation. It is postponed to the next chapter, § 6.

119. Properties of the function x^2 summarised. To return to our special function x^2, let us sum up our acquired knowledge:

(i) it is monotonely decreasing for all negative real values of x and monotonely increasing for all positive values of x;

(ii) it is bounded above and below (i.e. < a fixed number K and > a fixed number K') for all bounded values of x; (for, if $|x| < K$, $0 \leqslant x^2 < K^2$);

(iii) it is unbounded as x increases beyond all limit positively or negatively; (this is easily proved);

(iv) it is continuous for all values of x;

(v) it is differentiable for all values of x, and its differential coefficient is $2x$;

(vi) its differential coefficient is positive when x is positive and negative when x is negative;

(vii) it is integrable between 0 and 1, and in fact between any two values a and b;

(viii) the area bounded by the graph, the x axis, and the ordinate $x = 1$, is $1/3$; and in fact the area bounded by the graph, the x axis, and the ordinate x, is $x^3/3$; (this is easily proved as above);

(ix) it is of bounded variation in any bounded range.

We notice that the function is increasing wherever the differential coefficient is positive, and decreasing wherever it is negative. This fact is not a mere accident†.

We leave as an exercise to the student the proof that the only point at which the tangent is parallel to the x axis is where $x = 0$. Such a value of x is called a *turning value* (or a *maximum* or *minimum*), if, as in this case, the function is, on one side of the point increasing, and on the other decreasing. The function has at

* I.e. N increases indefinitely.
† See p. 157 and Chapter IV, § 5 below and Ex. 8, p. 133 above.

this point its least value. The curve which is the graph of the function x^2 is a *parabola*, with OY for "axis" and O for "vertex."

EXAMPLES XIII.

1. Prove that the tangent to the graph of x^2 becomes more and more nearly parallel to the y axis as x increases or decreases indefinitely but is nowhere actually parallel to it.

2. Prove the "fundamental property of continuous functions" for the function x^2 (p. 140).

3. Prove that the area cut off between the graph of x^2 and the x axis by the two ordinates $x=a$, $x=b$ $(b>a>0)$, is $b^3/3 - a^3/3$.

4. Draw the graph of the function x, and establish all the essential properties of this function. Apply the general process of the text to evaluate $\int_0^1 x\,dx$.
Shew that $\int_{-1}^1 x\,dx = 0$. The graph of the "function" 1 may be similarly discussed, but the results are trivial.

5. Draw the graph of x^3, shewing in particular that the function is increasing for all values of x and unbounded above and below, that it is continuous, differentiable and integrable for all values of x, that $Dx^3 = 3x^2$, and that the differential coefficient is zero when $x=0$ but that this is not a turning value.

6. The fundamental properties of the functions $\log_a x$ and a^x (where a is any positive number, $\neq 1$) have been established in Chapter II. Sketch the graphs of these functions for $a=2$, finding the slopes at all points and shewing that the area under the graph of a^x between the ordinates $x=0$, $x=1$ is $\log_2 e$.

7. Prove that a circle has a tangent at all points according to the definition of p. 142. (See also p. 143.)

8. Prove the "fundamental property of continuous functions" (p. 140).

[Suppose for definiteness, $a < b$, $f(a) < f(b)$. The set of numbers x for which $f(x') < k$ for all numbers x' from a to x inclusive, has an upper bound,—X say. Let $f(X) = K$.

Because $f(x)$ is continuous at X, $f(x)$ differs from K by an arbitrarily small amount for all values of x sufficiently near to X. Therefore if (i) $K > k$, then $f(x) > k$ for all values of x sufficiently near to X, and therefore, in particular, $f(x) > k$ for values of x less than X; and if (ii) $K < k$, similarly $f(x) < k$ for *all* values of x greater than X but sufficiently near to X.

Both these conclusions contradict the definition of X. Hence $f(X) = k$, and the theorem is proved.]

9. Prove that if $f(x)$ is continuous and monotone between a and b and $f(a) < 0 < f(b)$, then $f(x) = 0$ for only one value of x between a and b.

10. Prove that the definition of differentiability may be expressed as $[f(x+h_1) - f(x)]/h_1$, $[f(x+h_2) - f(x)]/h_2$, ... tends to a unique limit, the same for any sequence h_1, h_2, ... which tends to zero.

11. If $f(x)$ is differentiable at x it must be continuous there, but the converse theorem is not true.

12. The function y defined as $\sin 1/x$ when $x \neq 0$ and $=0$ when $x=0$ ($\sin 1/0$ being meaningless) is not continuous at $x=0$.

[If e.g. for the sequence $x_1, x_2, \ldots$, which $\searrow 0$, $1/\pi, 1/2\pi, 1/3\pi, \ldots$ be chosen, the corresponding sequence $y_1, y_2, \ldots$ is $0, 0, 0, \ldots$ and has the unique limit 0; if $2/\pi, 2/5\pi, 2/9\pi, \ldots$ be chosen for $x_1, x_2, \ldots$, the corresponding sequence for $y_1, y_2, \ldots$ is $1, 1, 1, \ldots$ and has the unique limit 1; if the sequence $2/\pi, 2/2\pi,$ $2/3\pi, 2/4\pi, 2/5\pi, \ldots$ be chosen, the sequence for $y_1, y_2, \ldots$ is $1, 0, -1, 0, 1, \ldots,$ which is not convergent. Sequences $x_1, x_2, \ldots$ can in fact be found for which the corresponding sequence $y_1, y_2, \ldots$ tends to any limit between ± 1.]

13. The function $|x|$ is continuous at all points and differentiable at all points except $x=0$. At $x=0$ it is differentiable on the right and differentiable on the left but not differentiable.

[The two "semi-differential coefficients" are $+1$ and -1.]

14. The (positive) difference $M_k - m_k$ between the upper and lower bounds of a bounded function $f(x)$ in an interval δ_k (x_{k-1}, x_k), is called the *oscillation* of $f(x)$ in the interval δ_k. If the sum-total of the oscillations of $f(x)$ in all the intervals into which (a, b) is divided (as in the text, p. 146) is bounded, i.e. is less than a fixed number, K say, no matter how the points of division, $x_1, x_2, \ldots x_{m-1}$, are chosen, the function $f(x)$ is said to be *of bounded variation* in the interval (a, b). If $f(x)$ is of bounded variation in (a, b) there must be an upper bound of the sum-totals of the oscillations of $f(x)$ in the intervals into which (a, b) is divided, i.e. a least number which is greater than or equal to all possible sum-totals of oscillations. This upper bound is called the *total variation* of $f(x)$ in the interval (a, b). Prove:

(i) If a bounded function $f(x)$ is monotone throughout (a, b), it is of bounded variation in (a, b) and its total variation is $|f(b)-f(a)|$.

(ii) If $f(x)$ is expressible as the sum or difference of two bounded monotone functions then $f(x)$ is of bounded variation.

[If $f(x)=u(x)-v(x)$, where u and v are non-decreasing, oscillation of $f(x)$ in any interval $\delta_k \leqslant$ sum of oscillations of u and v in δ_k; therefore total oscillations of $f(x)$ in the intervals dividing $(a, b) \leqslant$ sum of total oscillations of u and v in those intervals. Therefore total oscillations of $f(x) \leqslant |u(b)-u(a)| + |v(b)-v(a)| < K.$]

(iii) If $V(a, x)$ denotes the total variation in the interval (a, x) of a function of bounded variation $f(x)$, then $V(a, x)$ is a positive non-decreasing bounded function of x.

(iv) If $V(a, x)$ and $f(x)$ are as in (iii), the functions $V(a, x)+f(x)$ and $V(a, x)-f(x)$ are bounded non-decreasing functions of x.

(v) Any function of bounded variation can be expressed as the sum or difference of two bounded monotone functions.

15. Prove that if $u(x)$ and $v(x)$ are any two bounded monotone functions

of x, the product uv and the quotient u/v are functions of bounded variation, provided, in the case of the quotient, that the lower bound of $|v|$ is not zero.

[For the product, if u and v are non-decreasing, $u(x) = U(x) - A$ and $v(x) = V(x) - B$, where $U(x)$, $V(x)$ are positive non-decreasing bounded functions and A and B positive constants. $uv = (UV + AB) - (AV + BU)$. From laws of inequalities (Chapter I, § 5) UV, etc. are non-decreasing (positive) functions. The result follows. For the quotient put $u/v = -\{u(-1/v)\}$.]

16. Prove that, if u and v are functions of bounded variation in (a, b), then $u + v$, $u - v$, uv and u/v are also of bounded variation,—provided, in the case of u/v, that the lower bound of $|v|$ is not zero.

[Result for $u \pm v$ obvious. That for uv follows from Ex. 15. Result for quotient will follow if it is first proved that $\dfrac{1}{v} = \dfrac{1}{y - z}$ is of bounded variation, y and z being non-decreasing. To prove this (with obvious notation): oscillation of $1/(y - z)$ in $\delta_k = $ upper bound of

$$\left\{ \frac{1}{y(x_1) - z(x_1)} - \frac{1}{y(x_2) - z(x_2)} \right\} \leqslant \frac{1}{B_k{}^2} \cdot (\text{oscillation of } y + \text{oscillation of } z),$$

where $B_k = $ lower bound of $|y - z|$ in δ_k; result follows.]

§ 2. POLYNOMIALS

120. The function x^n. The properties of the function x^n, where n is any positive integer, are easily investigated similarly.

(i) Since, if $x_1 > x_2 > 0$ then $x_1{}^n > x_2{}^n > 0$, it follows that the function x^n is *monotonely increasing* for all positive values of x and *increases beyond all limit* as x increases indefinitely, whilst for negative values of x, as x decreases indefinitely the modulus of x^n increases indefinitely, but the sign of x will then be positive or negative according as n is even or odd.

(ii) x^n is bounded above and below in any bounded range.

(iii) x^n is *continuous* for all values of x, for, by inequality (iii), p. 29 above, $[(x + h)^n - x^n]/h$ lies between $n(x + h)^{n-1}$ and nx^{n-1}; therefore, as $h \searrow 0$ or as $h \nearrow 0$, $(x + h)^n - x^n$, which lies between $h \cdot n(x + h)^{n-1}$ and $h \cdot nx^{n-1}$, must $\to 0^*$.

(iv) x^n is *differentiable* for all values of x and $Dx^n = nx^{n-1}$, for $[(x + h)^n - x^n]/h$ lies between $n(x + h)^{n-1}$ and nx^{n-1}, and by property (iii), applied to the function x^{n-1}, we know $(x + h)^{n-1} \to x^{n-1}$ as $h \searrow 0$ or $h \nearrow 0$; whence the incrementary ratio

$$[(x + h)^n - x^n]/h \to nx^{n-1}.$$

---

* I.e. as h takes on any sequence of values h_1, h_2, ..., which $\searrow 0$ or $\nearrow 0$ as the case may be, the corresponding sequence of values of $(x + h)^n - x^n \to 0$. The abbreviation $h \to 0$ is also convenient.

(v) Dx^n is positive for x positive whether n be odd or even; it is also positive when x is negative if n be odd, but is negative when x is negative if n be even.

(vi) The only value of x giving a horizontal tangent is $x = 0$, which will be a turning value (in fact a minimum) if n is even but not if n is odd.

(vii) x^n is *integrable* between a and b,—any two real numbers,—for the proof given in the last section of the integrability of x^2 depends only on the boundedness and monotony of x^2; these properties have been proved,—(i) and (ii) above,—to be possessed also by the function x^n, and the proof of the last section will therefore apply here also.

(viii) x^n is "even" if n is even,—i.e. the values of $(-x)^n$ and of x^n are the same; whilst x^n is "odd" if n is odd,—i.e. the values of $(-x)^n$ and of x^n are numerically equal but are opposite in sign.

Typical graphs are drawn in Fig. 10.

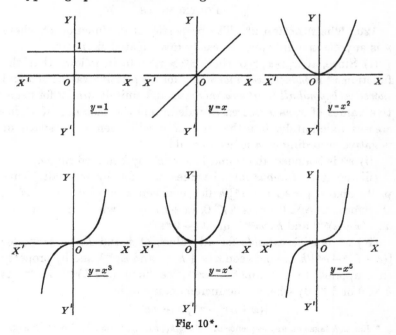

Fig. 10*.

* The graph of $y = x^0$ (or $y = 1$), though not of the type discussed, is added for completeness.

121. Example of a polynomial. Let us now consider the graph of a function defined by an expression consisting of two or more of the functions $1, x, x^2, x^3, \ldots$ combined by the operations of addition and of multiplication by "constants,"—i.e. fixed numbers independent of the variable x. For example let us consider the function $x^4 - 3x^2 + 2x + 4$.

We have firstly that if x increases indefinitely, the function (which we will call y) will also increase indefinitely, and, corresponding to any indefinitely increasing sequence for x, the sequence for y will be unbounded; for we have (for x positive) $y > x^4 - 3x^2$, but $x^4/2 > 3x^2$ if $x^2 > 6$ and therefore certainly if $x > 3$ say, hence for $x > 3$, $y > x^4/2$ and therefore as x increases indefinitely y is unbounded.

Also if x is negative and decreases indefinitely, y is similarly proved to be unbounded (above, as before), for

$$y > \frac{x^4}{2} \text{ if } x^4 - 3x^2 + 2x + 4 = x^4 - 4x^2 + x(2+x) + 4 > x^4/2,$$

which is certainly so if $x < -2$ and $x^2 > 8$.

Considering the question of continuity, we have, from the definition, that the function is continuous at a point x if and only if the sequence

$$y_1, y_2, y_3, \ldots \quad \ldots\ldots\ldots\ldots\ldots\ldots\ldots\ldots(1),$$

where $\quad y_1 = (x + h_1)^4 - 3(x + h_1)^2 + 2(x + h_1) + 4,$

$\qquad y_2 = (x + h_2)^4 - 3(x + h_2)^2 + 2(x + h_2) + 4,$ etc.,

tends to the unique limit $x^4 - 3x^2 + 2x + 4$ for all sequences $h_1, h_2, \ldots$ which $\to 0$.

Now we know that the function x^4 is continuous and therefore the sequence $(x + h_1)^4, (x + h_2)^4, \ldots \to x^4$;
and x^2 is continuous and therefore $(x + h_1)^2, (x + h_2)^2, \ldots \to x^2$, and therefore also the sequence $3(x + h_1)^2, 3(x + h_2)^2, \ldots \to 3x^2$;
and the function x is continuous and therefore the sequence $x + h_1, x + h_2, \ldots \to x$, and $2(x + h_1), 2(x + h_2), \ldots \to 2x$; and the sequence $4, 4, 4, \ldots$ clearly $\to 4$.

It is easily proved in general that if the sequence $s_1, s_2, \ldots \to a$ unique limit s, and the sequence $s_1', s_2', \ldots \to a$ unique limit s', then the sequences $(s_1 \pm s_1'), (s_2 \pm s_2'), \ldots$ are convergent and tend to $s \pm s'$ respectively*.

---

* By the condition for a unique limit (p. 45), for a certain value of n and all greater values $|s_n + s_n' - s - s'| < \epsilon$ because both $|s_n - s|$ and $|s_n' - s'|$ are less than $\frac{\epsilon}{2}$ say for all sufficiently large values of n.

Hence we deduce that the sequence $(1) \rightarrow x^4 - 3x^2 + 2x + 4$, and the continuity of the function for all values of x is established.

In precisely the same way, arguing from the sequences of incrementary ratios, we see that, since the component functions x^4, $3x^2$, $2x$, and 4 are each differentiable for all values of x, the compound function y is itself differentiable and we have

$$D\,(x^4 - 3x^2 + 2x + 4) = Dx^4 - D3x^2 + D2x + D4$$
$$= Dx^4 - 3Dx^2 + 2Dx + 4D1$$
$$= 4x^3 - 6x + 2.$$

122. Integrability of sum of two functions. Generalisation.

The integrability of this function y between any two values a and b rests on the same principles.

We first prove the general theorem:

If y_1 and y_2 are two functions which are integrable between a and b, then the function $y_1 + y_2$ is also integrable and its integral $\int_a^b (y_1 + y_2)\,dx$ *is the sum of the two separate integrals, viz.*

$$\int_a^b y_1\,dx + \int_a^b y_2\,dx.$$

The proof of this theorem is immediate from the definition of a definite integral; for if M_1 and M_2 are the upper bounds of y_1 and y_2 in any strip used in the definition of the definite integral, the upper bound of the compound function in that strip is clearly $\leqslant M_1 + M_2$. It therefore follows that at every stage of the division used in the definition the total area of the "large" rectangles relative to the compound function $y_1 + y_2$ is less than or equal to the sum of the total areas of the large rectangles relative to the two functions y_1 and y_2 separately; and similarly the total area of the "small" rectangles for $y_1 + y_2$ is greater than or equal to the sum of the total areas of the small rectangles for y_1 and y_2 separately.

That the conditions for integrability for $y_1 + y_2$ must be fulfilled if they are fulfilled for y_1 and y_2 separately is now evident. At the same time it is evident that the integral of the compound function $y_1 + y_2$, i.e. $\int_a^b (y_1 + y_2)\,dx$, = the sum of the two separate integrals, viz. $\int_a^b y_1\,dx + \int_a^b y_2\,dx.$ Q.E.D.

As a corollary to this theorem we have:

If $y_1, y_2, \ldots y_n$ are any functions which are each integrable between a and b, then $y_1 + y_2 + \ldots + y_n$ is also integrable between a and b, and

$$\int_a^b \cdot (y_1 + y_2 + \ldots + y_n)\, dx = \int_a^b y_1\, dx + \int_a^b y_2\, dx + \ldots + \int_a^b y_n\, dx.$$

We can in fact state a second corollary to include both the theorem and this first corollary as special cases.

Under the same circumstances, *if $k_1, k_2, \ldots k_n$ are any fixed numbers,—or "constants,"—the function $k_1 y_1 + k_2 y_2 + \ldots + k_n y_n$ is integrable and*

$$\int_a^b (k_1 y_1 + k_2 y_2 + \ldots + k_n y_n)\, dx$$

$$= k_1 \int_a^b y_1\, dx + k_2 \int_a^b y_2\, dx + \ldots + k_n \int_a^b y_n\, dx.$$

This follows from the evident fact that $\int_a^b ky\, dx$ exists and $= k \int_a^b y\, dx$ if y is integrable.

123. Applying the second corollary to our compound function $x^4 - 3x^2 + 2x + 4$, knowing that the functions x^4, x^2, x, 1 are integrable, we have that the function $y = x^4 - 3x^2 + 2x + 4$ is integrable between any two values a and b. The actual evaluation of the definite integral

$\int_a^b y\, dx$, which equals $\int_a^b x^4\, dx - 3 \int_a^b x^2\, dx + 2 \int_a^b x\, dx + 4 \int_a^b 1\, dx,$

depends on a knowledge of the separate integrals $\int_a^b x^4\, dx$, etc.

124. Question of monotony. We have seen that x^n is either monotone for all values of x or monotone for positive values of x and also monotone for negative values of x. Neither result is in general true for a polynomial, such as the function $x^4 - 3x^2 + 2x + 4$ under discussion. It can however be shewn that there are ranges in each of which the polynomial is monotone.

To handle this question in a general manner we need the theorem that where Dy is positive y is increasing (and where Dy is negative y is decreasing). This theorem is an immediate consequence of the theorems of pp. 119, 121. For let $x_1, x_2 (x_2 > x_1)$ be any two values of x in a range throughout which $Df(x)$, the derivative of the function $f(x)$, is positive, and let y denote the difference $f(x) - f(x_1)$. Then Dy is positive as long as $x_1 \leqslant x \leqslant x_2$, and therefore $f(x_2) > f(x_1)$ by the theorems cited. Similarly for a range where $Df(x)$ is negative.

For our polynomial $x^4 - 3x^2 + 2x + 4,$

$$Dy = 4x^3 - 6x + 2 = 4\left(x + \tfrac{1}{2} + \tfrac{1}{2}\sqrt{3}\right)\left(x + \tfrac{1}{2} - \tfrac{1}{2}\sqrt{3}\right)(x - 1),$$

(as is seen by noticing that when $x = 1$, $Dy = 0$, whence $x - 1$ is one factor of Dy, and the other two are obtained by division).

Thus $Dy > 0$ for x between $-\tfrac{1}{2} - \tfrac{1}{2}\sqrt{3}$ and $-\tfrac{1}{2} + \tfrac{1}{2}\sqrt{3}$ and for $x > 1$; whilst $Dy < 0$ for $x < -\tfrac{1}{2} - \tfrac{1}{2}\sqrt{3}$ and for x between $-\tfrac{1}{2} + \tfrac{1}{2}\sqrt{3}$ and 1.

Hence our function decreases as x increases up to $x = -\tfrac{1}{2} - \tfrac{1}{2}\sqrt{3}$, then increases up to $x = -\tfrac{1}{2} + \tfrac{1}{2}\sqrt{3}$, then decreases from there to $x = 1$, and, finally, increases as x increases beyond 1.

The function therefore, though not monotone, as were (e.g.) the functions x, x^3, is, we may say, monotone in stretches, as was the function x^2 (in two stretches, viz. up to $x = 0$, x^2 is decreasing and for $x > 0$, x^2 is increasing).

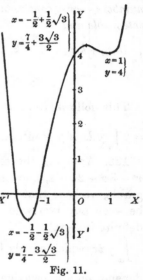

At each of the three points where the differential coefficient vanishes,—viz. the points

$$x = -\tfrac{1}{2} - \tfrac{1}{2}\sqrt{3}, \quad y = \tfrac{7}{4} - \tfrac{3\sqrt{3}}{2};$$

$$x = -\tfrac{1}{2} + \tfrac{1}{2}\sqrt{3}, \quad y = \tfrac{7}{4} + \tfrac{3\sqrt{3}}{2};$$

and $x = 1$, $y = 4$;—

the function changes from being increasing to decreasing or *vice versa*.

Our polynomial is also of bounded variation in any bounded range. (See p. 138 and Ex. 16, p. 153, above.)

The graph is sketched in Fig. 11.

In discussing the above and similar graphs the student will observe the great

Fig. 11.

utility of the differential coefficient as a means of arriving quickly and surely at the fundamental properties of the function,—properties which could only laboriously and doubtfully be discovered by mere plotting of points.

125. The general polynomial. We now proceed to discuss the general features of polynomials of any degree whatever.

Let us consider the general *polynomial*

$$y = a_n x^n + a_{n-1} x^{n-1} + \ldots + a_1 x + a_0,$$

where the coefficients $a_n, a_{n-1}, \ldots a_1, a_0$ are real numbers, with the proviso that the first (a_n) at least is not zero; the *degree* (n) of the polynomial is odd or even.

The value of y when $x = 0$ is a_0.

As x increases through positive values y may increase or decrease, but we can see, without much discussion, that as x increases, sooner or later the sign of y will be that of the first term $a_n x^n$; for evidently $|a_n x^n|$ will exceed $|na_{n-1} x^{n-1}|$ so soon as $x > |na_{n-1}/a_n|$, will exceed $|na_{n-2} x^{n-2}|$ so soon as $x^2 > |na_{n-2}/a_n|$, and so on, so that $|na_n x^n|$ will exceed $|n(a_{n-1} x^{n-1} + a_{n-2} x^{n-2} + \ldots + a_0)|$ so soon as x exceeds the greatest of the values

$$|na_{n-1}/a_n|, \; \sqrt{|na_{n-2}/a_n|}, \; \ldots \sqrt[n]{|na_0/a_n|},$$

i.e. the first term $a_n x^n$ will sooner or later exceed the sum of all the remaining terms in modulus and therefore the sign of the complete polynomial will be that of the first term.

Moreover, not only will the polynomial remain of one fixed sign for all values of x greater than a certain value, but its modulus will also increase beyond all limit; for, by similar reasoning, we can prove that $|a_n x^n|$ exceeds say *twice* the sum of the remaining terms, and therefore that the value of the polynomial certainly lies between $\frac{1}{2} a_n x^n$ and $\frac{3}{2} a_n x^n$; but each of these expressions increases indefinitely in the same sense, and therefore the polynomial increases likewise. By realising that if n is even x^n is positive whatever the sign of x, and that if n is odd x^n is positive or negative according as x is positive or negative, we see that, if n is even, for large positive and negative values of x the function is large and of the same sign as the first coefficient a_n, and that if n is odd, the function is large for large values of x and of opposite signs according as x is positive or negative.

The fundamental questions of continuity, differentiability, and integrability are easily disposed of:

If y_1 and y_2 are the values of the polynomial corresponding to the two values x_1 and x_2 of x, we have

$$y_1 - y_2 = (a_n x_1{}^n + \ldots + a_0) - (a_n x_2{}^n + \ldots + a_0)$$
$$= a_n (x_1{}^n - x_2{}^n) + \ldots + a_1 (x_1 - x_2)$$
$$= (x_1 - x_2) [a_n (x_1{}^{n-1} + x_1{}^{n-2} x_2 + \ldots + x_2{}^{n-1})$$
$$+ a_{n-1} (x_1{}^{n-2} + x_1{}^{n-3} x_2 + \ldots + x_2{}^{n-2}) + \ldots + a_1].$$

The expression in the square bracket is evidently bounded in magnitude (if x_1 and x_2 are bounded) and hence as $x_2 \to x_1$ the right-hand side tends to zero, i.e. as a sequence for x_2 tends to x_1

the corresponding sequence for y_2 tends to y_1; or *the function y is continuous* for all values x_1 of x.

The incrementary ratio

$(y_1 - y_2)/(x_1 - x_2)$
$= a_n (x_1{}^{n-1} + \dots + x_2{}^{n-1}) + a_{n-1} (x_1{}^{n-2} + \dots + x_2{}^{n-2}) + \dots + a_1$

itself is a continuous function of x_2 which tends to the limiting value $n a_n x_1{}^{n-1} + (n-1) a_{n-1} x_1{}^{n-2} + \dots + a_1$ as x_2 tends to x_1, i.e. *the original polynomial is differentiable* for all values of x, and its differential coefficient is

$$n a_n x^{n-1} + (n-1) a_{n-1} x^{n-2} + \dots + a_1.$$

The *integrability* of the polynomial follows immediately from the theorem and corollaries of pp. 156—157 above. We have

$$\int_a^b (a_n x^n + a_{n-1} x^{n-1} + \dots + a_1 x + a_0)\, dx$$

exists and equals

$$a_n \int_a^b x^n dx + a_{n-1} \int_a^b x^{n-1} dx + \dots + a_1 \int_a^b x dx + a_0 \int_a^b 1\, dx.$$

Once the separate integrals $\int_a^b x^n dx$, etc. are known the integral of the polynomial is known.

126. Let us now examine the differential coefficient

$$n a_n x^{n-1} + (n-1) a_{n-1} x^{n-2} + \dots + a_1.$$

We notice that it is itself a polynomial in x of the $(n-1)$th degree. It may be positive or negative for all values of x, or it may be positive for some and negative for other values of x. In virtue of the proved continuity of any polynomial and of the fundamental property of continuous functions*, the values (if any) of x for which it is positive must be separated from those for which it is negative by values for which it is zero.

Now there is an elementary proposition in the theory of equations to the effect that an equation of the mth degree cannot have more than m roots (unless it is an identity)†.

Applying this theorem to our differential coefficient, we see that it cannot vanish for more than $n-1$ values of x (unless it is identically zero for all values of x,—which can only happen if the "polynomial" reduces to the constant a_0) and thence that it cannot change sign more than $n-1$ times; or there are at most n ranges in each of which the differential coefficient is of constant sign, separated by points at which it vanishes.

* See p. 140 and Ex. 8, p. 151 above.
† See Ex. 4, p. 161 opposite.

Now, in ranges where the differential coefficient of a function is positive, the function is increasing, and where negative, decreasing. Hence we have proved that the graph of the general polynomial of degree n consists of at most n stretches in each of which it constantly increases or decreases, separated by points where the graph is stationary.

It may happen that the graph does not have as many as n such stretches, and it may happen that the differential coefficient may vanish for some values of x other than the turning points (e.g. the point $(0, 0)$ on the graph of x^3); but we know that it cannot have more than n such stretches nor more than $n-1$ such turning points, and that all the maxima and minima are included in the points where the differential coefficient vanishes. More than this cannot well be said. The details of the graph depend on the peculiarities of the particular polynomial concerned.

The polynomial is not, in general, monotone; but, since it is formed by the addition and subtraction of functions which are severally monotone (or expressible as the difference of two monotone functions), it follows that the polynomial is necessarily expressible as the difference of two non-decreasing functions, i.e. is *of bounded variation* over any bounded range (p. 138).

We have now seen that polynomials behave in all essential respects like the simple functions x, x^2, x^3, etc. They are the simplest class of function considered in analysis.

EXAMPLES XIV.

1. Trace the graph of the cubic polynomial $4x^3 + 9x^2 - 12x - 1$, marking the turning points.

2. Trace the graph of $x^3 + 3x^2 + 9x - 4$. Verify from first principles that this function is continuous and differentiable. Given that

$$\int_a^b x^3\,dx = (b^4 - a^4)/4, \quad \int_a^b x^2\,dx = (b^3 - a^3)/3,$$
$$\int_a^b x\,dx = (b^2 - a^2)/2, \quad \int_a^b 1\,dx = b - a,$$

find the area enclosed between the curve, the x axis, and the two ordinates $x=1$, $x=2$.

3. Sketch the graph of $12x - 4 - 4x^2$. Find its maximum value and the area enclosed in the loop above the x axis.

4. Prove that an equation of the mth degree cannot have more than m roots.

[If the equation $P_m = a_m x^m + \ldots + a_0 = 0$ could have more than m roots, a_1, a_2, etc., the relations $a_m a_1{}^m + \ldots + a_0 = 0$, etc. are all satisfied; therefore

$$a_m x^m + \ldots + a_0 = (a_m x^m + \ldots + a_0) - (a_m a_1{}^m + \ldots + a_0),$$

which $= (x - a_1) P_{m-1}$, where P_{m-1} is a polynomial of degree $m-1$. If a_2 be substituted for x in P_{m-1}, the result is zero, for with this substitution P_m is rendered zero and $a_2 - a_1$ is not zero. Therefore as before $P_{m-1} = (x - a_2) P_{m-2}$, P_{m-2} being a polynomial of degree $m-2$; so that

$$P_m = (x - a_1)(x - a_2) P_{m-2}.$$

And so on. Thus $P_m = (x - a_1)(x - a_2)...(x - a_m)P_0$, where P_0 is a constant $(P_0 = a_m)$. It is now evident that P_m cannot be zero for any value of x (e.g. a_{m+1}) different from the m values $a_1, ... a_m$ unless P_0 (and therefore also P_m) is itself identically zero for all values of x.]

5. Prove that any equation of odd degree $P_n(x) = 0$ necessarily has at least one real root and that any equation of even degree will necessarily have at least two real roots if the coefficient of the term of highest degree and the constant term in $P_n(x)$ are of opposite signs.

6. Prove that a polynomial is necessarily of bounded variation over any bounded range.

7. Express the polynomials of Exs. 1, 2, 3 as sums of monotone functions, (i) applicable to a range including only positive values of x, (ii) applicable to any bounded range.

8. Prove that any bounded function whose graph consists of a (finite) number of monotone stretches is of bounded variation over any bounded range.

9. Prove that any function which is of bounded variation over a range is integrable over that range.

§ 3. RATIONAL FUNCTIONS

127. Polynomials are the simplest kind of the larger class of functions known as *rational functions*.

Briefly, a rational function of the real variable x is any expression in x in which the only signs of operation employed on x are those of addition, subtraction, multiplication, and division. The constants which occur need not be rational numbers.

Thus $(3x + 2)/(5x^2 - 4) + 1/x - x^3$ and $(\sqrt{2} . x - x^3)/(x^2 + 3)$ are rational functions of x, while $\log_a x$, 2^x, e^x, $\sin x$, $\sqrt[3]{x}$, $\sqrt{(x^2 - 2x - 3)}$, and the sum of such an infinite series as $1 + x + x^2/2! + ...$ are irrational.

It may happen exceptionally that a function defined in irrational form is in fact a rational function. Thus the sum of the infinite series

$$1 + x + x^2 + ... (-1 < x < 1)$$

is the rational function $\dfrac{1}{1 - x}$.

Like a polynomial, a rational function is defined precisely by means of the elementary arithmetical operations, and its value is a rational number for all rational values of x if the coefficients are rational,—with the possible exception of isolated values of x where any denominator is zero.

We will now consider the nature of the graphs of one or two of the simpler rational functions (other than polynomials) and observe the resemblances and differences between such functions and polynomials.

128. The function 1/x. Consider the function $y = 1/x$.

If we try to plot the points of the graph corresponding to different values of x we find that this can be done without difficulty except for one value, $x = 0$. If we substitute $x = 0$ in $1/x$ we get $1/0$, which is meaningless. We cannot divide 1 by 0. In fact *the function is not defined* for this value of x.

We notice, however, that no matter how small we take x (not actually zero) there is a perfectly definite corresponding value of y, and the corresponding point on the graph can be plotted. Confining ourselves for the moment to positive values of x, we get in fact a succession of greater and greater values of y corresponding to smaller and smaller values of x; moreover these values of y increase beyond all limit as x is made to decrease indefinitely: corresponding to any positive sequence of values for x having 0 as limit, the sequence of values of y is unbounded and not convergent. We say the function is *unbounded* (above) in the neighbourhood of $x = 0$. It is also *not continuous* on the right at $x = 0$, independently of the fact that the function is not defined at that point. Similarly if we consider negative values of x tending to zero, the corresponding values of y decrease beyond all limit, and the function is unbounded below and not continuous on the left.

In other essential respects this function behaves like a polynomial. If $x \neq 0$, y is *continuous* at x, for

$$1/(x+h) - 1/x = - h/[x(x+h)],$$

and therefore $\to 0$ corresponding to any sequence for h which tends to 0.

If $x \neq 0$, y is *differentiable* and $Dy = -1/x^2$, for the incrementary ratio $= \frac{1}{h}\left(\frac{1}{x+h} - \frac{1}{x}\right) = -\frac{1}{x(x+h)}$, which clearly tends to $-1/x^2$. $Dy = -1/x^2 < 0$ for all values of x (except $x = 0$); and, as is seen independently, y decreases steadily for all negative values of x and for all positive values of x. As x increases positively y steadily decreases and $\searrow 0$ as $x \to \infty$; and if x is negative and decreases indefinitely, y is negative, increases and $\nearrow 0$.

The function is also *integrable* over any range (a, b), where a and b are both positive or both negative; for in any such range y is continuous and monotone and the proof of the integrability of x^2 above will apply. The function is *not integrable in any range which includes the point of discontinuity* $x = 0$.

The evaluation of the integral is postponed to the next chapter*. There are no maxima or minima. The graph is as drawn in Fig. 12. It is a *rectangular hyperbola*.

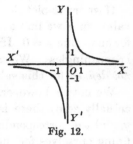

Fig. 12.

Similarly the graphs of the functions $1/(x-1)$ and $1/(x+2)$ are also rectangular hyperbolas. They have points of infinite discontinuity (similar to that of the graph $y = 1/x$ at $x = 0$) at the points $x = 1$ and $x = -2$ respectively.

129. The function $1/[(x-1)(x-2)]$. A slightly more complicated rational function is

$$1/[(x-1)(x-2)].$$

This function is undefined for two values of x, viz. $x = 1$ and $x = 2$; these are both points of infinite discontinuity of the same type as that in the case of $y = 1/x$.

The function is continuous at all other points, for the difference

$$1/[(x+h-1)(x+h-2)] - 1/[(x-1)(x-2)]$$

$$= \frac{x^2 - 3x + 2 - (x+h)^2 + 3(x+h) - 2}{(x+h-1)(x+h-2)(x-1)(x-2)}$$

$$= \frac{h(-2x-h+3)}{(x+h-1)(x+h-2)(x-1)(x-2)},$$

and the fraction last written is less than some fixed number, K say, no matter how small h be taken; the difference therefore $\rightarrow 0$ if h takes on a sequence of values which $\rightarrow 0$.

The function is also differentiable at all such points, for the incrementary ratio

$$\frac{1}{h}\left[\frac{1}{(x+h-1)(x+h-2)} - \frac{1}{(x-1)(x-2)} \right]$$

$$= \frac{-2x-h+3}{(x+h-1)(x+h-2)(x-1)(x-2)},$$

* See also Ex. 5, p. 166 below.

which, by argument similar to that by which the continuity has been established, $\to -\dfrac{2x-3}{(x-1)^2(x-2)^2}$ as h tends to zero, for all values of x other than 1 and 2. The differential coefficient is

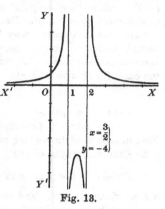

$-\dfrac{2x-3}{(x-1)^2(x-2)^2}$. This is clearly positive when $x < 3/2$, negative when $x > 3/2$, and zero when $x = 3/2$. Excluding the points of discontinuity, $x = 1$ and $x = 2$, the function is steadily increasing as x increases for $x < 3/2$ and decreasing for $x > 3/2$. At $x = 3/2$ the function has a maximum. It should be noted that a maximum does not mean the greatest value but only a value which exceeds all others *within a certain neighbourhood.* In

Fig. 13.

this example at $x = 3/2$ the function $= -4$, which is greater than its value for any other value of x between 1 and 2 but is not greater than the values of y outside these limits; we have seen in fact that y is unbounded.

As before, y is integrable over any range (a, b) which avoids the points of discontinuity.

The graph is drawn in Fig. 13.

130. Properties of rational functions in general. These two examples will suffice to shew the chief resemblances and differences between polynomials and rational functions which are not polynomials. The essential point to notice is that while polynomials are defined, continuous, and differentiable for all values of x and bounded and integrable in any bounded range for x, other rational functions may cease to be continuous, differentiable, or defined for some values of x, or bounded or integrable in some bounded ranges for x.

A rational function can always be expressed as the quotient of two polynomials, say $P_m(x)/Q_n(x)$, where $P_m(x)$ and $Q_n(x)$ are polynomials of degrees m and n respectively. It can be shewn that the number of points of discontinuity of the function will not exceed n. The points are in fact given by the roots of the equation of the nth degree $Q_n(x) = 0$. The function is continuous, differentiable and integrable throughout all ranges excluding these points, and it is monotone in stretches.

Rational functions are easily seen to be of bounded variation over any range (a, b) which includes no point of discontinuity, either interior to the range or at either end (a or b). (See Ex. 8 below.)

Non-rational functions may have other peculiarities. Thus $\sqrt{x}$ and $\log x$, though monotone, continuous, differentiable, integrable and bounded for all bounded *positive* values of x, are undefined for all negative values of x (and in the case of $\log x$ also for $x = 0$).

EXAMPLES XV.

1. Draw the graphs of

$1/(x+2)$, $x/(x-1)$, $(x^2-3x+2)/(x+1)$, and $(2x-3)/[(x-1)^2(x-2)^2]$.

Find in each case the turning points (if any).

2. Draw the graphs of $1/(1 \pm x)$, $1/(1 \pm x)^2$.

3. Prove that the definite integral $\int_k^1 \frac{1}{x} dx >$ any number K, however great, if k is sufficiently small (and positive).

4. Sketch the graphs of $1/x^2$ and $1/x^3$ and prove from the principles of section 1 above that

$$\int_a^b \frac{1}{x^2} dx = \frac{1}{a} - \frac{1}{b} \quad \text{and} \quad \int_a^b \frac{1}{x^3} dx = \left(\frac{1}{a^2} - \frac{1}{b^2}\right) \Big/ 2,$$

where a and b are both positive or both negative.

5. Prove from the principles of section 1 that

$$\int_a^b \frac{1}{x} dx = \log_e (b/a),$$

where a and b are both positive or both negative.

[Use the "fundamental inequality" of p. 110 above.]

6. Prove that the number of points of discontinuity of the function $P_m(x)/Q_n(x)$, where $P_m(x)$ and $Q_n(x)$ are polynomials of degrees m and n respectively, cannot exceed n.

7. Given that the differential coefficient of the rational function

$$f(x) = P_m(x)/Q_n(x) \text{ is } R_{m+n-1}(x)/[Q_n(x)]^2,$$

where $P_m(x)$, $Q_n(x)$, $R_{m+n-1}(x)$ are polynomials of degrees m, n, $m+n-1$ respectively, prove that, if the points of discontinuity be disregarded, the graph of $f(x)$ consists of at most $m+n$ stretches in each of which $f(x)$ is monotone.

8. Prove (i) from Ex. 7 and the definition of p. 138, and (ii) from Exs. 14—16, pp. 152–153, that a rational function is of bounded variation in any interval excluding the points of discontinuity.

9. Shew that the graph of the function $\log(x^2-1)$ consists of two separate infinite branches and that there is a strip of the coordinate plane (between the two lines $x = -1$, $x = 1$) where there is no point of the graph. Trace also the graph of the function $\log(1 - x^2)$.

§ 4. FUNCTIONS DEFINED BY POWER SERIES

131. Functions defined by convergent sequences and series of functions. In Chapter I we defined numbers by means of convergent sequences and series of numbers—and, in particular, irrational numbers by sequences and series of rational numbers. In the same way we can define *functions* by sequences and series of functions—and, in particular, irrational functions by sequences and series of rational functions or even simple polynomials. A series whose terms are functions of a real variable x may be convergent for certain ranges of values of x,—thus the series $1 - x + x^2 - x^3 + \ldots$, whose terms are simple powers of x with alternating signs, is convergent for all values of x between -1 and 1,—and the sum of the series will in general depend on x and be in fact a function of x,— viz. $1/(1 + x)$. There is only one obvious point of difficulty. If the series is convergent only for a restricted range of values of x, the function is thus defined only for values of x in that range. There is however nothing logically surprising or new about the idea of a function defined only for a restricted range of values of the variable. For example, $\log x$ is defined only for positive real values of x. In such a case then we shall content ourselves with the restricted definition.

The extension of the range of definition of such functions is in fact possible in some cases, but lies outside the scope of this course.

132. Approximation by sequences of polynomials. We can approach this question from a different point of view. We can endeavour to approximate to a given function which is not a polynomial by means of a set of polynomials. It is evident *a priori* that such a function as $1/(1 + x)$ has properties not possessed by any polynomial whatsoever, and therefore cannot be *accurately* represented as such. Just as it is possible however to find rational approximations to irrational numbers, it may here be possible to find polynomial *approximations*.

Let us consider the rational function $1/(1 + x)$.

By dividing out, we obtain, as successive quotients, the polynomials $1, 1 - x, 1 - x + x^2, \ldots$.

Drawing the graphs (C_1, C_2, etc.) of these successive polynomials, as well as that (C) of the rational function itself (see Fig. 14), we see that the successive polynomials form better and better approximations to the rational function within the limited range $-1 < x < 1$.

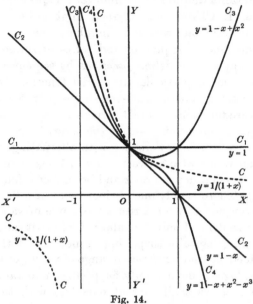

Fig. 14.

Outside this range, however, the polynomials seem to have no connection with the rational function.

In this case we know already that for any value of x between -1 and 1 the series $1 - x + x^2 - \dots$ is convergent and has $1/(1 + x)$ as its sum. This is clearly only another way of stating that the sequence of polynomials 1, $1 - x$, $1 - x + x^2$, $\dots$ gives indefinitely close approximations to the rational function in the range $-1 < x < 1$. It is in fact evident that *the representation of a function by the sum of a convergent infinite series of powers of x is equivalent to the indefinitely close approximation to the function by means of a sequence of polynomials of this type*[*].

133. Binomial theorem for negative integral index. Other

* I.e. a sequence of polynomials of steadily increasing degree, say $P_1(x), P_2(x), \dots$, where any polynomial differs from the preceding one only by the (possible) addition of a single power of x (i.e. $P_n(x) - P_{n-1}(x) = a_n x^n$, where a_n is a constant). The problem of the approximation to a function by sequences of polynomials in general lies quite beyond the scope of this course.

known rational functions may be discussed similarly. A particularly interesting set of such rational functions are the inverse powers of $1 + x$, the corresponding series being the *binomial series*.

We have proved that

$$\frac{1}{1+x} = 1 - x + x^2 - x^3 + \dots \text{ for } -1 < x < 1.$$

Dividing this infinite series in the ordinary way by $1 + x$, we obtain as successive quotients the sequence

$$1, \quad 1 - 2x, \quad 1 - 2x + 3x^2, \quad 1 - 2x + 3x^2 - 4x^3, \dots.$$

It is easy to see by direct multiplication that

$$(1 - 2x + 3x^2 - \dots \pm nx^{n-1})(1+x)^2 = 1 \pm [(n+1)x^n + nx^{n+1}].$$

If $-1 < x < 1$, $(n+1)x^n + nx^{n+1} \to 0$ as $n \to \infty$ *, and therefore

$$1 - 2x + 3x^2 - \dots \pm nx^{n-1} \to 1/(1+x)^2;$$

i.e. the infinite series $1 - 2x + 3x^2 - \dots$ is convergent for $-1 < x < 1$ and its sum is $1/(1+x)^2$.

Again in the same way

$$1/(1+x)^3 = 1 - 3x + \frac{3.4}{1.2}x^2 - \frac{3.4.5}{1.2.3}x^3 + \dots \text{ for } -1 < x < 1.$$

And so on.

By induction *we can shew in general that if n is any positive integer then*

$$(1+x)^{-n} = 1 - nx + \frac{n(n+1)}{1.2}x^2 - \frac{n(n+1)(n+2)}{1.2.3}x^3 + \dots$$

$$\dots\dots(1)$$

for $-1 < x < 1$.

Thus

$$\left[1 - (n+1)x + \frac{(n+1)(n+2)}{1.2}x^2 - \dots \right.$$
$$\left. + (-1)^r \frac{(n+1)\dots(n+r)}{1.2.\dots,r}x^r \right](1+x)$$

$$= 1 - (n+1)x + \frac{(n+1)(n+2)}{2!}x^2 - \dots + (-1)^r \frac{(n+1)\dots(n+r)}{r!}x^r$$

$$+ x - (n+1)x^2 - \dots - (-1)^r \frac{(n+1)\dots(n+r-1)}{(r-1)!}x^r$$

$$+ (-1)^r \frac{(n+1)\dots(n+r)}{r!}x^{r+1}$$

* See sequence (ii), p. 50 above.

$$= 1 - nx + \frac{n(n+1)}{2!} x^2 - \ldots + (-1)^r \frac{n(n+1)\ldots(n+r-1)}{r!} x^r$$
$$+ (-1)^r \frac{(n+1)\ldots(n+r)}{r!} x^{r+1} \quad \ldots\ldots\ldots\ldots\ldots(2).$$

The last term in this expression (2) can be shewn to tend to zero as $r \to \infty$ if $-1 < x < 1$. Thus, writing s_r for $\frac{(n+1)\ldots(n+r)}{r!} x^{r+1}$, we have $|s_r/s_{r-1}| = |(n+r)x/r| <$ a number θ less than 1 so soon as $1 + n/r < \theta/|x|$, i.e. so soon as $r > n|x|/(\theta - |x|) = m$ say; and such a number θ can be chosen, between $|x|$ and 1, to make this choice of m always possible. It will follow that

$$|s_{m+p}| = \left| s_{m-1} \frac{s_m}{s_{m-1}} \ldots \frac{s_{m+p}}{s_{m+p-1}} \right| < |s_{m-1}| \theta^{p+1} \to 0 \text{ as } p \to \infty.$$

Hence, as $r = m + p \to \infty$, $|s_r| \to 0$.

It follows now from (2) that if the sequence whose rth term is

$$1 - nx + \frac{n(n+1)}{2!} x^2 - \ldots + (-1)^r \frac{n(n+1)\ldots(n+r-1)}{r!} x^r$$

is convergent for $-1 < x < 1$ and has $(1+x)^{-n}$ for its unique limit, then the sequence whose rth term is

$$\left[1 - (n+1)x + \ldots + (-1)^r \frac{(n+1)\ldots(n+r)}{r!} x^r \right] (1+x)$$

is also convergent and has the same limit.

Hence *if* the series $1 - nx + \frac{n(n+1)}{2!} x^2 - \ldots$ is convergent for $-1 < x < 1$ and has $(1+x)^{-n}$ for its sum, the series

$$1 - (n+1)x + \frac{(n+1)(n+2)}{2!} x^2 - \ldots$$

is also convergent for $-1 < x < 1$ and has $(1+x)^{-(n+1)}$ for its sum.

We have already proved that $(1+x)^{-1} = 1 - 1x + \frac{1.2}{2!} x^2 - \ldots$;

hence $(1+x)^{-2} = 1 - 2x + \frac{2.3}{2!} x^2 - \ldots$, and, in general, the relation (1) is true.
$$\qquad\qquad\qquad\qquad\qquad\qquad\qquad\qquad\qquad\qquad\text{Q.E.D.}$$

The student will notice that the substitution of $-m$ say for n m being then a negative integer, gives the result

$$(1+x)^m = 1 + mx + \frac{m(m-1)}{2!} x^2 + \ldots \text{ for } -1 < x < 1,$$

formally identical with the binomial theorem for a positive integral index (p. 22 above). The binomial theorem is in fact true for all real values of the index, but the distinction between the case of a positive integral index,—where the " series " consists of only $m + 1$ terms and the expansion is valid for *all* values of x,—and the other cases,—where the series is an infinite series, convergent only for values of x between -1 and 1,—is essential*.

134. Power series. Expansions of this type can be used as a means of studying the known rational functions represented. We have in fact already, in obtaining the logarithmic series, essentially made use of the expansion $(1 + x)^{-1} = 1 - x + x^2 - \dots$ in this way. The most fundamental purpose of sequences and series of functions is however *to define new functions*. Various types of such sequences and series are possible, but sequences of polynomials and series of positive integral powers are the simplest and most generally useful. In this course we shall confine ourselves entirely to such series of powers,—or *power series*.

A power series in a real variable x is a series of the type

$$a_0 + a_1 x + a_2 x^2 + a_3 x^3 + \dots,$$

where the coefficients a_0, a_1, a_2, ... are real numbers independent of x, i.e. are real constants.

135. Monotony, continuity, etc. It is seen at once that, if the coefficients a_0, a_1, a_2, ... of a power series are all of the same sign, then, for all positive values of x, the function defined as the sum of the series is monotone.

In this case, for negative values of x, the function is not monotone, but it can be expressed as the sum of two monotone functions,—viz.

$$y_1 = a_0 + a_2 x^2 + a_4 x^4 + \dots, \quad \text{and} \quad y_2 = a_1 x + a_3 x^3 + a_5 x^5 + \dots,$$

for those values of x for which these two series are convergent (and therefore for those values of x for which the original power series is absolutely convergent).

It follows that the function so defined is *integrable* over any interval (a, b) which is such that the series is absolutely convergent for all values of x belonging to the interval (i.e. for $a \leqslant x \leqslant b$).

* The cases when the index is fractional or irrational are dealt with in Chapter iv, § 8 below.

For, if $b > a \geqslant 0$, the proof of the integrability of the function x^2, given in section 1 above,—which depends solely on the monotony of the function,—will apply to our function; and, if $a < b \leqslant 0$, the theorem of p. 156 above will apply. Recourse to the fundamental idea of the definite integral completes the proof, for the case when $a < 0 < b$.

In this connection it is easy to see that *the function defined by any power series is necessarily of bounded variation over any range throughout which the power series is absolutely convergent.*

For, the terms having positive coefficients will form a convergent series* and so define a function y_1 say, which will be monotone over any range including only positive values of x; and the terms having negative coefficients will form another convergent series, defining another function y_2 say, which will also be monotone over any such range. The function defined by the power series equals $y_1 + y_2$ and therefore is of bounded variation (p. 138) over any such range including only positive values of x.

The completion of the proof,—to apply to any bounded range throughout which the power series is absolutely convergent,—is left to the student.

From this result we can deduce, as above, the important result that *the function defined by any power series is necessarily integrable over any range throughout which the power series is absolutely convergent.*

It can be proved† that a power series, if convergent for any (non-zero) values of x whatever, is convergent for all values within a range $(-R, R)$,—called the *range of convergence,*—and may be convergent also for one or both of the extreme values, $-R$ and R; and that, in any case, the series is absolutely convergent for all values of x *within* the range, i.e. for $-R < x < R$. The results stated are therefore true throughout any interval wholly included within the range for which the function is defined.

It might appear that, just as the continuity and differentiability of the general polynomial follow immediately from the same properties of the simple powers, these properties will follow at once for the sums of power series in general. It is as a matter of fact true that these properties hold in general for power series, but the following considerations will shew that they are not immediately evident.

If $u_1(x) + u_2(x) + \dots$ is any convergent series of functions, having the properties of continuity, etc., we have that the sum function

$$f(x) = [u_1(x) + u_2(x) + \dots + u_n(x)] + [u_{n+1}(x) + \dots]$$
$$= s_n(x) + R_n(x) \text{ say.}$$

---

* See p. 78 above.   † See Ex. 13, p. 194 below.

We are entitled to argue that if the separate functions $u_1(x), u_2(x), \ldots$ are continuous and differentiable, then $s_n(x)$ is also continuous and differentiable, and, e.g. that the differential coefficient of $s_n(x)$ is the sum of the differential coefficients of the separate functions $u_1(x), \ldots u_n(x)$. But all we obviously know about $R_n(x)$ is that it $\rightarrow 0$ as $n \rightarrow \infty$. It is conceivable that $R_n(x)$ may be a function like $(\sin nx)/n$*. This function $\rightarrow 0$ as $n \rightarrow \infty$ for all values of x, is differentiable and $D \sin nx/n = \cos nx$; but this differential coefficient $\cos nx$ does not tend to 0 as $n \rightarrow \infty$, nor in fact does it tend to any unique limit. In this case it would be clearly unjustifiable to assume either that the infinite series of differential coefficients $Du_1(x) + Du_2(x) + \ldots$ is convergent, or that the function $f(x)$ is differentiable, or that the sum of the series of differential coefficients equals the differential coefficient of the sum function $f(x)$.

A closer examination of the properties of such series is needed. This will be best carried out directly when needed in dealing with special functions.

136. The series $1 + x + x^2/2! + x^3/3! + \ldots$. As an illustration of how power series may be used to define and discuss functions, let us consider the series

$$1 + x + x^2/2! + \ldots \quad \ldots\ldots\ldots\ldots\ldots\ldots(3),$$

which we have found (Chapter II, § 5) to be convergent for all values of x and to have for its sum the exponential function e^x. Ignoring for the moment that the sum of this series is the known function e^x, let us call the sum of (3) $E(x)$†.

We have

$$E(x + h) = 1 + (x + h) + (x + h)^2/2! + \ldots$$
$$= 1 + (x + h) + (x^2/2! + xh + h^2/2!) + \ldots \quad \ldots(4).$$

Now $E(x) = 1 + x + x^2/2! + \ldots$

and $E(h) = 1 + h + h^2/2! + \ldots,$

and these series are both absolutely convergent‡ for all values of x and h.

---

* The series in this case would not of course be a *power* series.

† The direct proof that (3) is convergent for all values of x is easy,—by application of the "ratio test" (p. 80 above) or otherwise.

‡ See Chapter I, § 7, p. 77 etc. above.

Therefore the two series may be "multiplied"* and we have

$$E(x) . E(h) = 1 + (x + h) + (x^2/2! + xh + h^2/2!) + \ldots \ldots .(5).$$

The typical terms of (4) and (5) are identical, being both equal to

$$x^n/n! + x^{n-1}h/(n-1)! + x^{n-2}h^2/2!(n-2)! + \ldots + h^n/n!.$$

That is, the two series (4) and (5) are identical and we have the *addition theorem* for the exponential function

$$E(x + h) = E(x) . E(h) \ldots \ldots \ldots \ldots (6).$$

From this relation we can at once deduce that, since the sum of the series $1 + 1 + 1/2! + 1/3! + \ldots$ is the number e†, so that $e = E(1)$, therefore $E(x) = e^x$ for all positive integral values of x, because

$$E(x) = E(1 + 1 + \ldots + 1),$$

there being x 1's in the bracket,

$$= E(1) . E(1) \ldots E(1),$$

there being x factors $E(1)$,

$$= [E(1)]^x$$
$$= e^x.$$

The same is true if x is any positive rational number, p/q say, for

$$[E(p/q)]^q = E\left(\frac{p}{q}q\right) = E(p) = e^p,$$

and therefore $E(x) = E(p/q) = e^{p/q} = e^x.$ Q.E.D.

If x is any positive rational number,

$$E(x) . E(-x) = E(x - x) = E(0) = 1,$$

whence $E(-x) = 1/E(x) = e^{-x}$, and the identity of the function $E(x)$ with the exponential function e^x for all rational values of x, positive and negative, is established.

The identity holds for all real values of x‡.

The continuity, etc., of the function $E(x)$ may be deduced from the addition theorem, thus:

The incrementary ratio

$$= [E(x + h) - E(x)]/h$$
$$= E(x) . \frac{E(h) - 1}{h} .$$

---

* See p. 80 above.   † See p. 73 above.
‡ See Ex. 4, p. 175 opposite.

Now
$$[E(h) - 1]/h = (1 + h + h^2/2! + \dots - 1)/h$$
$$= 1 + h/2! + h^2/3! + \dots$$
$$\to 1 \text{ as } h \to 0^*,$$

for the difference between the last expression and 1 is certainly less than $\dfrac{h}{2} \dfrac{1}{1 - \dfrac{h}{3}}$ if $h < 3$. Hence the incrementary ratio $\to E(x)$

and the function is differentiable and $DE(x) = E(x)$.

The continuity is implied in the differentiability.

The integrability follows from the monotony,—which is easily established from the equation (6).

Thus we have established by direct consideration of the series all the essential properties of the exponential function $E(x)$ or e^x.

137. General power series. It is not of course to be expected that the functions defined by *all* power series will be so easily discussed as the very special exponential function.

With the exception of the properties proved on pp. 171—172 above and another general property of power series, which will be proved in § 6, pp. 191—192 below, we omit all discussion of general power series. In the next two sections we shall discuss in detail the nature of the functions defined by certain important power series of a type resembling the exponential series.

EXAMPLES XVI.

1. Draw the graphs of $(1 + x)^{-2}$ and of the functions $1 - 2x$, $1 - 2x + 3x^2$, etc. formed from the first few terms of the expansion. Compare the values of the functions considered for $x = \pm\frac{1}{2}$, $\pm\cdot 1$, $\pm\cdot 01$; and for $x = \pm 2$.

2. Draw the graphs of $1/(x + 2)$ and $1/2 - x/4 + x^2/8 - x^3/16$. Expand $1/(x + 2)$ in an infinite series of powers of x. For what values of x is the expansion valid?

3. Use the binomial theorem to evaluate $1/(1\cdot 1)^5$, $1/(\cdot 99)^{10}$, $(99/101)^{10}$ correct to three decimal places.

4. Prove that if a is any real number, $E(a) = e^a$.

[Let $a = (x|y)$, x and y being rational. Then $e^a = (e^x|e^y)$ by Chapter II, § 1, $= (E(x)|E(y))$ by the rational case proved in the text, $= E(a)$ by the same kind of argument by which the fact that $(e^x|e^y)$ is a Dedekindian classification defining e^a was established in Chapter II.]

* See footnote, p. 153 above.

5. Prove directly from the series for $E(x)$ and $E(-x)$ that $E(x)E(-x)=1$.

6. Prove directly from the series and Ex. 5 that the function $E(x)$ is monotonely increasing for all values of x.

7. Shew that the series $1+\frac{1}{2}x+\frac{1.3}{2.4}x^2+\frac{1.3.5}{2.4.6}x^3+...$ is convergent for $-1<x<1$, and that its sum $=(1-x)^{-\frac{1}{2}}$.

8. Deduce from Ex. 7 that

$$1-\frac{1}{2}x-\frac{1}{2.4}x^2-\frac{1.3}{2.4.6}x^3-...=(1-x)^{\frac{1}{2}}$$

for $-1<x<1$.

9. Prove that if in any power series the error (or "remainder") after n terms $=R_n(x)<Kx^n$ numerically, where K is a fixed constant, independent of x and n, then the function defined by the power series is continuous for all values of x numerically less than 1.

[If the function is $f(x)$, the difference

$$f(x+h)-f(x)=s_n(x+h)-s_n(x)+R_n(x+h)-R_n(x),$$

where s_n denotes the sum of the first n terms.

Then $\qquad |R_n(x+h)-R_n(x)|<K|x+h|^n+K|x|^n<2K\theta^n,$

where θ is some number between $|x|$ and 1 and between $|x+h|$ and 1, $|x+h|$ being supposed <1. This $\to 0$ as $n\to\infty$. n can be chosen sufficiently great to make $|R_n(x+h)-R_n(x)|$ as small as we please. When this is done, the difference $s_n(x+h)-s_n(x)\to 0$ as $h\to 0$, because $s_n(x)$ is a continuous polynomial.]

10. Prove that if the coefficients of the power series $a_0+a_1x+a_2x^2+...$ are bounded, i.e. $|a_n|<K$ for all values of n, the series is absolutely convergent for $|x|<1$ and that the function defined by the sum of the series is differentiable for $|x|<1$.

[With the notation of Ex. 9, the incrementary ratio of $f(x)$ is the sum of the incrementary ratios of $S_n(x)$ and $R_n(x)$. Argue about these separately.]

§ 5. THE TRIGONOMETRICAL FUNCTIONS*

138. The functions C (x) and S (x) defined by the series

$$1-\frac{x^2}{2!}+\frac{x^4}{4!}-... \text{ and } x-\frac{x^3}{3!}+\frac{x^5}{5!}-....$$

Consider the two series

$$1-\frac{x^2}{2!}+\frac{x^4}{4!}-\frac{x^6}{6!}+... \qquad ....(1),$$

$$x-\frac{x^3}{3!}+\frac{x^5}{5!}-\frac{x^7}{7!}+... \qquad ....(2).$$

These series have their terms alternately positive and negative;

---

* The definitions and main results of this section are applicable also if x is any complex number. See Appendix.

from and after the Nth term, where N is the integer next greater than $\frac{1}{2}(|x|+1)$, the terms of both series certainly steadily decrease in modulus and they tend to zero, whatever real number x may be.

The series are therefore convergent and their sums define functions of x for all real values of x.

We will denote these functions by $C(x)$ and $S(x)$ respectively.

We have $C(0)=1$, $S(0)=0$, and, to two places of decimals,

$$C(1)=1-\frac{1}{2!}+\frac{1}{4!}-\ldots=\cdot54,\qquad C(-1)=\cdot54,$$

$$S(1)=1-\frac{1}{3!}+\frac{1}{5!}-\ldots=\cdot84,\qquad S(-1)=-\cdot84,$$

$$C(2)=1-\frac{2^2}{2!}+\frac{2^4}{4!}-\ldots=-\cdot42,\qquad S(2)=2-\frac{2^3}{3!}+\ldots=\cdot91.$$

Other values may be obtained.

$C(x)$ is an *even* function,—i.e. $C(-x)=C(x)$; and $S(x)$ is *odd*, —i.e. $S(-x)=-S(x)$.

139. The addition theorems. Let us try to establish *addition theorems*,—expressing $C(x+y)$ and $S(x+y)$ in terms of $C(x)$, $C(y)$, $S(x)$, and $S(y)$.

We shall find that, not only do these functions possess addition theorems of this kind, but also the theorems themselves will give us very valuable information about the functions.

The series (1) and (2) are not only convergent for all values of x;—they are *absolutely convergent**.

For, arguing from first principles, if the terms of the series (1), for example, be replaced by their moduli, the terms from the Nth onwards, N being the integer next greater than $\frac{1}{2}(|x|+1)$, are less than (or equal to)

$$\frac{x^{2N-2}}{(2N-2)!},\quad \frac{x^{2N-2}}{(2N-2)!}\theta,\quad \frac{x^{2N-2}}{(2N-2)!}\theta^2,\ldots,$$

where $\qquad\qquad \theta=\left(\frac{x}{2N-1}\right)^2<1.$

These are the terms of a convergent geometrical progression, of common ratio θ, and the sum of any number of them

$$<\frac{x^{2N-2}}{(2N-2)!}\frac{1}{1-\theta}.$$

* See p. 77 above.

Hence the sum of any number of terms of the series (1) as modified is bounded,—being less than the fixed number

$$1 + \frac{x^2}{2!} + \dots + \frac{x^{2N-4}}{(2N-4)!} + \frac{x^{2N-2}}{(2N-2)!} \frac{1}{1-\theta}.$$

The modified series of positive terms is therefore convergent and the original series (1) is absolutely convergent for all values of x. Similarly the series (2) is also absolutely convergent.

An appeal to d'Alembert's ratio test for absolute convergence (p. 80 above) would dispense with this proof, for in each case $|u_{n+1}/u_n| \leqslant$ a fixed number k which < 1, for all values of n beyond a certain value.

Let, now, x and y be any two real numbers.

Then

$$C(x+y) = 1 - \frac{(x+y)^2}{2!} + \frac{(x+y)^4}{4!} - \dots$$

$$= 1 - \left(\frac{x^2}{2!} + xy + \frac{y^2}{2!}\right) + \left(\frac{x^4}{4!} + \frac{x^3 y}{3!} + \frac{x^2 y^2}{2!\,2!} + \frac{xy^3}{3!} + \frac{y^4}{4!}\right) - \dots$$

$$\dots\dots\dots\dots\dots(3)$$

and $S(x+y) = (x+y) - \frac{(x+y)^3}{3!} + \dots$

$$= (x+y) - \left(\frac{x^3}{3!} + \frac{x^2 y}{2!} + \frac{xy^2}{2!} + \frac{y^3}{3!}\right) + \dots \dots\dots\dots(4).$$

Also the series for $C(x)$, $C(y)$, $S(x)$, $S(y)$ are absolutely convergent and therefore any two of these four series may be "multiplied*."

We have in fact

$$C(x)\,C(y) = \left(1 - \frac{x^2}{2!} + \frac{x^4}{4!} - \dots\right)\left(1 - \frac{y^2}{2!} + \frac{y^4}{4!} - \dots\right)$$

$$= 1 - \left(\frac{x^2}{2!} + \frac{y^2}{2!}\right) + \left(\frac{x^4}{4!} + \frac{x^2 y^2}{2!\,2!} + \frac{y^4}{4!}\right) - \dots \dots\dots\dots(5),$$

$$S(x)\,S(y) = \left(x - \frac{x^3}{3!} + \dots\right)\left(y - \frac{y^3}{3!} + \dots\right)$$

$$= xy - \left(\frac{x^3 y}{3!} + \frac{xy^3}{3!}\right) + \dots \dots\dots\dots\dots\dots(6),$$

$$S(x)\,C(y) = \left(x - \frac{x^3}{3!} + \dots\right)\left(1 - \frac{y^2}{2!} + \dots\right)$$

$$= x - \left(\frac{x^3}{3!} + \frac{xy^2}{2!}\right) + \left(\frac{x^5}{5!} + \frac{x^3 y^2}{3!\,2!} + \frac{xy^4}{4!}\right) - \dots \dots\dots\dots(7),$$

---

* See p. 80 (theorem B) above.

$$C(x)\,S(y) = \left(1 - \frac{x^2}{2!} + \frac{x^4}{4!} - \cdots\right)\left(y - \frac{y^3}{3!} + \cdots\right)$$

$$= y - \left(\frac{x^2 y}{2!} + \frac{y^3}{3!}\right) + \left(\frac{x^4 y}{4!} + \frac{x^2 y^3}{2!\,3!} + \frac{y^5}{5!}\right) - \cdots \quad \cdots\cdots\cdots(8),$$

and these series are all absolutely convergent.

Comparing the series (3) with series (5) and (6) we notice that the terms of (3) are precisely the differences of corresponding terms of series (5) and (6), i.e. the nth term (or bracket) of series (3)

$$= (-1)^{n+1}\left[\frac{x^{2n-2}}{(2n-2)!} + \frac{x^{2n-3}y}{(2n-3)!} + \frac{x^{2n-4}y^2}{(2n-4)!\,2!} + \cdots \right.$$
$$\left. + \frac{xy^{2n-3}}{(2n-3)!} + \frac{y^{2n-2}}{(2n-2)!}\right]$$

$$= (-1)^{n-1}\left[\frac{x^{2n-2}}{(2n-2)!} + \frac{x^{2n-4}y^2}{(2n-4)!\,2!} + \cdots + \frac{y^{2n-2}}{(2n-2)!}\right]$$
$$- (-1)^n\left[\frac{x^{2n-3}y}{(2n-3)!} + \frac{x^{2n-5}y^3}{(2n-5)!\,3!} + \cdots + \frac{xy^{2n-3}}{(2n-3)!}\right]$$

= the nth term of series (5) − the $(n-1)$th term of series (6).

This is true for all values of n greater than 1; and for $n = 1$ we have that the first term of (3) = 1 = the first term of (5).

Hence* the sum of the series (3) = the sum of (5) − the sum of (6), or

$$C(x+y) = C(x)\,C(y) - S(x)\,S(y) \quad \cdots\cdots\cdots\cdots(9),$$

the required *addition theorem for* $C(x+y)$.

Comparing similarly series (4) with series (7) and (8), we obtain the *addition theorem for* $S(x+y)$, viz.

$$S(x+y) = S(x)\,C(y) + C(x)\,S(y) \quad \cdots\cdots\cdots(10).$$

The student will notice that the two functions $C(x)$ and $S(x)$ are intimately connected, both addition theorems involving all the four functions $C(x)$, $C(y)$, $S(x)$, $S(y)$.

140. Consequences of addition theorems. We can at once deduce from the addition theorems various important algebraical consequences:

Putting $y = x$, we have

$$C(2x) = [C(x)]^2 - [S(x)]^2$$
and
$$S(2x) = 2S(x)\,C(x) \qquad \left.\right\} \quad \cdots\cdots\cdots\cdots(11).$$

* P. 80 (theorem A) above.

Putting $y = -x$,
$$1 = C(0) = C(x) C(-x) - S(x) S(-x)$$
and therefore $$[C(x)]^2 + [S(x)]^2 = 1 \quad \dots\dots\dots\dots(12).$$

This relation (12) reveals still more clearly the intimate connection between the two functions $C(x)$ and $S(x)$, for either can now be expressed directly in terms of the other; thus
$$C(x) = \pm \sqrt{\{1 - [S(x)]^2\}}, \quad S(x) = \pm \sqrt{\{1 - [C(x)]^2\}}.$$
An important deduction from this relation (12) is that $|C(x)|$ *and* $|S(x)|$ *are less than or equal to 1 for all values of* x ....(13).

Again, since $x = \dfrac{x+y}{2} + \dfrac{x-y}{2}$ and $y = \dfrac{x+y}{2} - \dfrac{x-y}{2}$, we have
$$C(x) + C(y) = C\left(\frac{x+y}{2}\right) C\left(\frac{x-y}{2}\right) - S\left(\frac{x+y}{2}\right) S\left(\frac{x-y}{2}\right)$$
$$+ C\left(\frac{x+y}{2}\right) C\left(\frac{x-y}{2}\right) + S\left(\frac{x+y}{2}\right) S\left(\frac{x-y}{2}\right).$$
Hence
$$C(x) + C(y) = \quad 2C\left(\frac{x+y}{2}\right) C\left(\frac{x-y}{2}\right)$$
and similarly
$$C(x) - C(y) = -2S\left(\frac{x+y}{2}\right) S\left(\frac{x-y}{2}\right)$$
$$S(x) + S(y) = \quad 2S\left(\frac{x+y}{2}\right) C\left(\frac{x-y}{2}\right)$$
$$S(x) - S(y) = \quad 2C\left(\frac{x+y}{2}\right) S\left(\frac{x-y}{2}\right)$$
$$\left. \right\} \dots\dots\dots\dots\dots\dots(14),$$
expressing the sums and differences of the two functions $C(x)$ and $C(y)$ (and $S(x)$ and $S(y)$) as products of two C and S functions.

Other formulae will occur to the student of elementary trigonometry when he realises that our addition theorems are formally identical with the formulae for the sines and cosines of the sum of two angles, x and y, the function $S(x)$ replacing the sine and $C(x)$ the cosine. *All the formulae of elementary trigonometry which can be deduced algebraically from these addition formulae can be established for our functions* $S(x)$ *and* $C(x)$.

141. Continuity and differentiability of C (x) and S (x). With the help of the relations (14) (deduced from the addition theorems), or direct from the addition theorems, we can prove that *the functions* $C(x)$ *and* $S(x)$ *are continuous and differentiable for all values of* x and $DC(x) = -S(x)$, $DS(x) = C(x)$.

We have in fact from (14), if x is any number and h any number other than zero,
$$C(x+h) - C(x) = -2S(x+h/2)\,S(h/2)$$
and
$$S(x+h) - S(x) = 2C(x+h/2)\,S(h/2).$$

From (13), $S(x+h/2)$ and $C(x+h/2)$ are numerically less than 1; and therefore each of these differences is less numerically than $2S(h/2)$.

But $2S(h/2) = 2\left[\dfrac{h}{2} - \dfrac{1}{3!}\left(\dfrac{h}{2}\right)^3 + \ldots\right]$ and is less numerically than h provided only that $\dfrac{1}{3!}\left(\dfrac{h}{2}\right)^2 < 1$ (i.e. certainly if $|h| < 4$), for then the series last written is of the "alternating" type*.

Hence if h takes on any sequence of values tending to the limit 0, the corresponding sequence for $2S(h/2)$ also tends to 0, and the sequences for $C(x+h) - C(x)$ and $S(x+h) - S(x)$ therefore both tend to zero. The continuity of $C(x)$ and $S(x)$ for all values of x is established.

To establish the differentiability, we have that the incrementary ratios for $C(x)$ and $S(x)$ are respectively
$$\frac{C(x+h) - C(x)}{h} = -S(x+h/2)\frac{S(h/2)}{h/2}$$
and
$$\frac{S(x+h) - S(x)}{h} = C(x+h/2)\frac{S(h/2)}{h/2} \quad\Bigg\rbrace \ldots\ldots(15).$$

In virtue of the proved continuity, if h takes on a sequence of values tending to zero, $S(x+h/2) \to S(x)$ and $C(x+h/2) \to C(x)$. But
$$[S(h/2)]/(h/2) = 1 - \frac{(h/2)^2}{3!} + \frac{(h/2)^4}{5!} - \ldots \quad\ldots\ldots(16).$$

The series (16),—certainly if $|h| < 4$,—is of the alternating type and its sum differs from the first term (viz. 1) by less than the second term $(h/2)^2/3!$. As $h \to 0$†, $(h/2)^2/3! \to 0$. Therefore the sum of the series (16) $\to 1$.

Hence, from (15), if h takes on a sequence of values tending to the unique limit 0, the corresponding sequences of the values of the incrementary ratios for $C(x)$ and $S(x)$ are convergent and tend to the unique limits $-S(x)$ and $C(x)$ respectively, no matter what such sequence for h be taken.

* Pp. 74—76 above.   † See footnote p. 153 above.

That is, the functions $C(x)$ and $S(x)$ are differentiable for all values of x and $DC(x) = -S(x)$, $DS(x) = C(x)$.   Q.E.D.

142. Monotony of C (x) for x positive and less than 2. It also follows from the formulae (14), or direct from the addition theorems, that the function $C(x)$ *is monotonely decreasing at least for positive values of x less than 2.*

For, if $0 < x < x+h < 2$,

$$C(x+h) - C(x) = -2S(x+h/2)\, S(h/2);$$

and $S(x+h/2) = S(x')$ say

$$= x' - x'^3/3! + \ldots,$$

which is positive since $x'^3/3! = x'(x'^2/3!) < x'$
and the series for $S(x')$ is of the alternating type; while

$$S(h/2) = \frac{h}{2}\left[1 - \frac{(h/2)^2}{3!} + \ldots\right] > 0 \quad \text{because} \quad \frac{(h/2)^2}{3!} < 1 \quad \text{and} \quad h > 0;$$

and therefore the difference $C(x+h) - C(x)$ is negative and $C(x)$ is decreasing for $0 < x < 2$ *.

143. Existence of a least positive number ($\varpi/2$) for which C ($\varpi/2$) = 0. We now know that

$$C(0) = 1 \quad \text{and} \quad C(2) = 1 - 2^2/2! + 2^4/4! - \ldots < -1/3;$$

and the function $C(x)$ is continuous and monotonely decreasing as x increases from 0 to 2.

By an appeal to the fundamental property of continuous functions† we could deduce at once, firstly that *for at least one value of x between 0 and 2, $C(x) = 0$*, and secondly, in virtue of the monotony of $C(x)$, that *there is only one such value of x in this range.* Or we can prove this directly thus:

If x is any number between 0 and 2, $C(x)$ is either positive, negative, or zero. Let y denote any number between 0 and 2 for which $C(y) \geqslant 0$ and z denote any number between 0 and 2 for which $C(z) < 0$.

Then every number $y <$ every number z, because if a number $y \geqslant$ a number z, $C(y)$ would $\leqslant C(z)$, which would be incompatible with $C(y) \geqslant 0 > C(z)$.

* This fact also follows immediately from the fact that the differential coefficient of $C(x)$, viz. $-S(x)$, is negative for $0 < x < 2$. We do not imply that $C(x)$ ceases to decrease as soon as x is as large as 2. It does not.

† P. 140 above.

The classification $(y \mid z)$ is therefore a Dedekindian classification of the system of real numbers if to the class y be added all numbers less than or equal to 0 and to the class z all numbers $\geqslant 2$; for then every real number is included in one of the two classes. Let us call the real number defined by this classification α, i.e. $(y \mid z) = \alpha$.

Now the function $C(x)$ is continuous at the point α and therefore, if an increasing sequence of numbers $y_1, y_2, \ldots$ be taken, having α as upper bound and unique limit, the corresponding sequence $C(y_1), C(y_2), \ldots$ is also convergent and has $C(\alpha)$ for unique limit. But these numbers $y_1, y_2, \ldots$, being less than α, belong to the lower class (y) of the classification $(y \mid z)$; so that

$$C(y_1) \geqslant 0, \ C(y_2) \geqslant 0, \ \ldots;$$

and it follows that $C(\alpha) \geqslant 0$.

Similarly, by taking a decreasing sequence of numbers $z_1, z_2, \ldots$, tending to the unique limit α, we prove $C(\alpha) \leqslant 0$.

Since $C(\alpha) \geqslant 0$ and $\leqslant 0$ it follows that $C(\alpha) = 0$.

That α is the only such number between 0 and 2 is evident from the monotony. The theorem stated is completely proved.

This number α,—thus defined as *the least positive number for which* $C(\alpha) = 0$,—is destined to play an important part in the theory of these functions $C(x)$ and $S(x)$ and elsewhere. We shall find later that $\alpha = \pi/2$, where π is the ratio of the circumference to the diameter of a circle. We will call it $\varpi/2$.

144. So far, all we know as to the value of this number $\varpi/2$ is that it is positive and less than 2. However, we know that $C(1)$ is positive and $C(2)$ negative and therefore (by the fundamental property of continuous functions, or by the argument of the last paragraph) this number $\varpi/2$, for which $C(\varpi/2) = 0$, must lie between 1 and 2. By evaluating to a sufficient degree of accuracy $C(1\cdot5)$ and $C(1\cdot6)$, which are respectively positive and negative, $\varpi/2$ is proved to lie between $1\cdot5$ and $1\cdot6$. It is possible to evaluate ϖ to any desired degree of accuracy in this way, but the work involved in thus obtaining any considerable degree of accuracy would be prohibitive. Special series can be found for this purpose*. The value of the number ϖ correct to six places of decimals is $3\cdot141593$.

145. Periodicity formulae. We have proved that $C(\varpi/2) = 0$.

* See Exs. 16 and 19, p. 189 below.

By the relation $[C(x)]^2 + [S(x)]^2 = 1$ it follows that $S(\varpi/2) = \pm 1$. Since $S(x)$ is known to be positive for all positive values of x less than 2, it follows that $S(\varpi/2) = 1$.

Thus $$C(\varpi/2) = 0 \text{ and } S(\varpi/2) = 1 \dots\dots\dots\dots(17).$$

From (17) and the addition theorems we get

$$C(\varpi) = -1, \ S(\varpi) = 0; \quad C(3\varpi/2) = 0, \ S(3\varpi/2) = -1;$$
$$C(2\varpi) = 1, \ S(2\varpi) = 0; \quad \text{etc.}$$

Also, if x is any number,

$$C(\varpi/2 \mp x) = C(\varpi/2)\,C(x) \pm S(\varpi/2)\,S(x),$$

whence $\quad C(\varpi/2 \mp x) = \pm S(x)$

and similarly $S(\varpi/2 \mp x) = C(x)$ $\qquad \Big\}$ $\dots\dots\dots\dots\dots\dots(18).$

Also similarly, $\quad C(\varpi \pm x) = -C(x)$

$\qquad\qquad\qquad\quad S(\varpi \pm x) = \mp S(x)$ $\Big\}$ $\dots\dots\dots\dots(19),$

$\qquad\qquad\qquad\quad C(2\varpi + x) = C(x)$

and $\qquad\qquad\quad S(2\varpi + x) = S(x)$ $\Big\}$ $\dots\dots\dots\dots(20).$

These formulae (18)—(20),—called briefly the *periodicity formulae*—shew that *the functions $C(x)$ and $S(x)$ are both completely known for all values of x if either $C(x)$ or $S(x)$ is known in the restricted range* 0 to $\varpi/2$.

If, for example, x lies between $6\varpi - \varpi/2$ and 6ϖ,

$$S(x) = S(2\varpi + 2\varpi + \varpi + \varpi/2 + y)$$

where y is some number between 0 and $\varpi/2$; and therefore

$$S(x) = S(\varpi + \varpi/2 + y) \text{ by applying (20) twice}$$
$$= -S(\varpi/2 + y) \text{ by (19)}$$
$$= -C(y) \text{ by (18)}$$

or $\qquad\qquad\qquad = -S(\varpi/2 - y);$

and $S(x)$ is expressed in terms of $C(y)$ where $0 < y < \varpi/2$ or alternatively in terms of $S(\varpi/2 - y)$ where $0 < \varpi/2 - y < \varpi/2$.

146. The graphs of C (x) and S (x). The complete graphs of the functions $C(x)$ and $S(x)$ can now be drawn. We have seen that $C(x)$ decreases steadily from 1 to 0 as x increases from 0 to $\varpi/2$. When x lies between $\varpi/2$ and ϖ, $-C(x) = C(\varpi - x)$ and therefore $-C(x)$ decreases from 1 to 0 as x decreases from ϖ to $\varpi/2$ in precisely the same way as $C(x)$ decreases as x increases from 0 to $\varpi/2$; or as x increases from $\varpi/2$ to ϖ, $C(x)$ is negative and decreases from 0 to -1 precisely as $-C(x)$ decreases from 0 to -1 as x decreases from $\varpi/2$ to 0. When x lies between ϖ and 2ϖ, $C(x) = -C(\varpi + x)$ and therefore $-C(x)$ behaves in the range ϖ to 2ϖ precisely as $C(x)$ behaves in the range 0 to ϖ.

Between 2ϖ and 4ϖ, $C(x)$ behaves precisely as between 0 and 2ϖ, because $C(2\varpi + x) = C(x)$; and the graph repeats itself between 2ϖ and 4ϖ, again between 4ϖ and 6ϖ, and so on.

Since $C(x)$ is an even function, the part of the graph to the left of the origin is precisely similar to that part to the right of the origin.

Also, since $S(x + \varpi/2) = C(x)$, and therefore $S(x) = C(x - \varpi/2)$, it follows that in any range, e.g. that from 0 to $\varpi/2$, $S(x)$ behaves precisely as $C(x)$ behaves in the range distant $\varpi/2$ to the left of this range, e.g. that from $-\varpi/2$ to 0; so that the graph of $S(x)$ is identical with that of $C(x)$ except that it is displaced a distance $\varpi/2$ to the right.

The graphs consist of endless successions of "waves" of length 2ϖ; they are drawn in Fig. 15.

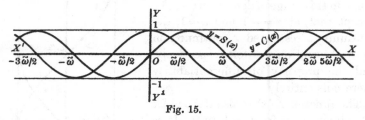

Fig. 15.

147. Integrability of C(x) and S(x). The *integrability* of $C(x)$ and $S(x)$ in the interval $(0, \varpi/2)$ (or any interval included in this) is a direct consequence of the proved monotony of these functions*. The integrability of these functions in any interval whatever follows immediately. For, from the nature of a definite integral, since e.g. $C(x)$ is integrable in the interval $(0, \varpi/2)$ and also in the interval $(\varpi/2, \varpi)$, it is also integrable in any interval (a, b), including parts of both these intervals; and similarly for any interval; and for the function $S(x)$.

---

* The proof of pp. 144—148 above of the integrability of x^2 is applicable with no vital change.

148. The inverse functions, $\bar{C}(x)$ and $\bar{S}(x)$. In the light of our knowledge of the functions $C(x)$ and $S(x)$ our proof above of the existence of one and only one number $\varpi/2$ between 0 and 2 for which $C(\varpi/2) = 0$, can,—by an obvious and very slight modification,—be made to cover the important "existence theorem for the inverse C function":

If x is any number from -1 to 1 inclusive, there exists one and only one number y from 0 to ϖ inclusive for which $C(y) = x$.

We have similarly also the companion theorem for S:

If x is any number from -1 to 1 inclusive, there exists one and only one number y from $-\varpi/2$ to $\varpi/2$ inclusive for which $S(y) = x$.

We will denote the inverse C and S functions thus established by $\bar{C}$ and $\bar{S}$. $\bar{C}(x)$ *is defined as that number y from 0 to ϖ inclusive for which $C(y) = x$, and $\bar{S}(x)$ is defined as that number y from $-\varpi/2$ to $\varpi/2$ inclusive for which $S(y) = x$.*

The function $\bar{C}(x)$ is defined for all values of x from -1 to 1. It is clearly monotonely decreasing and easily proved to be continuous, differentiable, and integrable throughout any range in this range of definition $(-1, 1)$, except that, at $x = -1$ and $x = 1$, the continuity is restricted to continuity on the right and on the left respectively, and the property of differentiability there fails entirely.

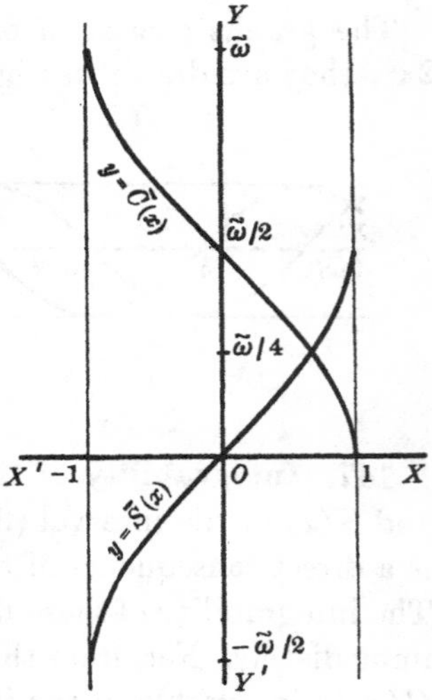

The function $\bar{S}(x)$ is also defined for all values of x from -1 to 1. It is likewise monotonely increasing, continuous, differentiable, and integrable throughout any interval included in the interval $(-1, 1)$, with the same exception.

The graphs of the functions $\bar{C}(x)$, $\bar{S}(x)$ are drawn in Fig. 16.

Fig. 16.

It should be particularly noted that, in the definition of these inverse functions $\bar{C}(x)$ and $\bar{S}(x)$, the restriction that y must lie in the range from 0 to ϖ and $-\varpi/2$ to $\varpi/2$ respectively, is essential; but it must not be thought, for example, that $\bar{C}(x)$ is the only number y,—of any magnitude,—for which $C(y) = x$. As we know, from the periodicity of $C(x)$ and $S(x)$, this is far from being the case.

149. Resemblance to the functions of trigonometry. The student conversant with elementary trigonometry cannot have failed to notice the remarkable similarity between the functions $C(x)$, $S(x)$ and the trigonometrical ratios, or circular functions, $\cos x$, $\sin x$. The properties considered above are shared in every particular by the two pairs of functions; and this agreement will be complete if the unit of angle used is such that $180°$ is represented by the number ϖ,—which will be the case if radian measure be used if (as seems probable) ϖ is identical with the number π representing the ratio of the circumference of a circle to its diameter. In spite of the apparent total lack of connection of the sums of the infinite series (1) and (2) with ratios of the sides of right-angled triangles (or even with circles), the student will probably feel fairly confident that the functions $C(x)$ and $S(x)$ actually are the same functions as $\cos x$ and $\sin x$;—and his feeling will be correct. It is not indeed difficult,—once we have satisfactory analytical definitions of the geometrically-defined trigonometrical ratios,—to prove, on lines similar to those adopted in Chapter II in obtaining the logarithmic and exponential series, that $C(x) = \cos x$ and $S(x) = \sin x$ for all values of x, the angle x being measured in radians. The proof is postponed to the next chapter[*].

We call the functions $C(x)$ and $S(x)$ *the trigonometrical functions*, and the functions $\bar{C}(x)$ and $\bar{S}(x)$ the *inverse* trigonometrical functions. We do not of course assume any identity between these functions and the functions $\cos x$ and $\sin x$, arc $\cos x$ (or $\cos^{-1} x$) and arc $\sin x$ (or $\sin^{-1} x$); or between the numbers ϖ and π.

EXAMPLES XVII.

1. Verify the addition theorems for $C(x)$ and $S(x)$ if x and y are so small that all terms occurring of degree exceeding three in x and y are negligible.

2. Evaluate correct to two decimal places $S(\cdot 1)$, $S(1)$, $S(2)$, $S(10)$, $S(20)$.

3. Calculate $\log C(\cdot 1)$ correct to three decimal places.

4. Prove
$$C(\varpi/4) = S(\varpi/4) = (\sqrt{2})/2,$$
$$C(\varpi/3) = S(\varpi/6) = 1/2,$$
$$C(\varpi/6) = S(\varpi/3) = (\sqrt{3})/2.$$

---

[*] See p. 277. For the identity of π and ϖ see p. 272. See also Exs. 7, p. 231; 9 and 10, p. 260; 9, p. 266; and 8, 10 and 13, pp. 279, 280.

5. Prove the "triple angle" formulae:
$$C(3x) = 4[C(x)]^3 - 3C(x), \quad S(3x) = 3S(x) - 4[S(x)]^3.$$

6. Prove the "half angle" formulae:
$$C\left(\frac{x}{2}\right) = \pm \sqrt{\frac{1 + C(x)}{2}}, \quad S\left(\frac{x}{2}\right) = \pm \sqrt{\frac{1 - C(x)}{2}}.$$

7. Prove (by induction) the formulae for "multiple angles":
$$C(nx) = C^n - \frac{n(n-1)}{2!} S^2 C^{n-2} + \frac{n(n-1)(n-2)(n-3)}{4!} S^4 C^{n-4} - \dots,$$
$$S(nx) = nSC^{n-1} - \frac{n(n-1)(n-2)}{3!} S^3 C^{n-3} + \dots,$$

where n is any positive integer, and C and S denote $C(x)$ and $S(x)$ respectively.

8. Prove $C(x)\,C(y) = \frac{1}{2}[C(x+y) + C(x-y)]$ direct from the series. Establish similar formulae for $S(x)\,S(y)$, $C(x)\,S(y)$, $S(x)\,C(y)$.

9. Prove that $S(1)$ is an irrational number.

[See Ex. 17, p. 60 above, where the irrationality of e is established. The method there adopted will apply here also.]

10. Shew that the graphs of $C(x)$ and $S(x)$ cut the x axis at an angle of 45° at all points of intersection with the axis.

Prove also that the slope of either of the graphs at any point of intersection of the two graphs is $\pm 1/\sqrt{2}$.

11. Prove that $C(x)$ has maxima at all the points $0, \pm 2\varpi, \pm 4\varpi, \dots$ and minima at all the points $\pm \varpi, \pm 3\varpi, \dots$; the values of $C(x)$ at the maxima and minima being respectively 1 and -1.

Shew also that the maximum and minimum values of $S(x)$ are 1 and -1 and that they occur at the points $\varpi/2, 5\varpi/2, 9\varpi/2, \dots$; $-3\varpi/2, -7\varpi/2, \dots$; and $3\varpi/2, 7\varpi/2, \dots$; $-\varpi/2, -5\varpi/2, \dots$ respectively.

12. Prove that
$$1 + C(a) + C(2a) + C(3a) + \dots + C[(n-1)a] = \frac{1}{2} + \frac{S[(n-\frac{1}{2})a]}{2S(a/2)}$$
and
$$S(a) + S(2a) + \dots + S[(n-1)a] = \frac{C(a/2) - C[(n-\frac{1}{2})a]}{2S(a/2)},$$

n being any positive integer, and a any real number.

Deduce $\int_0^a C(x)\,dx = S(a)$ and $\int_0^a S(x)\,dx = 1 - C(a)$.

Hence prove that the area of the "wave" of the graph of $C(x)$ between $-\varpi/2$ and $\varpi/2$ is 2.

[The summation formulae come by induction. For the integration cut up the range into n equal parts; the total areas of the "small rectangles" are then the above sums multiplied by a with $a = a/n$.]

13. Prove the existence theorem for $\overline{S}(x)$,—the inverse S function.

14. If x is any real number from -1 to 1 inclusive, shew that

(i) $C(y) = x$ if and only if y is one of the numbers $2n\varpi \pm \overline{C}(x)$, where n has all positive and negative integral values (and zero), and

(ii) $S(y) = x$ if and only if y is one of the numbers $n\varpi + (-1)^n \overline{S}(x)$.

15. Prove that the inverse trigonometrical functions $\bar{C}(x)$ and $\bar{S}(x)$ are continuous and differentiable for $-1 < x < 1$; and shew
$$D\bar{C}(x) = -1/[\surd(1-x^2)], \quad D\bar{S}(x) = 1/[\surd(1-x^2)].$$
[If $\bar{C}(x) = y$, so that $x = C(y)$, $S(y) = \surd[1-(C(y))^2] = \surd(1-x^2)$. Use the method of pp. 128—129.]

16. Establish the series
$$\bar{S}(x) = x + \frac{1}{2}\frac{x^3}{3} + \frac{1.3}{2.4}\frac{x^5}{5} + \frac{1.3.5}{2.4.6}\frac{x^7}{7} + \cdots$$
for $-1 \leqslant x \leqslant 1$.

By taking $x = \frac{1}{2}$ deduce the value of ϖ correct to two decimal places.

[Use the results of Ex. 15 and of Ex. 7, p. 176 and the general method used in the deduction of the logarithmic series in Chapter II.]

17. Defining a function $T(x)$ as the ratio $[S(x)]/[C(x)]$, obtain its fundamental properties and draw its graph.

Prove in particular that $T(x+y) = \dfrac{T(x) + T(y)}{1 - T(x)T(y)}$ and $DT(x) = 1/[C(x)]^2$ for all values of x which are not positive or negative odd multiples of $\varpi/2$.

Establish also the properties of the inverse T function, defined thus,—$\bar{T}(x)$ is that number y between $-\varpi/2$ and $\varpi/2$ for which $T(y) = x$, this definition being valid for all values of x. Prove in particular that $D\bar{T}(x) = 1/(1+x^2)$.

Discuss also the functions $1/S(x)$, $1/C(x)$, $C(x)/S(x)$ and the corresponding inverse functions.

18. Deduce from Ex. 17 the power series for $\bar{T}(x)$:
$$\bar{T}(x) = x - \frac{x^3}{3} + \frac{x^5}{5} - \cdots,$$
valid for all values of x from -1 to 1 inclusive. (Gregory's series.)

19. Shew that, if $T(x) = \dfrac{1}{5}$, $T\left(4x - \dfrac{\varpi}{4}\right) = \dfrac{1}{239}$; and thence deduce the equality $\dfrac{\varpi}{4} = 4\bar{T}\left(\dfrac{1}{5}\right) - \bar{T}\left(\dfrac{1}{239}\right)$. From this formula and the series of Ex. 18 shew that, to six places of decimals, $\varpi = 3\cdot141593$.

§ 6. THE HYPERBOLIC FUNCTIONS

150. The functions cosh x and sinh x. A pair of series similar to the trigonometrical series and defining functions of importance are
$$1 + x^2/2! + x^4/4! + \cdots \quad \ldots\ldots\ldots\ldots\ldots\ldots(1)$$
and
$$x + x^3/3! + x^5/5! + \cdots \quad \ldots\ldots\ldots\ldots\ldots(2).$$
These series are absolutely convergent for all values of x. We will call their sums $\cosh x$ and $\sinh x$ respectively. These functions are the *hyperbolic cosine* and *hyperbolic sine*. Alternative definitions are the equalities of Ex. 1 below.

We can calculate the values and establish the properties of these functions in just the same way as before. We have in fact

cosh $0 = 1$, sinh $0 = 0$; cosh x is an *even* function; sinh x is *odd*.

sinh x is monotonely increasing for all values of x.

cosh x is monotonely decreasing for negative values of x and monotonely increasing for positive values of x.

Both cosh x and sinh x increase beyond all limit.

sinh x takes on all real values; cosh x takes on all real values greater than 1.

Both functions are integrable throughout any range.

The addition theorems are

$$\cosh (x + y) = \cosh x \cosh y + \sinh x \sinh y,$$
$$\sinh (x + y) = \sinh x \cosh y + \cosh x \sinh y.$$

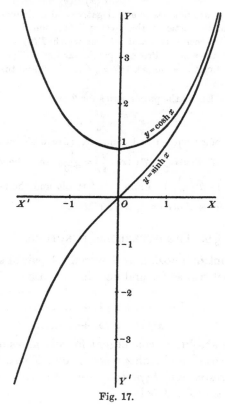

Fig. 17.

We have the relation $(\cosh x)^2 - (\sinh x)^2 = 1$ for all values of x*.

The functions are continuous and differentiable and
$$D \cosh x = \sinh x, \quad D \sinh x = \cosh x.$$

The proofs of these properties are left to the student.

The hyperbolic functions are not periodic and there is no real number x for which $\cosh x = 0$.

The graphs of these functions are as drawn in Fig. 17.

From the functions $\sinh x$ and $\cosh x$ are derived other hyperbolic functions, defined thus:

$$\tanh x = \frac{\sinh x}{\cosh x}, \quad \operatorname{sech} x = \frac{1}{\cosh x}, \quad \operatorname{cosech} x = \frac{1}{\sinh x},$$

$$\coth x = \frac{\cosh x}{\sinh x}.$$

The properties of these functions can be deduced from those of $\sinh x$ and $\cosh x$.

151. The inverse hyperbolic functions. The existence of *inverse hyperbolic functions* follows from the continuity and monotony of the functions $\cosh x$, $\sinh x$. They are defined thus:

arg $\cosh x$ = that *positive* number y for which $\cosh y = x$.

This function,—the *inverse hyperbolic cosine*,—is defined for all values of x greater than (or equal to) 1, but *for no values of x less than* 1.

arg $\sinh x$ = that number y (positive or negative) for which
$$\sinh y = x.$$

This function,—the *inverse hyperbolic sine*,—is defined *for all real values of x*†.

The functions inverse to the other hyperbolic functions, $\tanh x$, etc., are defined similarly.

The properties of the inverse functions can be deduced from our knowledge of the direct functions.

152. Differentiability of function defined by general power series. In this and the preceding section we have established in detail the properties of four functions defined by special power series. As mentioned in the last section, it is not possible to discuss so easily the functions defined by *every* power series. It is however possible to prove, for example, that functions so defined necessarily

* $(\cosh x)^2$ and $(\sinh x)^2$ are usually written $\cosh^2 x$ and $\sinh^2 x$.
† arg $\cosh x$ and arg $\sinh x$ are also denoted by $\cosh^{-1} x$ and $\sinh^{-1} x$.

possess (within certain limitations) the important property of differentiability.

Thus if the power series

$$a_0 + a_1x + a_2x^2 + \ldots \quad\ldots\ldots\ldots\ldots\ldots\ldots(3)$$

is convergent for all values of x and has its coefficients $a_0, a_1, a_2, \ldots$ all positive, then the function $f(x)$ which is the sum of this series is differentiable for all positive values of x, and

$$Df(x) = a_1 + 2a_2x + 3a_3x^2 + \ldots \quad\ldots\ldots\ldots\ldots(4);$$

this series being itself convergent for all values of x.

To prove this, we have, if x and h are any positive numbers,

$$f(x+h) - f(x) = a_0 + a_1(x+h) + a_2(x+h)^2 + \ldots$$
$$- a_0 - a_1x - a_2x^2 - \ldots$$
$$= a_1h + a_2(2hx + h^2) + a_3(3hx^2 + 3h^2x + h^3) + \ldots,$$

by subtracting the two convergent series; and therefore the incrementary ratio $[f(x+h) - f(x)]/h$ equals

$$a_1 + a_2(2x + h) + a_3(3x^2 + 3hx + h^2) + \ldots \quad\ldots\ldots\ldots(5)$$
$$= a_1 + a_22x + a_2h + a_33x^2 + a_33hx + a_3h^2 + \ldots \quad\ldots\ldots(6),$$

(because all the terms of (6) are positive and the series (6) converges because (5) converges)

$$= a_1 + 2a_2x + 3a_3x^2 + \ldots$$
$$+ h[a_2 + 3a_3x + a_3h + 6a_4x^2 + 4a_4xh + a_4h^2 + \ldots] \quad\ldots\ldots(7),$$

by separating the terms which do not contain h,—this derangement being valid and the two series in (7) being convergent because all the terms are positive *.

If $h < 1$ say, the sum of the series in the bracket in (7) is less than

$$a_0 + a_1(x+1) + a_2(x+1)^2 + a_3(x+1)^3 + \ldots,$$

(because this series includes, besides other terms, all the terms of the series concerned with h replaced by 1), and therefore is less than $a_0 + a_1R + a_2R^2 + \ldots$, i.e. $f(R)$, where R is any number exceeding $x + 1$. Hence the incrementary ratio tends to $a_1 + 2a_2x + 3a_3x^2 + \ldots$ as $h \searrow 0$, and the differentiability of $f(x)$ on the right is established. The modification of the proof needed to establish the complete differentiability is slight and is left to the student.

* See Ex. 12, p. 84 above.

By further modifications of the proof we can prove the more general theorem :

If the power series $a_0 + a_1 x + a_2 x^2 + \dots$ *is absolutely convergent for all values of x between* $-R$ *and R, then the sum-function* $f(x)$ *is continuous and differentiable for all values of x between* $-R$ *and R, and in that range*

$$Df(x) = a_1 + 2a_2 x + 3a_3 x^2 + \dots.$$

A proof of this important general theorem is given in Ex. 14 below.

In general (see Ex. 13 below) the range of convergence of a power series consists of all values of x between $-R$ and R where R is some number (called the radius of convergence). The series may or may not be convergent also for $x = -R$ and $x = R$, but on the one hand it is not convergent for any value of x such that $|x| > R$ and on the other hand it is absolutely convergent for all values of x between $-R$ and R (i.e. such that $|x| < R$). The above theorem therefore applies to any power series over its entire range of convergence ($-R$, R) except the extreme values $x = -R, x = R$. As to the behaviour of the series and the function for the extreme values, we say nothing. Not only may the series be not convergent for these values but the function may be unbounded in the complete interval ($-R, R$) and have other peculiarities for the extreme values. The discussion of these questions belongs to the theory of functions.

In virtue of the above theorem, and the theorems of pp. 170—171 above, we see the essential amenability of the sum-functions of power series to the fundamental processes of analysis. Other types of series*, though often of very great utility and importance, do not as a rule possess these general properties.

EXAMPLES XVIII.

1. Prove that $\cosh x = (e^x + e^{-x})/2$ and $\sinh x = (e^x - e^{-x})/2$.

2. Evaluate to four decimal places $\cosh 1$, $\sinh 1$, $\cosh 10$, $\sinh 10$.

3. Prove that
$$\cosh x + \cosh y = 2 \cosh \frac{x+y}{2} \cosh \frac{x-y}{2},$$
$$\cosh x - \cosh y = 2 \sinh \frac{x+y}{2} \sinh \frac{x-y}{2},$$
$$\sinh x + \sinh y = 2 \sinh \frac{x+y}{2} \cosh \frac{x-y}{2},$$
$$\sinh x - \sinh y = 2 \cosh \frac{x+y}{2} \sinh \frac{x-y}{2}.$$

4. Prove the addition theorems and other relations stated in the text. Prove also that
$$\cosh 2x = (\cosh x)^2 + (\sinh x)^2 \quad \text{and} \quad \sinh 2x = 2 \sinh x \cosh x.$$

5. Prove that the functions $\cosh x$ and $\sinh x$ are continuous, differentiable, and integrable for all values of x and monotone in positive or negative ranges ; and that they are bounded in any bounded range.

6. Prove that the graph of $\cosh x$ has one turning point,—a minimum at $x = 0$, where $\cosh x = 1$: and that the graph of $\sinh x$ has no turning point.

* E.g. Fourier series.

7. Prove that the graph of $\sinh x$ cuts the x axis at the origin at an angle of $45°$.

8. Prove that

$$1 + \cosh a + \cosh 2a + \cosh 3a + \ldots + \cosh (n-1)\,a = \frac{\sinh \frac{a}{2} + \sinh \left(n - \frac{1}{2}\right) a}{2 \sinh \frac{a}{2}}$$

and

$$\sinh a + \sinh 2a + \sinh 3a + \ldots + \sinh (n-1)\,a = \frac{\cosh \left(n - \frac{1}{2}\right) a - \cosh \frac{a}{2}}{2 \sinh \frac{a}{2}}.$$

Deduce that

$$\int_0^a \cosh x\,dx = \sinh a \quad \text{and} \quad \int_0^a \sinh x\,dx = \cosh a - 1 \, ;$$

and thence

$$\int_a^b \cosh x\,dx = \sinh b - \sinh a \quad \text{and} \quad \int_a^b \sinh x\,dx = \cosh b - \cosh a.$$

[The first part by induction. For the integrals, which are known to exist, divide up $(0, a)$ into n equal intervals $a/n = a$ say. The total area of the "small" rectangles of the definition then $= a[1 + \cosh a + \ldots + \cosh (n-1)\,a]$, and the result for $\int_0^a \cosh x\,dx$ follows.]

9. Solve the equation $\sinh x = 2x$.

[A graphical solution can be obtained by drawing the graphs of $\sinh x$ and $2x$ and finding the abscissae of the points of intersection. Analytically, the function $\sinh x - 2x$ is continuous; it is negative when $x = 1$ and positive when $x = 4$; whence there is a root between 1 and 4. And so on.]

10. Shew that if x is sufficiently great, $\cosh x$ and $\sinh x$ exceed x^n, no matter how great the index n.

11. Discuss the properties of $\tanh x$, and the inverse hyperbolic tangent $\operatorname{arg\,tanh} x$, which is defined for all values of x between ± 1. Shew in particular $D \tanh x = (\operatorname{sech} x)^2$; $D \operatorname{arg\,tanh} x = \dfrac{1}{1 - x^2}$. Discuss also $\operatorname{sech} x$ and $\operatorname{cosech} x$.

12. Discuss the properties and sketch the graphs of the inverse hyperbolic functions. Prove that

$$\operatorname{arg\,sinh} x = \log_e [x + \sqrt{(x^2 + 1)}], \quad \operatorname{arg\,cosh} x = \log_e [x + \sqrt{(x^2 - 1)}],$$
$$\operatorname{arg\,tanh} x = \tfrac{1}{2} \log_e [(1 + x)/(1 - x)].$$

13. If a power series is convergent for $x = a$, it is absolutely convergent for all values of x numerically less than a.

[Use the "comparison theorem" for convergence of series of positive terms, taking a geometrical progression as series of comparison.]

14. Prove the theorem of the text that if a power series is absolutely convergent for all values of x numerically less than R then its sum is a continuous and differentiable function for all such values of x and the differential coefficient is the sum of the series whose terms are the differential coefficients of the terms of the original power series.

[Use the notation of the text. If $|x| + |h| < R$, the series

$$a_0 + a_1 x + a_1 h + a_2 x^2 + 2a_2 xh + a_2 h^2 + a_3 x^3 + 3a_3 x^2 h + 3a_3 xh^2 + a_3 h^3 + \dots \quad \dots (1)$$

is absolutely convergent because, when the terms are replaced by their numerical values, the series is equivalent to the series of positive terms

$$|a_0| + |a_1|(|x| + |h|) + |a_2|(|x| + |h|)^2 + \dots,$$

known to be convergent. The series (1) can therefore be deranged. On the one hand $(1) = a_0 + a_1(x+h) + \dots = f(x+h)$. On the other hand

$$(1) = a_0 + a_1 x + a_2 x^2 + \dots + h(a_1 + 2a_2 x + 3a_3 x^2 + \dots)$$
$$+ h^2(a_2 + 3a_3 x + a_3 h + \dots).$$

The series in the last bracket is bounded. Therefore

$$\frac{f(x+h) - f(x)}{h} \to a_1 + 2a_2 x + \dots \text{ as } h \to 0,$$

and the theorem is proved.]

15. Prove that if the series

$$|a_0| + |a_1|(|x| + |h|) + |a_2|(|x| + |h|)^2 + \dots$$

is convergent and if the sum-function of the series $a_0 + a_1 x + a_2 x^2 + \dots$ is $f(x)$,

then
$$f(x+h) = f(x) + h\,Df(x) + \frac{h^2}{2!}\,D^2 f(x) + \frac{h^3}{3!}\,D^3 f(x) + \dots,$$

where $Df(x)$ denotes the differential coefficient of $f(x)$, $D^2 f(x)$ the differential coefficient of the differential coefficient, and so on. (Taylor's theorem.)

13—2

CHAPTER IV

DIFFERENTIAL AND INTEGRAL CALCULUS

153. The most important practical application of the ideas of limits of sequences is to be found in the differential and integral calculus. The calculus is a very powerful instrument in geometry, dynamics and almost all branches of mathematics and applications of mathematics to the sciences. The main results of the calculus were discovered independently of any theory of convergent sequences, or of irrational numbers, on which the theory of sequences depends, as set out above. Historically the same order of progress has been maintained in the case of the calculus as has been the case with the rational numbers and with most instruments of mathematical thought,—the practical use has been fully and advantageously understood long before the principles on which the instrument logically rests have been at all clear. For some time after the introduction of the calculus by Newton and Leibniz in the seventeenth century, there was considerable controversy as to the validity of the new process, philosophers, such as Bishop Berkeley, maintaining that it was without foundation and that all results obtained by its aid were worthless. In so far as the attacks were restricted to the lack of logical basis, these criticisms of the philosophers were sound, but the great and obviously true results obtained by using the calculus were sufficient justification for the mathematicians' scepticism of the philosophers' arguments. This controversy, including attempts by mathematicians such as Maclaurin and Lagrange to put the calculus on a basis not liable to the attacks of the philosophers, may be said to have lasted throughout the eighteenth century. At the beginning of the nineteenth century,—the great rush of new results consequent on the introduction of the calculus having somewhat abated,—a new movement for rigour and criticism of principles in mathematics began, initiated largely by Abel and Cauchy. It is this movement which has resulted, towards the end of the last century, in the complete clarification of the principles of the calculus, as also of the ideas of irrational numbers and infinite sequences. As a result of this movement we are now in a position to build up the calculus on the logically safe foundation of the ideas of limits.

154. We have already, in Chapters II and III, in discussing the behaviour of functions, been compelled to introduce some of the special kinds of limits which are the special province of calculus. In this chapter we shall discuss systematically the fundamental notions of the calculus and illustrate these notions by simple applications and developments in analysis, geometry, and applied mathematics.

§ 1. GENERAL IDEA OF A LIMIT

155. The fundamental idea of a limit is that of an infinite sequence (Chapter I, § 6) where we have an unending set of numbers tending to a unique limit. Limits may occur in other forms. In one form or another they are of great importance in the application of mathematics to physical problems, because many phenomena are measurable only in terms of limits. In geometry, for example, the notions of the length of a curve, the area of a region, the volume of a solid, and the tangent to a curve are of considerable importance and yet none of these notions can be defined satisfactorily (except in special simple cases) other than in terms of limits*. And again in mechanics, velocity, acceleration, centres of mass, and moments of inertia necessitate the consideration of limits. And generally whenever we consider mathematically any continuous change of any measurable quantity,—i.e. briefly throughout all mathematical physics,—the use of limits is unavoidable.

156. Limit of a function of an integral variable. The simple notion of the limit of a sequence may be considered as the *limit of a function of an integral variable*. The nth term of a sequence is a number depending on the integer n, and is therefore a function of n, defined for all positive integral values of n. If the sequence is convergent, then we know that this function, $f(n)$ say, tends to coincide with the unique limit as the variable n increases indefinitely; and in fact the convergence of such a sequence is in essence a description of the behaviour of the function for indefinitely large values of n. It is therefore appropriate to consider the limit of the sequence as the limit of the function $f(n)$ as the variable n

* Or bounds.

increases indefinitely. The notation $\lim\limits_{n \to \infty} f(n) = L$ or $f(n) \to L$ as $n \to \infty$, introduced in Chapter I as a convenient abbreviated statement, is natural and appropriate, as directing attention to the function $f(n)$ and its behaviour as n increases indefinitely.

157. Limits of a function of a continuous variable. We have already met, in our study of the properties of polynomials and other functions, what may be looked on as another kind of limit,—*the limit of a function of a continuous variable.* We have seen that for some functions of the continuous variable x, any decreasing sequence of values of x which $\searrow a$ gives rise to a sequence of corresponding values of the function (y) which tends to a unique limit, and that this limit (L_1 say) is the same for all such sequences. In such a case we say that *the function y has a unique limit on the right at the point $x = a$, viz. L_1.*

Similarly a function y has *a unique limit on the left at $x = a$*, viz. L_2, if the sequence of the values of y corresponding to any increasing sequence for x which $\nearrow a$ is convergent and has the unique limit L_2.

If a function has a unique limit on the right and also a unique limit on the left at $x = a$ and if these two limits are equal $(= L)$, it is said to have a unique limit (L) at $x = a$[*].

Denoting the function by $f(x)$, we use the notation $\lim\limits_{x \to a} f(x) = L$, or alternatively, $f(x) \to L$ as $x \to a$[†].

Another way of defining the limit of a function of a continuous variable which in some ways appears simpler is the following:

The function $f(x)$ of the continuous variable x tends to a unique limit L as x tends to a (i.e. $\lim\limits_{x \to a} f(x)$ exists and $= L$) if, corresponding to any positive number ϵ however small, it is possible to choose another positive number δ so small that for all values of x between $a - \delta$ and $a + \delta$ (with the possible exception of the value $x = a$ itself) the difference between $f(x)$ and the fixed number L is less than the number ϵ, i.e.

$$|f(x) - L| < \epsilon \quad \text{for} \quad a - \delta < x < a + \delta \, (x \neq a).$$

---

[*] Corresponding to the more general limiting numbers of a sequence (p. 44) there are non-unique limiting numbers (also often called limits) of a function at a point. They lie beyond the scope of this course. The word limit in what follows means unique limit.

[†] We have in fact already used this notation as an abbreviation. (See e.g. the footnote to p. 153 above.)

That these two definitions are equivalent will be obvious after a little thought; for if $f(x)$ has the limit L as $x \to a$ in accordance with the first definition, it will follow that, for *all* values of x sufficiently near a (i.e. within some distance δ of a) on either side of a, $f(x)$ differs from L by less than any arbitrarily small number; for if, no matter how small we took δ, there were always, for x within the distance δ of a, values of $f(x)$ differing from L by more than ϵ, we could then find a sequence of values of x tending to a as limit for which $|f(x) - L| > \epsilon$ and therefore not having L as the limit of the sequence,—contrary to the conditions of the first definition. The conditions of the first definition therefore imply those of the second. The converse is even more obvious.

It is with limits of this kind,—limits of functions of a continuous variable,—that differential calculus is mainly concerned.

Examples of limits of functions are:

$\lim_{x \to 0} x^2 = 0$; $\lim_{x \to 1} x^2 = 1$; $\lim_{x \to 0} (\sinh x/x) = 1$; $\lim_{x \to 0} x \sin(1/x) = 0$; $\lim_{x \to 0} \sin(1/x)$ does not exist; $\lim_{x \to 1} \log x = 0$; $\lim_{x \to 0} \log x$ does not exist; $\lim_{x \to \frac{1}{2}} [x]$, where $[x]$ stands for the greatest integer not exceeding x, exists and $= 0$; $\lim_{x \to 1} [x]$ does not exist, though if we restrict ourselves to values of x greater than 1 the limit ("on the right") exists and $= 1$ and if we restrict ourselves to values of x less than 1 the limit ("on the left") also exists but $= 0$.

The notions of the *limit* of a function at a point and the *value* of the function at that point are, of course, quite distinct and must not be confused.

The student will notice that the definition of continuity (Chapter III, p. 139) may now be restated in the form:

The function $f(x)$ is continuous at the point $x = a$ if $\lim_{x \to a} f(x)$ *exists and* $= f(a)$.

158. First definition of the length of a curve. The *length of a curve* provides an example of another type of limit,—similar in essentials to that needed in the definition of the definite integral or area of a plane region. The measuring of a length is essentially performed by means of straight lines, length having originally no meaning except along a straight line. In practice it is true one can often determine the length of a curve by replacing it by a flexible string and measuring the string when straightened out; but such a method will not answer the mathematical problem of the measurement of the length of the curve, for in such a method we have to presuppose that the string has the same length in the two positions (curved and straight),—a supposition which is entirely meaningless unless the length of the curve has been independently defined. We can proceed thus: On the curve AB (supposed continuous) take a number of points, $P_1, P_2, \ldots P_n$ say, and join them up in pairs in order, so that a "polygon" is formed of the straight lines $P_1P_2, P_2P_3, \ldots P_{n-1}P_n$ all of whose vertices lie on the curve. (See Fig. 18.) If all these chords P_1P_2 etc. are small the polygon will appear (if the curve be a simple one) very much like the curve. The perimeter

$$P_1P_2 + P_2P_3 + \ldots + P_{n-1}P_n$$

of this polygon is known, being

Fig. 18.

merely the sum of a number of measurable straight lines. Suppose now, each of the arcs of the curve cut off by the points $P_1, P_2, \ldots P_n$ is subdivided by taking other intermediate points on the curve, and all the points so obtained are joined in the same way,—thus forming another polygon $(P_1'P_2'P_3'P_4' \ldots P_m'$ say) of a greater number of sides than the first polygon. Since this polygon will, in general, lie closer to the curve than the first, it will naturally be considered to give a better approximation to the length of the curve than the first. This process of interpolation of new vertices may be repeated indefinitely. (The sequence of the perimeters of these polygons, being increasing, will necessarily be convergent if bounded.) It may happen (as it does in the case of most simple curves like circles, conics, etc.) that the perimeters of the successive approximating polygons not only tend to a definite limit as the length of the greatest of the chords forming the sides is indefinitely diminished in this way, but also have the same limit for all possible choices of such a sequence of polygons (provided that the maximum chord is made to tend to zero). *If this is the case, the curve is said to be rectifiable, or to have a length, and the limit of the perimeters of the inscribed polygons is called the length of the curve.* If no such unique limit exists the curve is not rectifiable, i.e. has no length.

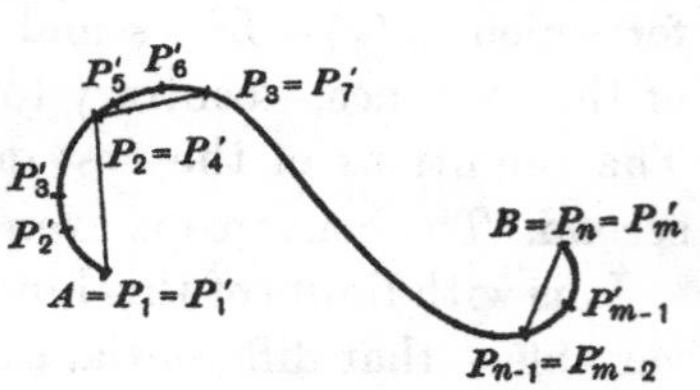

159. In this definition of the length of a curve it will be clearly understood that we do not confuse the curve with any of the polygons. A polygon will always consist of straight sides and angular vertices, and a smooth curve like a circle has neither straight sides nor vertices. The object of the construction of polygons of large numbers of sides is not to obtain a polygon "indistinguishable" from the curve and then to assert that any property of such a polygon applies to the curve. Such an assertion would be very like asserting, for example, that in an ideally constructed American city with a large number of streets running all North and South or East and West the path of a man who travels from the S.W. to the N.E. corner of the city along the streets closest to the diagonal is practically indistinguishable from the diagonal itself and therefore is practically indistinguishable from this diagonal in length,—the truth being of course quite different, the ratios of the two distances being $\sqrt{2}$ or $1\cdot 4$....

160. Proof of rectifiability of a circle. Using this definition we will now prove that *any arc, or the whole circumference, of a circle is rectifiable.* We know (from the first book of Euclid) that if we take any inscribed polygon, say $P_1P_2P_3P_4P_5$ (Fig. 19), and interpolate, on the arcs cut off, other points, with which to form vertices of an inscribed polygon of an extended number of sides, and then interpolate further points, and so on, the perimeter of the polygons steadily increases at each stage. But it is also easy to prove that the perimeter of any inscribed polygon is less than (say) that of a square circumscribed to the circle,—viz. $8R$ if R is the radius of the circle. Hence the

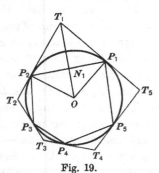

Fig. 19.

sequence of perimeters of inscribed polygons obtained by successive interpolation of vertices on the circumference is a steadily increasing sequence which is also bounded. Therefore this sequence has a unique limit, say L_1, which $\leqslant 8R$.

If we circumscribe polygons to the circle by drawing the tangents at the vertices of these inscribed polygons, we shall similarly obtain a monotonely decreasing sequence of perimeters which is bounded below. This sequence has a lower bound and unique limit, say L_2.

Further, supposing that all the chords which are the sides of the inscribed polygons tend to zero, these two limits are identical. For the difference between the perimeters of an inscribed polygon (e.g. $P_1P_2P_3P_4P_5$) and the corresponding circumscribed polygon ($T_1T_2T_3T_4T_5$) equals the sum of such differences as

$$(P_1T_1 + T_1P_2) - P_1P_2;$$

and this difference equals

$$P_1P_2(OP_1/ON_1 - 1).$$

The difference of perimeters therefore $< (P_1P_2 + \ldots + P_5P_1)\alpha$ if α is the greatest of the numbers $OP_1/ON_1 - 1$ etc.; i.e. $< 8R\alpha$.

But, by choosing the chords P_1P_2 etc. sufficiently small, α (and therefore also $8R\alpha$) can be made as small as we like, i.e. the perimeters of the circumscribed and inscribed polygons can be made to differ by less than any pre-assigned positive number (however small). The two limits L_1 and L_2 of these perimeters therefore cannot be unequal, i.e. $L_1 = L_2 = L$ say.

It now follows easily that this common limit L is the same no matter what systems of polygons may have been chosen (provided the maximum side tends to zero). For if $L_1' = L_2' = L'$ are the limits corresponding to any other system of polygons, we have easily (since *any* circumscribed perimeter $>$ *any* inscribed perimeter*) $L_1 \leqslant L_2'$, $L_2 \geqslant L_1'$; i.e. $L \leqslant L'$, $L \geqslant L'$; whence $L = L'$.

The limit of any sequence of inscribed perimeters is thus independent of the particular set of polygons chosen, provided only that the greatest side $\rightarrow 0$. The circumference of the circle has therefore a length in accordance with the above definition. Slight modifications will make the proof applicable to any arc of a circle.

The actual determination of the length of the circumference of the circle can be carried out by calculating the perimeters of successive inscribed and circumscribed polygons (preferably regular);

---

* If $P_1P_2 \ldots P_n$ is any inscribed polygon, and $Q_1, Q_2, \ldots Q_m$ the points of contact with the circle of any circumscribed polygon and if we form two more polygons, inscribed and circumscribed, having all the points $P_1, P_2, \ldots P_n, Q_1, Q_2, \ldots Q_m$ as vertices and points of contact with the circle, it follows (from Euclid i, 20) that if p, q, r_i, r_c are the perimeters of the four polygons $p \leqslant r_i < r_c \leqslant q$; which proves the statement.

the length of the circumference being intermediate to the inscribed
and circumscribed perimeters. By taking polygons of 96 sides, as
was done by Archimedes in the third century B.C., the value of the
ratio of the circumference of any circle to its diameter (π) can be
calculated correct to two places of decimals.

The evaluation of lengths of curves and limits of similar type
depends in general on integration.

161. Second definition of length of a curve. The above
definition of the length of a curve is that which has up to the
present time been most established by custom. Such lengths may
however be defined in another way which is simpler and more
convenient in theory and practice.

Let us consider the set of all possible inscribed polygons (the
vertices being taken in order along the curve). The set of numbers
representing the perimeters of these polygons will not be a finite
set or even a sequence (as it is impossible to set them all down in
order as the terms of a sequence). But it is quite possible that all
these numbers are less than some fixed number (K say); i.e. that
the set is *bounded* in the same sense as a sequence or a finite set
may be bounded. *If the set of perimeters of inscribed polygons is
bounded in this way, the curve is rectifiable.* If this is the case, the
system of real numbers can be divided into two classes, x and y,
such that y typifies any number (such as K) exceeding all possible
inscribed perimeters and x is any number less than all such numbers
y; and the number $(x|y)$ defined by this classification is the least
number not exceeded by any perimeter. That is, the set of peri-
meters has an *upper bound* $(x|y)$. *This upper bound of the perimeters
of inscribed polygons is the length of the curve.*

It is easy to prove that a curve, rectifiable according to our first
definition, is rectifiable also according to this second definition and
that the lengths so obtained agree. The converse is also true but
the proof of this is more difficult.

Using this second definition, the rectifiability of the circle is
an immediate consequence of the fact that the perimeter of any
inscribed polygon is less than that of the circumscribed square.

**162. Definitions of circular measure and the circular func-
tions.** Having defined the length of a curve, and proved that any

arc of a circle has a length, we are now in a position to measure angles in radian measure and to define and discuss the circular functions.

The *radian measure* of an angle $A\hat{O}P$ (Fig. 20) is the length of the arc AP which the angle cuts off the circumference of a circle, of centre O and unit radius $OA = OP = 1$. Thus if the length of the arc $AP = x$, the radian measure of the angle $A\hat{O}P = x$.

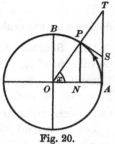

Fig. 20.

We suppose in the first instance $0 \leqslant x < 2\pi$ and the arc AP measured from A in a definite direction, say that indicated in the figure by the arrow. Through the centre O erect OB, perpendicular to OA, in the direction indicated in the figure. Considering OA and OB as a pair of rectangular Cartesian axes, let the coordinates of P be X, Y; so that, with due attention to sign, $X = ON$, $Y = NP$. *The circular functions* $\cos x$ *and* $\sin x$ *are defined as* $\cos x = X$, $\sin x = Y$.

If $x \geqslant 2\pi$, an arc AP of length x will $\geqslant$ the complete circumference. In the measurement of such an arc the circumference is followed a second time, and so on if necessary. Corresponding to any positive angle x, there will always be a point P which marks the extremity of the arc AP of length x. If x is negative, the arc AP is measured similarly, but in the reverse direction. In all cases there is a definite point P corresponding in this way to any real number x. The above definition of $\cos x$ and $\sin x$ will therefore apply to angles of any magnitude as so defined.

The derived circular functions are defined as:
$$\tan x = \sin x / \cos x, \quad \mathrm{cosec}\, x = 1/\sin x,$$
$$\sec x = 1/\cos x, \quad \cot x = \cos x/\sin x;$$
these definitions being valid for all values of x except those for which the denominators are zero.

We notice that if x is positive and $< \pi/2$,
$$\cos x = ON, \quad \sin x = NP, \quad \tan x = AT,$$
and are all positive.

From these definitions the usual results of elementary trigonometry are deduced. In particular the circular functions have

addition theorems and periodicity formulae which are identical
with those of pp. 179, 184 above, with the substitution of
cos for C, sin for S, and π for ϖ.

Inverse circular functions arc sin x, arc cos x, arc tan x (also written
$\sin^{-1} x$, $\cos^{-1} x$, $\tan^{-1} x$) can be defined in a manner similar to that in
which the inverse trigonometrical functions $\bar{S}(x)$, $\bar{C}(x)$, $\bar{T}(x)$ were
defined (p. 186 and Ex. 17, p. 189); thus arc sin x *is defined for all
values of x from* -1 *to* 1 *inclusive as that angle y, measured in radians,
restricted to lie in the interval* $-\pi/2$ *to* $\pi/2$ (i.e. $-\pi/2 \leqslant y \leqslant \pi/2$), *for
which* sin $y = x$.

In theoretical work angles are invariably measured in radians.
In what follows the angles are all so measured.

163. Important inequalities and limits. We have the useful
and important *inequalities*:
$$\sin x < x < \tan x \quad \text{if} \quad 0 < x < \pi/2.$$
For the angle x is less than a right angle and we have from Fig. 20
that $NP < \text{arc } AP < AT$, because NP and AT are respectively less
and greater than inscribed and circumscribed polygons (viz. AP
and $AS + SP$ respectively) belonging to sequences of which the
length of the arc AP is respectively the upper and the lower bound.

We deduce from these inequalities the important *limits*:
$$\lim_{x \to 0} \frac{\sin x}{x} = 1, \quad \lim_{x \to 0} \frac{\tan x}{x} = 1.$$
For $\dfrac{\sin x}{\tan x} = \cos x = \dfrac{ON}{OA} \to 1 \text{ as } x \searrow 0$

and therefore, *a fortiori*, $\dfrac{\sin x}{x} \to 1$ and $\dfrac{\tan x}{x} \to 1$ as $x \searrow 0$; and
as $x \nearrow 0$ these functions behave in the same way.

164. Areas and tangents. The *area of a portion of a plane*
bounded by curved lines is another example of a similar kind of
limit,—the area, as with the special case of Chapter III, being
defined as the limit, if it exists, of the areas of any set of rectangles
which tends to coincide with the area concerned.

The important case of the definition and determination of
tangents to curves has been considered before (in Chapter III) in
discussing the continuity and differentiability of polynomials and
other functions.

165. Velocity. Let us take a kinematical illustration:—What is meant by the *velocity* at a certain time of a body moving along a straight line? The *average velocity* over a specified period of time is simply the ratio of the distance covered to the time taken to cover the distance; but the body may have been moving at different velocities in different parts of its motion. Our problem is to define the velocity *at* any particular instant.

Consider the average velocities over different intervals of time beginning or ending at the instant under consideration;—these simply being the ratios of the distances covered by the intervals of time. It may happen that these average velocities all approach more and more nearly to some fixed velocity as the intervals of time are taken smaller and smaller. If we represent the distance travelled by the body in the time t (measured algebraically) by $y(t)$, y being a function of t, the average velocity between the times t and $t+h$ will be $[y(t+h)-y(t)]/h$. If this average velocity,—which is a function of the continuous variable h,—tends to a unique limit as $h \to 0$, *this limit is called the velocity of the body at the time t*. If there is no such limit the term velocity has in this case no meaning. The naturalness of this definition as the mathematical interpretation of our preconceived physical ideas of velocity will be sufficiently apparent to the student.

166. Of the special types of limits, that of a function of a continuous variable is much the most important. Throughout the remainder of this course we shall use the term limit solely in this sense except where there is occasion to use it in its original sense as the limit of a sequence (or as a limiting number).

EXAMPLES XIX.

1. The following are examples of limits of functions of an integral variable: $\lim_{n \to \infty} (\frac{1}{2})^n = 0$, $\lim_{n \to \infty} (1+1/n)^n = e$, $\lim_{n \to \infty} S(n)/n = 0$ (where $S(n)$ is the trigonometrical function defined in Chapter III, § 5), $\lim_{n \to \infty} S(n)$ does not exist.

2. The following are examples of limits of functions of a continuous variable: $\lim_{x \to 0} x = 0$, $\lim_{h \to 0} [(x+h)^2 - x^2]/h = 2x$, $\lim_{h \to 0} [S(h)]/h = 1$.

3. The curvature of a plane curve is the limit of the "average curvature" over an arc of the curve as the arc tends to zero in length, the average curvature being the ratio of the angle between the tangents at the two extremities of the arc to its length.

4. Evaluate the ratio of the circumference to the diameter of a circle (π) correct to one decimal place by calculating the perimeters of polygons of a sufficient number of sides inscribed and circumscribed to the circle.

5. The angle x being measured in radians, establish the inequalities
$$2/\pi < (\sin x)/x < 1, \text{ where } 0 < x < \pi/2.$$

6. Prove $\lim\limits_{x \to 0} \sin x = 0$, $\lim\limits_{x \to 0} \cos x = 1$, $\lim\limits_{x \to 0} (1 - \cos x)/x = 0$,
$$\lim\limits_{x \to 0} (1 - \cos x)/x^2 = \tfrac{1}{2}.$$

7. Prove the addition theorems for $\sin x$ and $\cos x$. Hence prove that $\sin x$ and $\cos x$ are continuous for all values of x.

8. The area of a circle may be defined as the limit of the areas of inscribed polygons as in the text. With this definition shew that a circle has an area and that it $= \pi R^2$ if $R =$ the radius.

9. The circular measure of an angle $A\hat{O}P$ (Fig. 20) may be defined as half the area of the sector AOP. Prove, from this definition, the inequalities and limits of p. 205.

10. The notation $\lim\limits_{x \to \infty} f(x)$ is self-explanatory. Prove
$$\lim\limits_{x \to \infty} x^n/e^x = 0, \quad \lim\limits_{x \to \infty} [S(x)]/x = \lim\limits_{x \to \infty} [C(x)]/x = 0, \quad \lim\limits_{x \to \infty} (\log x)/x = 0.$$

11. Prove that any part of the graph of a bounded function is rectifiable, (i) if the function is monotone in the range considered, or (ii) if the range can be divided into portions in each of which the function is monotone.

[Form polygons as in the definition. If the extremities of one of the chords are (x_1, y_1) and (x_2, y_2) the length of the chord is clearly less than the sum of the two positive numbers $x_2 - x_1$, $y_2 - y_1$ (supposing the function increasing for definiteness). Adding up for all the chords we see that the perimeter of any of the polygons is less than the total sum $(b-a)+(B-A)$, where the two extremities of the arc are denoted by (a, A) and (b, B). The set of perimeters is therefore bounded. The second part is obvious.]

12. Any part of the graph of any of the following functions is rectifiable: x^2, x^n, $\log x$ (both extremities of the arc lying to the right of the origin), e^x, $C(x)$, $S(x)$, $\overline{C}(x)$, $\overline{S}(x)$, $\cosh x$, $\sinh x$, $\arg\cosh x$, $\arg\sinh x$, any polynomial, any rational function (the arc avoiding all points of discontinuity), $\sin x$, $\cos x$, $\tan x$ (both extremities lying between $x = -\pi/2$ and $\pi/2$), $\arcsin x$, $\arccos x$, $\arctan x$, any sum-function of a power series (the extremities in the interior of the range of convergence).

[For the last see pp. 171—172.]

13. The curve defined by the Cartesian equation $x^2/a^2 + y^2/b^2 = 1$ (the ellipse) is rectifiable.

[Since $y = \pm b \sqrt{(1 - x^2/a^2)}$ the curve is seen to consist of four monotone continuous arcs.]

14. Prove that any part of the graph of a function formed by the addition and subtraction of any number of monotone functions is rectifiable.

[Any such function can be expressed as the difference of two increasing functions, say $f'(x) - f''(x)$. The length of any chord which is a side of the defining polygons, as in Ex. 11, is less than the sum of the four positive numbers $x_2 - x_1$, $x_2 - x_1$, $y_2' - y_1'$, $y_2'' - y_1''$ corresponding to the two functions. Adding up we get that the perimeters of the polygons are bounded by the number $(b-a) + (b-a) + (B' - A') + (B'' - A'')$, and the result follows.]

15. The function y, defined as $x \sin(1/x)$ [or $x S(1/x)$] when $x \neq 0$ and $= 0$ when $x = 0$, is continuous for all values of x. Prove that any arc of the graph of this function which passes through the origin is not rectifiable.

[We can, by taking the extremities of the chords at the points $2/(4n \pm 1)\pi$, obtain for the perimeter of a polygon a number exceeding the sum of any number of terms of the series $\dfrac{1}{\pi}\left[\dfrac{1}{1 + \frac{1}{4}} + \dfrac{1}{2 + \frac{1}{4}} + \dfrac{1}{3 + \frac{1}{4}} + \dots\right]$ which is known to be divergent. Hence the set of perimeters is not bounded.]

16. In order that the graph of a function should be rectifiable it is necessary and sufficient that the function should be of bounded variation in the range concerned.

17. If a bounded function $f(x)$ is monotone, or of bounded variation, in a range, the limits of $f(x)$ on the right and left at any point a of the range necessarily exist.

18. Let $f(x)$ be any bounded function, h any positive number, and let $u(h)$ denote the upper bound of $f(x)$ in the range a to $a+h$ (a excluded), i.e. the least number not exceeded by the value of $f(x)$ for any value of x such that $a < x \leqslant a+h$. Shew that $u(h)$ is a non-decreasing function of h; that the correspondingly-defined lower bound $l(h)$ is a non-increasing function of h; and that these functions $u(h)$ and $l(h)$ have (unique) limits on the right at $h = 0$. [These limits,—which are the lower bound of $u(h)$ and the upper bound of $l(h)$,—are called respectively *the upper and lower limits on the right* at $x = a$ of the function $f(x)$.]

19. Shew that if the upper and lower limits on the right at $x = a$ of a bounded function $f(x)$ coincide and $= L$, the (unique) limit of $f(x)$ on the right at $x = a$ exists and equals L; and conversely.

[For the first part, with obvious notation, $u(h) \searrow L$ as $h \searrow 0$ and $l(h) \nearrow L$ as $h \searrow 0$. Also if $a < x \leqslant a+h$, $l(h) \leqslant f(x) \leqslant u(h)$. The result is now evident. For the converse, call the upper limit L_1. We have

$f(x) \leqslant u(h)$,—the *upper* bound of $f(x)$,—for all values of x concerned,

    $< L_1 + \epsilon$,—any number greater than L_1 the lower *bound* of $u(h)$,—for some values of x indefinitely close to a. ....(1)

But $f(x) > u(h) - \epsilon$,—any number less than $u(h)$ the upper *bound* of $f(x)$,—for some values of x indefinitely close to a,

$\geqslant L_1 - \epsilon$,—any number less than L_1 (the *lower* bound of $u(h)$),—for values of x sufficiently close to a. ....(2)

By hypothesis, $f(x) \to L$. Relations (1) and (2) shew that $L = L_1$. Similarly $L = L_2$ and the result is proved.]

20. Shew that if the upper and lower limits of a function $f(x)$ on the right at $x = a$ and the correspondingly-defined upper and lower limits on the left all coincide and $= f(a)$ then $f(x)$ is continuous at $x = a$; and conversely.

§ 2. PROPERTIES OF LIMITS

167. There is no general method by which limits of functions of a continuous variable can be evaluated. From the nature of the case the evaluation will depend on the mode of definition of the particular function whose limit is sought. When we reflect that the determination of such limits may be regarded as the general problem of which the evaluation of the sums of convergent series (or limits of convergent sequences) is but a special case, and that there is no general method of determining even such limits, the difficulty of the general problem is realised. It is fortunately true however that the most common types of function are so defined that it is usually possible to determine their limits (when they exist).

168. Limits of sums, products, etc. We consider first the laws dealing with the limit of the sum or product or other combinations of two or more functions whose limits are known (or known to exist)*. They are:

If u and v are two functions of x such that $\lim\limits_{x \to a} u$ and $\lim\limits_{x \to a} v$ both exist (a being some real number) then

$$\lim_{x \to a} (u + v) \text{ exists and } = \lim_{x \to a} u + \lim_{x \to a} v,$$

$$\lim_{x \to a} (u \times v) \text{ exists and } = \lim_{x \to a} u \times \lim_{x \to a} v,$$

$$\lim_{x \to a} (u/v) \text{ exists and } = (\lim_{x \to a} u)/(\lim_{x \to a} v)$$

provided $\lim\limits_{x \to a} v$ is not zero.

In the case of division, if $\lim\limits_{x \to a} v$ is zero whilst $\lim\limits_{x \to a} u$ is not zero,

---

* These laws are also applicable to limits of simple sequences; $\lim\limits_{x \to a}$ has only to be replaced by $\lim\limits_{n \to \infty}$ in the symbolic statements.

the limit of the ratio, $\lim_{x \to a} (u/v)$, cannot exist,—excluding, as we do, "infinity" from our number system; but if $\lim_{x \to a} u$ and $\lim_{x \to a} v$ are both zero we cannot say without further information whether the limit of the ratio exists or not, nor, if it does exist, what is its value.

To prove these rules is necessary, though simple.

To prove the second law for example, let $\lim_{x \to a} u = U$ and $\lim_{x \to a} v = V$.

The existence of the limits implies that $|u|$ and $|v|$ are bounded (i.e. $<$ some fixed number K) for all values of x concerned.

Therefore
$$|uv - UV| = |(u - U)v + (v - V)U|$$
$$\leqslant |u - U||v| + |v - V||U|$$
$$< \frac{\epsilon}{2K} K + \frac{\epsilon}{2K} K$$

if x is sufficiently near to a, K being some number exceeding all values of $|u|$ and $|v|$ concerned and ϵ any positive number whatever; i.e. $|uv - UV| < \epsilon$, or $uv \to UV$.   Q.E.D.

The third law may be proved similarly. The first is simpler.

It should be noticed that none of these laws is reciprocal,—i.e. the converse is not true. For example it may happen that neither $\lim_{x \to a} u$ nor $\lim_{x \to a} v$ exists but that $\lim_{x \to a} (u + v)$ does exist [e.g. if $u = \sin(1/x)$,—or $S(1/x)$, the trigonometrical function of Chapter III,— $v = -u = -\sin(1/x)$,—or $-S(1/x)$,—and $a = 0$].

The laws however imply that if any *two* of the three limits occurring in the statement of any one of the laws exist, then the third limit also exists and the equality holds; subject to the restriction that in the third law as stated $\lim_{x \to a} v$ must not be zero, and in the second law if, for example, the existence and value of $\lim_{x \to a} u$ are to be inferred from knowledge of the other two limits, $\lim_{x \to a} v$ must not be zero.

169. We can evidently extend and restate these laws as one law:

If a function is defined as a combination of functions all of whose limits (as the variable x tends to some limiting value a) exist, then the function has a limit (as $x \to a$) which is the same function of the *limits as the function is of the component functions,—provided the functions are combined only by the four cardinal operations (addition, etc.) and no divisions by zero are involved.*

Or symbolically, using $A\,(u,\,v,\,w,\,...)$ to denote any such arithmetical function of a number of functions $u,\,v,\,w$, etc., then $\lim\limits_{x\to a} A\,(u, v, w, ...) = A\,(\lim\limits_{x\to a} u, \lim\limits_{x\to a} v, \lim\limits_{x\to a} w, ...)$ provided the limits $\lim\limits_{x\to a} u, \lim\limits_{x\to a} v, \lim\limits_{x\to a} w$, etc. exist and no divisions by zero are involved in the right-hand side.

Other combinations which are not of this arithmetic type suggest themselves, such as powers like u^v or $\log_u v$, and it is natural to ask if it is still true that (e.g.) $\lim\limits_{x\to a} u^v$ exists and $= U^V$ if the limits $\lim\limits_{x\to a} u$ and $\lim\limits_{x\to a} v$ exist and equal U and V respectively.

As a matter of fact this particular relation is true (if $\lim\limits_{x\to a} u > 0$), for if $\lim\limits_{x\to a} u = U$ and $\lim\limits_{x\to a} v = V$, we have ($u$ being positive for all values of x sufficiently near a), $u^v - U^V = (u^v - u^V) + (u^V - U^V)$; but $u^v - u^V \to 0$ as $x \to a$, in virtue of the continuity of the exponential function (Ex. 4, p. 91), because as $x \to a$, $v \to V$; and $u^V - U^V$ also $\to 0$ as $x \to a$, in virtue of the continuity of the power function x^n (Ex. 6, p. 92), because as $x \to a$, $u \to U$.

It will in fact be seen in general that the truth or otherwise of the law $\lim\limits_{x\to a} F\,(u, v, w, ...) = F\,(\lim\limits_{x\to a} u, \lim\limits_{x\to a} v, \lim\limits_{x\to a} w, ...)$,—where $F\,(u, v, w, ...)$ is used to denote any mathematical combination (or function) of $u,\,v,\,w$, etc.,—will depend on what may be termed the continuity of the defining function F. It is out of place here to make precise the meaning of this type of continuity.

An important special case of this general relation is the case when only one component function occurs. For example, we may want the limit of say u^2 or a^u or $\log_a u$, when we know the limiting value of the function u. It is here seen plainly that the question is simply that of the continuity or otherwise of the defining function $(x^2,\,a^x,\,\log_a x)$. As such, this question is that of the existence and evaluation of limits of special functions, and belongs properly to the study of the functions concerned. If we know that a function $F(x)$ is continuous for a certain value of $x\,(x = U$ say) then we know that

14—2

the relation $\lim\limits_{x \to a} F(u)$ exists and $= F(U)$ is true,—U being the limit, $\lim\limits_{x \to a} u$, supposed to exist.

170. In virtue of these properties of limits and of the simple knowledge that the single function x has a unique limit (a) as x tends to any limiting value a, we can conclude that any power x^n, or any polynomial $P_n(x)$, has such a unique limit, a^n, or $P_n(a)$, as x tends to any value a; any rational function $R(x)$ has a unique limit $R(a)$ as $x \to a$ if a is any number other than a value for which the function is not defined.

In fact the student will easily verify the general theorem:

If $F(x)$ is any function of x defined by means of a combination of symbols representing continuous functions, the function $F(x)$ is continuous. If any of the component functions is undefined or discontinuous for a value $x = a$, the function will in general be undefined and discontinuous at $x = a$ and $\lim\limits_{x \to a} F(x)$ may not exist; otherwise $\lim\limits_{x \to a} F(x)$ *exists and equals* $F(a)$*.

If the functions concerned are all defined and continuous for all values of x there is no difficulty in applying this theorem; but in the general case it may not always be easy to determine the exceptional values (or ranges of values) of x where the component functions are discontinuous or not defined.

Thus for example $x^2 + \log(2 - \sqrt{e^x})$ is continuous for all values of x less than $2\log_e 2$ and if $a < 2\log_e 2$, $\lim\limits_{x \to a} [x^2 + \log(2 - \sqrt{e^x})]$ exists and equals $a^2 + \log(2 - \sqrt{e^a})$. If $a \geqslant 2\log_e 2$, $\sqrt{e^a} \geqslant 2$, $2 - \sqrt{e^a} \leqslant 0$, and $\log(2 - \sqrt{e^x})$ is not defined for values of x near to a; the limit cannot then exist.

It is not true conversely that the function $F(x)$ will be not continuous if any of the defining functions is not continuous.

An important class of limits which exist but are not covered by the above theorem is that class known as *indeterminate forms*, which includes all differential coefficients. For example, the limits

$$\lim_{x \to a} \frac{x^2 - a^2}{x - a} \ (= 2a), \qquad \lim_{h \to 0} \frac{S(h)}{h} \ (= 1), \qquad \lim_{x \to 0} (1 + x)^{\frac{1}{x}} \ (= e)$$

---

* Such a function as $x/(x - a)$ will not be continuous at $x = a$ because the symbol of division by $x - a$ defines a function which is not continuous.

cannot be dealt with directly by the theorem, though the limits are known to exist.

171. Distinction between combinations of functions and their limits. We can conclude from the above theorem that any ordinary function (being a combination of a number of the functions studied above and similar functions) has a limit at all ordinary points; and we can usually see at once any exceptional points*. But in this statement *we cannot include under the term "ordinary function" any function defined as a limit,*—e.g. the sum-function of a convergent series,—unless we have specially established the continuity of such function. It is quite possible for the sum-function of a convergent series of continuous functions to be not continuous†. Using for the moment the words "finite" and "infinite" in a loose manner, we observe that this essential difference between the above "finite combinations" of functions and "infinite combinations," or series of functions, is not really surprising;—but it will serve as a reminder that such "infinite combinations" are admitted to analysis only in an arbitrary (and well defined) sense (as limits), and that the notion of the "sum" of an infinite series is arbitrary and different from the notion of the sum of a finite number of terms.

EXAMPLES XX.

1. Prove from first principles that the existence of two of the three limits in the relation $\lim_{x \to a} (u+v) = \lim_{x \to a} u + \lim_{x \to a} v$ is sufficient to ensure the existence of the third limit and the truth of the relation. Prove also the other laws stated in the text.

2. Determine the limits:

$$\lim_{x \to 1} (x+x^2),\ \lim_{x \to 0} (x\cos x),\ \lim_{x \to 0} [x\,C(x)],\ \lim_{x \to 1} \left(\frac{x^2-1}{x^2-2x}\right),\ \lim_{x \to 0} [x\sin(1/x)].$$

[The last limit $=0$ because $\sin(1/x)$ lies between ± 1 and therefore $x\sin(1/x)$ lies between $\pm x$; but $x \to 0$.]

3. Shew that $\lim_{n \to \infty} (1-1/n^2)^n = 1$, and that $\lim_{n \to \infty} (1-1/n)^{-n^2}$ does not exist.

4. Determine the limits $\lim_{x \to 0} (1-\sin x)^{\operatorname{cosec} x}$, $\lim_{x \to 0} 2^{x^2}$, $\lim_{x \to 0} \log\cos x$.

5. Shew that the function

$$\frac{\sqrt{[x+\log(1-e^{-x})]} - \log(10-x)}{(x-2)(x-1)}$$

is continuous for all values of x between $\log_e 2$ and 10 except at $x=1$ and $x=2$.

* See Ex. 5 below.   † See Exs. 8 and 9 below.

6. Shew that $\lim\limits_{x \to a} (\log_u v)$ exists and equals $\log_U V$ if U and V are the limits of u and v as $x \to a$ (supposed to exist). Is this true for all values of the letters concerned?

[See Ex. 5, p. 99.]

7. It is possible for the product of two functions, neither of which has a limit as $x \to a$, nevertheless to have a unique limit. The product yz, where the function y is defined to be 1 for all rational values of x and 0 for all irrational values, whilst z is zero for all rational values of x and 1 for all irrational values, is such a function; for neither of the limits $\lim\limits_{x \to a} y$ and $\lim\limits_{x \to a} z$ exists but the product is identically zero for all values of x.

8. The terms of the series
$$|\arc\tan 2x - \arc\tan x| + |\arc\tan 3x - \arc\tan 2x| + \ldots$$
(where arc tan x is the inverse tangent, i.e. that angle y between $-\pi/2$ and $\pi/2$ in radian measure for which $\tan y = x$) are continuous and the series is convergent for all values of x. The sum-function is however not continuous at the point $x = 0$.

[If $s(x)$ is the sum-function, $\lim\limits_{x \to 0} s(x)$ exists and equals $\pi/2$ but $s(0) = 0$.]

9. The series $\dfrac{x}{x^2+1} + \left(\dfrac{x}{x^2+\frac{1}{2}} - \dfrac{x}{x^2+1}\right) + \left(\dfrac{x}{x^2+\frac{1}{3}} - \dfrac{x}{x^2+\frac{1}{2}}\right) + \ldots$ has all its terms continuous and is convergent for all values of x. The sum-function $s(x)$ however is discontinuous at $x = 0$ and in fact $\lim\limits_{x \to 0} s(x)$ does not exist.

10. Prove the laws of p. 209 by using the ideas of Exs. 17—20 of the last section, taking first the case of monotone functions and deducing the general case.

11. The laws of limits (p. 209) will not apply to upper and lower bounds of functions or sequences; but the following properties hold:

If $B(y)$ and $b(y)$ denote the upper and lower bounds of the function y, then

$\qquad$ (i) $\quad b(u) + b(v) \leqslant b(u+v) \leqslant B(u+v) \leqslant B(u) + B(v)$,

and $\qquad$ (ii) $\quad b(u) b(v) \leqslant b(uv) \leqslant B(uv) \leqslant B(u) B(v)$;

where $b(u)$, $b(v)$, $B(u)$, $B(v)$ are supposed to exist, and, in (ii), $b(u)$ and $b(v)$ are not negative.

12. Shew that if $u \to 0$ through positive values as $x \to a$ and if $v \to V$, then

$\qquad$ (i) $\quad$ if $V > 0$, $\lim\limits_{x \to a} u^v$ exists and $= 0$,

$\qquad$ (ii) $\quad$ if $V < 0$, $\lim\limits_{x \to a} u^v$ does not exist,

$\qquad$ (iii) $\quad$ if $V = 0$, $\lim\limits_{x \to a} u^v$ may or may not exist.

§ 3. DIFFERENTIAL COEFFICIENTS

172. We have seen in Chapters II and III that if the graph $y = f(x)$ of a function be drawn, the inclination of the chord joining two points $P(x, y)$ and $Q(x+h, y+k)$ on the curve may be measured by the ratio k/h, i.e. $[f(x+h) - f(x)]/h$;—which is the trigonometrical tangent of the angle of inclination of PQ to the axis OX. Considering the behaviour of this ratio as the increment h is diminished, we saw also that in certain cases the limit $\lim_{h \to 0} [f(x+h)-f(x)]/h$ exists, and in those cases measures in the same way the *inclination of the tangent to the graph at P*, or the *slope* of the graph at P.

Again, if the variable x be taken to represent the time and y, or $f(x)$, is some measurable physical quantity depending on the time (e.g. the distance moved through in time x by a moving body), the same ratio $[f(x+h)-f(x)]/h$ will measure the average rate of increase of the quantity y in the interval of time from x to $x+h$ (e.g. the average speed of the body). And, considering in such a case the behaviour of this ratio as the increment h is diminished, we saw, in § 1 of this chapter, that in cases where the limit $\lim_{h \to 0} [f(x+h)-f(x)]/h$ exists it will measure the *rate of increase of the quantity y at the time x* (e.g. the velocity of the body at the time x).

Limits of this kind are the primary concern of the differential calculus.

173. Definitions and notation. For the sake of completeness we restate the definitions.

Let $f(x)$ be a function of the continuous (real) variable x. The *incrementary ratio* of the function $f(x)$ with respect to the variable x and the increment h is defined to be the ratio $\dfrac{f(x+h)-f(x)}{h}$, h being any real number whatever other than zero. (If the function is not defined for all real values of x we suppose x and h are such that $f(x)$ is defined for the values x, $x+h$, and all intermediate values.)

This incrementary ratio depends on x and h. If we consider x to be fixed and h to vary, it may be regarded as a function of the continuous variable h, defined for all real values of h (sufficiently small if $f(x)$ is not defined for all real values), except $h = 0$. If the unique limit $\lim_{h \to 0} \dfrac{f(x+h)-f(x)}{h}$ exists we say that the function $f(x)$ is *differentiable* for the value of x concerned, and the limit is called the *differential coefficient* or *derivative* of the function $f(x)$ with regard to the variable x.

As a rule we shall use the term derivative only when we look on the limit as itself a function of x, i.e. when we are concerned with the values of the differential coefficient over a whole range of values of x rather than for a single value. Otherwise we use the term differential coefficient.

It is convenient in some connections to use the notation $\dfrac{dy}{dx}$ to denote the differential coefficient of a function y with regard to x, and this notation is the most customary. We shall use this notation in places where convenient. We shall often, however, continue to use the notation Dy or $Df(x)$. In using the notation $\dfrac{dy}{dx}$ it must be clearly understood that the fractional form is merely symbolic of the definition of the differential coefficient as the limit of the ratio of the increments of x and of y. It must be used as a whole and not in any sense looked on as a quotient of two numbers dx and dy. We shall also occasionally use the notation $f'(x)$.

It is to be noticed that in the evaluation of the limit of the incrementary ratio, the exclusion of the limiting value $(h = 0)$ is essential, as the function concerned,—the incrementary ratio,—is not defined for that value of h*. We observe also that for the same reason the existence and value of the differential coefficient cannot in general be deduced directly from the rules of the last section.

174. Non-differentiable functions. Differentiable functions necessarily continuous. It should be noticed that the definition does not presuppose that any functions actually are differentiable. In order to use differential coefficients it will be necessary for us to prove their existence in the cases which concern us. We know already (see Chapter III) that some simple functions have differential coefficients; but we must not therefore jump to the conclusion that all functions have this property.

It is indeed evident that *in order to be differentiable at a point x (i.e. for a value x) a function must be continuous at that point.*

For the ratio $[f(x+h)-f(x)]/h$ cannot tend to any number as unique limit as the denominator $h \to 0$ unless the numerator also $\to 0$, i.e. unless $f(x)$ is continuous at the point x.

Consider for example the function y defined thus: $y = \sin(1/x)$ for all values of x except $x = 0$, and $y = 0$ for $x = 0$,—this value being assigned to complete the

---

* This exclusion agrees with the definition of limit (p. 198).

definition of the function (sin (1/0) having of course no meaning). This function is discontinuous and not differentiable at the point $x=0$. In fact

$$\lim_{h \to 0} y(h) = \lim_{h \to 0} \sin(1/h)$$

and does not exist because $\sin(1/h)$ takes on all values between -1 and 1 for values of h indefinitely close to 0. [The upper and lower limits of y at $x=0$,—(see Ex. 18, p. 208),—are 1 and -1.]

Also the incrementary ratio $=[\sin(1/h) - 0]/h$ and has all real numbers for its values for values of h indefinitely close to 0 and therefore has no unique limit (even on the right or left). [The function $[\sin(1/x)]/x$ is unbounded in any neighbourhood $0 < x \leqslant h$.]

It is not even true that all continuous functions are differentiable.

For example, if $y = |x|$, y is continuous for all values of x but not differentiable at the point $x=0$; for the incrementary ratio there $=|h|/h = \pm 1$ according as h is positive or negative and therefore has no unique limit as $h \to 0$,—though the limits on the right and left, i.e. the differential coefficients on the right and left, exist. (They are $+1$ and -1 respectively.)

Or again, the function defined thus:

$$y = x \sin(1/x) \text{ for all values of } x \text{ except } x=0, \; y=0 \text{ for } x=0,$$

is continuous at $x=0$ because $|y| \leqslant |x| \to 0$ as $x \to 0$.

But this function is not differentiable at $x=0$ (nor differentiable on the right or left) because the incrementary ratio, viz.

$$[h \sin(1/h) - 0]/h = \sin(1/h),$$

does not tend to a unique limit as $h \to 0$.

In view of the existence of non-differentiable functions it would be very rash to assume that all the functions met in the complex reality of nature are differentiable. In analysis of course we assume nothing whatever as to the amenability of our functions to such operations as differentiation. The cases in which such operations are possible will be the most useful in practical applications, though the other cases are of importance in the fundamental theory of functions. We are concerned here with the more practical class of functions.

175. Differentiation of sums, products and quotients. There are certain *general rules* whereby the differentiability and differential coefficients of functions which are combinations of simpler functions can be deduced from knowledge of the simpler functions.

Let u and v be any two functions of x which are differentiable for a certain value of x, and let their differential coefficients with respect to x be Du and Dv.

Then *the functions* $u + v$, $u - v$, $u \times v$, u/v *are differentiable and their differential coefficients are given by*

$$D(u + v) = Du + Dv, \quad D(u - v) = Du - Dv,$$
$$D(u \times v) = uDv + vDu, \quad D(u/v) = (vDu - uDv)/v^2,$$

provided only that in the case of the quotient u/v, the denominator v must not be zero for the value of x concerned.

The first two of these rules follow directly from the corresponding rules of the last section for limits of sums and differences, for the incrementary ratio of $u \pm v$ = the incrementary ratio of $u \pm$ the incrementary ratio of v.

To prove the third, the incrementary ratio of uv

$$= [u(x + h)v(x + h) - u(x)v(x)]/h$$
$$= u(x + h)\frac{v(x + h) - v(x)}{h} + v(x)\frac{u(x + h) - u(x)}{h}$$
$$\rightarrow u(x)Dv(x) + v(x)Du(x),$$

by using the rules for the limit of a product and of a sum of two functions (p. 209)*; and the rule is established.

Similarly for the quotient, the incrementary ratio of u/v

$$= \frac{1}{h}\left[\frac{u(x + h)}{v(x + h)} - \frac{u(x)}{v(x)}\right] = \frac{u(x + h)v(x) - u(x)v(x + h)}{hv(x)v(x + h)}$$
$$= \left[v(x)\frac{u(x + h) - u(x)}{h} - u(x)\frac{v(x + h) - v(x)}{h}\right]\Big/[v(x)v(x + h)]$$
$$\rightarrow [v(x)Du(x) - u(x)Dv(x)]/[v(x)]^2,$$

from the rules for limits, the denominator $v(x)v(x + h)$ having the unique limit $[v(x)]^2$ (which $\neq 0$) as $h \rightarrow 0$. The rule is established.

176. Differentiation of a function of a function. These rules may evidently be extended to cover similar combinations of several functions, but in practice it is best to make repeated applications of these rules. A fifth rule, dealing with the differential coefficient of a *function of a function*, is however of importance.

If $u(x)$ is any function of x, differentiable with respect to x for a certain value of x, and $y(u)$ is any function of u, differentiable with respect to u for the value of u corresponding to this value of x, then y may be regarded as a function of x; y is differentiable with regard to x for the value concerned; and, if u' is the differential coefficient of u with respect to x and y' the differential coefficient

---

* $u(x + h) \rightarrow u(x)$ because the function $u(x)$ is necessarily continuous.

of y with respect to u, then the differential coefficient of y with respect to $x = y' \times u'$. Expressed in the usual notation

$$\frac{dy}{dx} = \frac{dy\,(u)}{du} \times \frac{du\,(x)}{dx} = \frac{dy}{du}\cdot\frac{du}{dx} \text{ for brevity.}$$

The proof of this is simple:

The incrementary ratio of y regarded as a function of x

$$= [y\,\{u\,(x+h)\} - y\,\{u\,(x)\}]/h$$

$$= \frac{y\,\{u\,(x+h)\} - y\,\{u\,(x)\}}{u\,(x+h) - u\,(x)} \cdot \frac{u\,(x+h) - u\,(x)}{h} \to y'u' \text{ as } h \to 0$$

because $u\,(x+h) - u\,(x) \to 0$ as $h \to 0$ and therefore the incrementary ratio $\dfrac{y\,\{u\,(x+h)\} - y\,\{u\,(x)\}}{u\,(x+h) - u\,(x)} \to$ the differential coefficient y'.

As an example of this theorem consider the function $y = \log_e x^2$. The differential coefficient of $u\,(= x^2)$ with regard to $x = dx^2/dx = 2x$, and the differential coefficient of $y\,(= \log_e u)$ with regard to $u = d\log_e u/du = 1/u$ (supposing $u > 0$, i.e. $x \neq 0$). Therefore the differential coefficient of $y\,(= \log_e x^2)$ with regard to x (for $x \neq 0$)

$$= \frac{dy}{dx} = \frac{dy}{du}\frac{du}{dx} = \frac{1}{u}\,2x = 2x/x^2 = 2/x.$$

The student may verify from first principles that the function $\log_e x^2$ is differentiable for all positive and negative values of x and that its differential coefficient $= 2/x$.

177. Differentiation of inverse functions. Existence theorem.
As a corollary of this theorem we have the following theorem, useful in the *differentiation of inverse functions*:

If a pair of variables x and y are related in such a way that each may be regarded as a function of the other (e.g. $y = x^2$, $x = \sqrt{y}$; or $y = a^x$, $x = \log_a y$) then *if y is differentiable with regard to x* for a value x and the differential coefficient $= dy/dx$, *and if x is differentiable with regard to y* for the corresponding value of y and the differential coefficient $= dx/dy$, *then* $dx/dy = 1/(dy/dx)$.

For $1 = dx/dx = (dx/dy)\,(dy/dx)$, i.e. $dx/dy = 1/(dy/dx)$.

Regarding the conditions under which a functional relation $y = f(x)$,—the function $f(x)$ being for example continuous and differentiable,—leads to an inverse functional relation of the same character, it is easy to prove the following theorem:

If a function $y = f(x)$ is steadily increasing (or steadily decreasing), continuous and differentiable in a certain range, then there is an inverse function of y, viz. $x = F(y)$, defined in the corresponding range, which is also steadily increasing (or decreasing), continuous and differentiable throughout its range, and the relation $dx/dy = 1/(dy/dx)$ holds.

The analytical proof is left to the student.

Geometrically also this theorem is simple:

Draw (Fig. 21) the graph of $y = f(x)$.

By interchanging the x and y axes this graph may be considered as the graph of x as a function of y,—for to

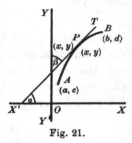

every value of y in the range (c, d) there corresponds a definite value of x. The graph is continuous and has at all points a tangent PT which is not parallel to either axis,— and this is clearly independent of whether the graph is looked on as the graph of y as a function of x or that of x as a function of y. Therefore x, considered as a function of y, is continuous and differentiable at all points.

Fig. 21.

Also if α and β are the angles PT makes with OX and OY respectively, $\tan \alpha = dy/dx$, $\tan \beta = dx/dy$.

But the angles α and β are complementary and therefore
$$\tan \beta = \cot \alpha = 1/\tan \alpha.$$
Therefore $$dx/dy = 1/(dy/dx).$$

It will be noticed that for the truth of this theorem it is essential that $f(x)$ should be definitely increasing (or decreasing) (and that dy/dx must not be zero). Unless we are prepared to consider "multiple valued functions" and "infinite differential coefficients" we cannot dispense with this condition.

EXAMPLES XXI.

1. Prove directly from the rules for the differentiation of products and quotients and the facts $Dx = 1$, $D1 = 0$, that any positive integral power of x, viz. x^n, is differentiable and $Dx^n = nx^{n-1}$ for all values of x; and that any negative integral power of x (x^n where n is a negative integer) is differentiable for all values of x other than zero and $Dx^n = nx^{n-1}$.

$[Dx^2 = D(x.x) = xDx + xDx = x + x = 2x;$
$Dx^3 = D(x.x^2) = xDx^2 + x^2Dx = x2x + x^21 = 3x^2$, etc.

And $\qquad D(1/x) = (xD1 - 1Dx)/x^2 = -1/x^2,$

$D(1/x^2) = D\left(\dfrac{1}{x} \bigg/ x\right) = [xD(1/x) - (1/x)Dx]/x^2 = (-1/x - 1/x)/x^2 = -2/x^3$, etc.

The proofs are completed by induction.]

2. Prove from the rules for differentiation of sums etc. that any polynomial $P_n(x)$ is differentiable for all values of x and that any rational function $P_n(x)/Q_m(x)$ is differentiable for all values of x which are not roots of the equation $Q_m(x) = 0$.

3. Prove that if $u_1, u_2, \ldots u_n$ are differentiable functions of x and $a_1, a_2, \ldots a_n$ are constants, then $a_1u_1 + a_2u_2 + \ldots + a_nu_n$ is differentiable and
$$D(a_1u_1 + \ldots + a_nu_n) = a_1Du_1 + \ldots + a_nDu_n.$$

4. If $u_1(x) + u_2(x) + \ldots$ is a convergent series of differentiable functions, it is not in general true that the sum-function is differentiable. If the series is a power series however,
$$a_0 + a_1x + a_2x^2 + \ldots,$$
it is true that $\qquad D(a_0 + a_1x + a_2x^2 + \ldots)$
exists and equals
$$Da_0 + Da_1x + Da_2x^2 + \ldots = a_1 + 2a_2x + 3a_3x^2 + \ldots.$$
[See Ex. 14, p. 195.]

5. Given $Dx^n = nx^{n-1}$ for all values of x if n is any positive integer, prove that the inverse function $\sqrt[n]{x}$ is also differentiable and $D\sqrt[n]{x} = \dfrac{1}{n}x^{\frac{1}{n}-1}$. For what values of x is the result valid?

6. Using the results of Ex. 1 and that $De^x = e^x$, shew that
$$De^{x^2} = 2xe^{x^2}; \quad D(e^x)^2 = 2e^{2x}.$$
[Function of a function.]

7. Given $\qquad D\log_e x = 1/x$ (for $x > 0$), $DS(x) = C(x)$,
and $\qquad\qquad\qquad DC(x) = -S(x)$,
shew that $\qquad D\log_e S(x) = C(x)/S(x)$, $D\log_e C(x) = -S(x)/C(x)$,
$$DS(\log_e x) = C(\log_e x)/x, \quad DC(\log_e x) = -S(\log_e x)/x;$$
the functions $S(x)$ and $C(x)$ being the trigonometrical functions of Chapter III. State in each case the ranges of values of x for which the result is valid.

8. Given $De^x = e^x$, $De^{-x} = -e^{-x}$ and the differential coefficients in the last example, shew that
$$De^x S(x) = e^x[C(x) + S(x)], \quad De^x C(x) = e^x[C(x) - S(x)],$$
$$De^{-x} S(x) = e^{-x}[C(x) - S(x)], \quad De^{-x} C(x) = -e^{-x}[C(x) + S(x)],$$
$$De^{-ax} S(bx) = e^{-ax}[-aS(bx) + bC(bx)],$$
$$De^{-ax} C(bx) = -e^{-ax}[aC(bx) + bS(bx)];$$
a and b being any constants.

9. Shew $\dfrac{df(x+a)}{dx} = \dfrac{df(x+a)}{d(x+a)}$, $\dfrac{df(ax+b)}{dx} = a\dfrac{df(ax+b)}{d(ax+b)}$;

a and b being any constants; the differential coefficients figuring on the right being supposed to exist.

10. Using the results of the above examples, shew that

$$De^{ax+b} = ae^{ax+b}; \quad Dxe^{ax+b} = e^{ax+b}(ax+1);$$
$$De^{(x^2+\log_e x)} = e^{(x^2+\log_e x)}(2x+1/x) \text{ for } x>0;$$
$$Dxe^{x_2} = xe^{x^2}\left(2x+\frac{1}{x}\right).$$

11. If y is the distance moved through in time x by a moving body, then the velocity of the body at the end of any time x is $v = dy/dx$ and the acceleration (the rate of increase of velocity) $= f = dv/dx$.

12. It can be proved experimentally that the distance (y) moved through by a falling body (under gravity) in any time x is given approximately by the formula $y = 16x^2$,—(x and y being measured in seconds and feet). Shew that the velocity acquired in time x is $32x$ feet per second, and that the acceleration is constant ($=32$ feet per second per second).

§ 4. DIFFERENTIATION OF ELEMENTARY FUNCTIONS

178. Standard forms. In this section we will obtain the differential coefficients of the simple functions which are of importance. By using the general rules of the last section we shall then be in a position to differentiate all functions which arise out of ordinary combinations of these known functions.

We have the following standard forms:

I. *Powers of x.*

$Dx^n = nx^{n-1}$ if n is any real number;

valid for all values of x if n is a positive integer,

for all values of x except $x = 0$ if n is a negative integer or zero,

and for all positive values of x if n is any non-integral number.

II. *Logarithmic function.*

$D\log_e x = 1/x$, $D\log_a x = \log_a e/x$ if a is any positive number other than 1;

valid for all positive values of x.

III. *Exponential function.*

$De^x = e^x$, $Da^x = a^x . \log_e a$ if a is any positive number;

valid for all real values of x.

IV. *Circular and trigonometrical functions.*

(a) $D\sin x = \cos x$, $D\cos x = -\sin x$;

valid for all real values of x.

$D \tan x = \sec^2 x$, $D \sec x = \sin x \sec^2 x$;
valid for all values of x except odd multiples of $\pm \pi/2$.
$D \cot x = - \operatorname{cosec}^2 x$, $D \operatorname{cosec} x = - \cos x \operatorname{cosec}^2 x$;
valid for all values of x except 0 and multiples of $\pm \pi$.

(b) $DS(x) = C(x)$, $DC(x) = - S(x)$;
valid for all values of x.

$$DT(x) = \frac{1}{[C(x)]^2}, \quad D \frac{1}{C(x)} = \frac{S(x)}{[C(x)]^2};$$

valid for all values of x except odd multiples of $\pm \pi/2$.

$$D \frac{1}{T(x)} = - \frac{1}{[S(x)]^2}, \quad D \frac{1}{S(x)} = - \frac{C(x)}{[S(x)]^2};$$

valid for all values of x except 0 and multiples of $\pm \pi$.

V. *Inverse circular and trigonometrical functions.*

(a) $D \operatorname{arc} \sin x = 1/\sqrt{1 - x^2}$, $D \operatorname{arc} \cos x = - 1/\sqrt{1 - x^2}$;
valid for all values of x interior to the range of definition, i.e. for
$-1 < x < 1$.
$D \operatorname{arc} \tan x = 1/(1 + x^2)$, $D \operatorname{arc} \cot x = - 1/(1 + x^2)$;
valid for all values of x.

$$D \operatorname{arc} \sec x = \frac{1}{x \sqrt{x^2 - 1}}, \quad D \operatorname{arc} \operatorname{cosec} x = - \frac{1}{x \sqrt{x^2 - 1}};$$

valid for all values of x interior to the range of definition, i.e. for
which $|x| > 1$.

(b) $D \bar{S}(x) = \dfrac{1}{\sqrt{1 - x^2}}$, $D \bar{C}(x) = - \dfrac{1}{\sqrt{1 - x^2}}$;

valid for all values of x interior to the range of definition, i.e. for
which $|x| < 1$.

$D \bar{T}(x) = \dfrac{1}{1 + x^2}$, valid for all values of x.

VI. *Hyperbolic and inverse hyperbolic functions.*
$D \sinh x = \cosh x$, $D \cosh x = \sinh x$, $D \tanh x = \operatorname{sech}^2 x$,

$D \operatorname{sech} x = - \dfrac{\sinh x}{\cosh^2 x}$; valid for all values of x.

$D \coth x = - \operatorname{cosech}^2 x$, $D \operatorname{cosech} x = - \dfrac{\cosh x}{\sinh^2 x}$; valid for all values
of x except $x = 0$.

$D \arg \sinh x = \dfrac{1}{\sqrt{x^2 + 1}}$, valid for all values of x.

$D \arg \cosh x = \dfrac{1}{\sqrt{x^2 - 1}}$, valid for $x > 1$.

$D \arg \tanh x = \dfrac{1}{1 - x^2}$, valid for $|x| < 1$.

$D \arg \coth x = \dfrac{1}{1 - x^2}$, valid for $|x| > 1$.

$D \arg \operatorname{sech} x = -\dfrac{1}{x\sqrt{1 - x^2}}$, valid for $0 < x < 1$.

$D \arg \operatorname{cosech} x = -\dfrac{1}{x\sqrt{x^2 + 1}}$, valid for all values of x except $x = 0$.

VII. *Any power series.*

$D\,(a_0 + a_1 x + a_2 x^2 + \ldots) = a_1 + 2a_2 x + 3a_3 x^2 + \ldots$, a_0, a_1, a_2, $\ldots$ being constants; valid for all values of x interior to the range of convergence of the series.

179. Proofs. Powers, exponentials and logarithms. For the sake of completeness we append (or recall) proofs:

I. If n is a positive integer the incrementary ratio $[(x+h)^n - x^n]/h$ can be divided out by h and so reduced to

$$(x + h)^{n-1} + (x + h)^{n-2} x + \ldots + (x - h) x^{n-2} + x^{n-1},$$

which $\to nx^{n-1}$ as $h \to 0$, by the rules of § 2 of this chapter for the evaluation of limits. Similarly if n is a negative integer, $= -m$ say, the incrementary ratio

$$= \frac{1}{h}\left[\frac{1}{(x+h)^m} - \frac{1}{x^m}\right] = -\frac{(x+h)^m - x^m}{hx^m(x+h)^m}$$

$$= -\frac{(x+h)^{m-1} + (x+h)^{m-2} x + \ldots + x^{m-1}}{x^m(x+h)^m}$$

$$\to -\frac{mx^{m-1}}{x^{2m}} = -\frac{m}{x^{m+1}} = nx^{n-1},$$

provided $x \neq 0$ (it being then always possible to choose h so small that x and $x + h$ have the same sign).

In the general case we use the inequality (iii) of p. 29 as completed in Ex. 2, p. 91, viz. $(a^n - b^n)/(a - b)$ lies between na^{n-1} and nb^{n-1} if n is any real number and a and b any two numbers of the same sign.

Taking x positive (x^n being defined only for $x > 0$ if n is not integral), we can take $|h|$ so small that $x + h$ is also positive. The incrementary ratio $[(x+h)^n - x^n]/h$ therefore lies between $n(x+h)^{n-1}$ and nx^{n-1}.

But the power function x^n is continuous,—as follows immediately, as in the case of the integral power (p. 153 above). Therefore $n(x+h)^{n-1} \to nx^{n-1}$ as $h \to 0$.

Hence the incrementary ratio also $\to nx^{n-1}$ and the result is established.

Note. If $n \geqslant 1$, the differential coefficient on the right at $x = 0$ exists and $= 0$ whether n be integral or not.

II. The inequality of p. 110 above, viz.

$$\frac{\log_a(x+h) - \log_a x}{h} \text{ lies between } \frac{\log_a e}{x} \text{ and } \frac{\log_a e}{x+h},$$

holds if x and $x + h$ are any positive numbers and a any positive number other than 1. Hence the incrementary ratio of $\log_a x$, viz.

$$[\log_a(x+h) - \log_a x]/h,$$

tends to the common value of

$$\lim_{h \to 0} \frac{\log_a e}{x+h} \text{ and } \lim_{h \to 0} \frac{\log_a e}{x}, \text{ i.e. } (\log_a e)/x.$$

Therefore

$$D\log_a x = (\log_a e)/x \text{ and } D\log_e x = (\log_e e)/x = 1/x.$$

III. If $a \neq 1$, let $y = a^x$, so that $x = \log_a y$,—and this pair of inverse functions has been proved (Chapter II, §§ 1 and 2) to exist and be both monotone and continuous for all values of x and for $y > 0$.

By II above, $dx/dy = d\log_a y/dy = (\log_a e)/y$.

Therefore by the theorem of p. 220, or directly as on pp. 128—129, dy/dx exists and

$$= 1/(dx/dy) = y/(\log_a e) = a^x \log_e a;$$

and this result is true for all positive values of y and all values of x. When $a = 1$, $y = 1$ for all values of x and $Dy = 0 = a^x \log_e 1$.

$$De^x = e^x \log_e e = e^x.$$

180. The circular and inverse circular functions. IV. To differentiate the circular functions $\sin x$ and $\cos x$ we may use the addition theorems for the circular functions and the method of p. 181, or we may proceed independently thus:

Let $0 < x < \pi/2$ and let (Fig. 22) the point P on a unit circle represent the angle x, so that the arc $AP = x$.

Let Q be any point on the arc AB between P and B, and let the arc $PQ = h$.

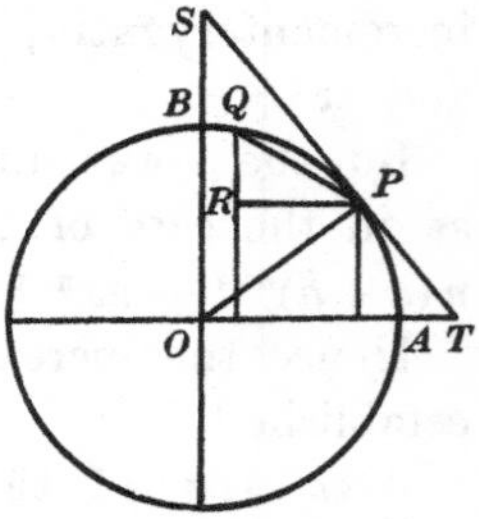

All distances concerned will be positive.

Let SPT be the tangent at P.

The incrementary ratios of $\cos x$ and $\sin x$ with respect to the increment h are respectively

$$- RP/\text{arc } PQ \text{ and } RQ/\text{arc } PQ,$$

i.e. $$- \frac{RP}{PQ} \frac{PQ}{\text{arc } PQ} \text{ and } \frac{RQ}{PQ} \frac{PQ}{\text{arc } PQ}.$$

Fig. 22.

Now as $Q \to P$ along the arc,

$$RP/PQ \searrow OT/ST = \sin O\hat{S}T = \sin x$$

and $$RQ/PQ \nearrow OS/ST = \cos O\hat{S}T = \cos x,$$

from the simple properties of the circle and of similar triangles. Also

$$PQ/\text{arc } PQ = [\sin (h/2)]/(h/2) \to 1$$

as $h \to 0$, because $\lim\limits_{x \to 0} \dfrac{\sin x}{x} = 1$ (p. 205).

Therefore, as $h \searrow 0$, the incrementary ratios for $\cos x$ and $\sin x$ tend to $- \sin x$ and $\cos x$ respectively.

This is true also as $h \nearrow 0$, and therefore

$$D \cos x = - \sin x, \quad D \sin x = \cos x.$$

This result,—thus proved for $0 < x < \pi/2$,—is easily extended to cover all cases.

To find $D \tan x$, $D \sec x$, $D \operatorname{cosec} x$ we use the rule for the differentiation of a quotient (p. 218), thus:

$$D \tan x = D \left(\frac{\sin x}{\cos x} \right) = \frac{\cos x\, D \sin x - \sin x\, D \cos x}{\cos^2 x}$$

$$= \frac{\cos^2 x + \sin^2 x}{\cos^2 x} = \frac{1}{\cos^2 x} = \sec^2 x;$$

valid for all values of x which do not make $\cos x = 0$, i.e. for all values of x except odd multiples of $\pm \pi/2$.

The differential coefficients of the trigonometrical functions $C(x)$ and $S(x)$ (defined in Chapter III, § 5) have been obtained before (pp. 180—182). $DT(x)$, etc. are derived from these precisely as $D \tan x$, etc. are derived from $D \sin x$ and $D \cos x$.

V. The inverse circular functions arc sin x and arc cos x are defined thus:

arc sin x *is that angle y, measured in radians, belonging to the interval* $\left(-\dfrac{\pi}{2}, \dfrac{\pi}{2}\right)$ (*i.e.* $-\pi/2 \leqslant y \leqslant \pi/2$), *for which* $\sin y = x$;

arc cos x *is that angle y, measured in radians, such that* $0 \leqslant y \leqslant \pi$, *for which* $\cos y = x$.

Both these functions are defined for all values of x between -1 and 1 and for $x = -1$ and $x = 1$, but for no other values. If the restrictions $(-\pi/2 \leqslant y \leqslant \pi/2$ or $0 \leqslant y \leqslant \pi)$ were removed the functions would be multiple-valued. The restrictions are arbitrary, but desirable. We retain them*.

To differentiate these inverse functions, the theorem on p. 220 will apply.

$y = $ arc sin x gives $x = \sin y$.

Therefore $dx/dy = \cos y = \sqrt{1 - \sin^2 y} = \sqrt{1 - x^2}$,

and this $(= \cos y)$ is positive (or zero) because $-\pi/2 \leqslant y \leqslant \pi/2$.

Therefore $dy/dx = +1/\sqrt{1 - x^2}$.

Similarly, from $y = $ arc cos x, $x = \cos y$,

$dx/dy = -\sin y = -\sqrt{1 - x^2}$ and $dy/dx = -1/\sqrt{1 - x^2}$.

The inverse tangent arc tan x *is defined as that angle y between* $-\pi/2$ *and* $\pi/2$ *for which* $\tan y = x$. It is defined for all real values of x.

Using the theorem of p. 220, if $y = $ arc tan x, $x = \tan y$,

$$dx/dy = \sec^2 y = 1 + \tan^2 y = 1 + x^2,$$

and therefore $dy/dx = 1/(1 + x^2)$.

The other inverse circular functions are defined and differentiated similarly.

The inverse trigonometrical functions $\overline{S}(x)$, $\overline{C}(x)$, $\overline{T}(x)$ are defined and differentiated in precisely the same way as the inverse circular functions, with the replacement of π by ϖ, sin by S, cos by C, and tan by T.

181. The hyperbolic and inverse hyperbolic functions. VI. To differentiate $\sinh x$, the incrementary ratio

$$= [\sinh (x + h) - \sinh x]/h = [\cosh (x + h/2) \sinh (h/2)]/(h/2),$$

---

* When multiple values are admitted these functions are termed the *principal values* of the functions arc sin x, etc. See also Ex. 4, p. 230 below.

by Ex. 3, p. 193. This tends to cosh x as $h \to 0$ by considerations analogous to those used in the differentiation of $S(x)$ and $C(x)$.

Similarly for cosh x: the incrementary ratio

$$= [\cosh(x+h) - \cosh x]/h = \sinh(x+h/2)\left(\frac{\sinh h/2}{h/2}\right)$$
$$\to \sinh x \text{ as } h \to 0.$$

To differentiate the hyperbolic tangent tanh x we have

$$D \tanh x = D(\sinh x/\cosh x) = \frac{\cosh x\, D \sinh x - \sinh x\, D \cosh x}{\cosh^2 x}$$
$$= \frac{\cosh^2 x - \sinh^2 x}{\cosh^2 x} = \frac{1}{\cosh^2 x},$$

for all values of x, because cosh x is never zero.

The other hyperbolic functions are differentiated similarly.

The inverse function theorem of p. 220 applies to all the inverse hyperbolic functions and establishes their ranges of definition and the differential coefficients, as given. These functions are necessarily single-valued*.

182. Sum-function of a power series. VII. The theorem that the sum-function of a power series $a_0 + a_1 x + a_2 x^2 + \dots$ is continuous and differentiable at all points interior to the range of convergence has been proved above (p. 193 and Ex. 14, p. 195). Some series are convergent for all values of x, but usually the values of x for which a power series is convergent consist of all the values between $-R$ and R, R being some positive number, called the "radius of convergence." The series may or may not converge also for $x = -R$ and $x = R$, but it is not convergent for any value of x for which $|x| > R$ and it is absolutely convergent for all values of x *between* $-R$ and R. (See Ex. 13, p. 194.) If the terms of the power series are differentiated (term by term) a second series, $= a_1 + 2a_2 x + 3a_3 x^2 + \dots$, is obtained. The theorem of p. 193 shews at once that the radius of convergence of this differentiated series is at least R; it is in fact precisely R†.

If a power series is convergent for *all* values of x (e.g. the series of Chapter III, §§ 5, 6) the differentiated series will also be convergent for all values of x and its sum will be the differential coefficient of the sum-function. By using this result the differential coefficients

---

* If complex numbers are introduced the hyperbolic functions are, like the circular functions, periodic and the inverse hyperbolic functions many-valued.

† By the comparison theorem (I), p. 79 above.

of the functions $C(x)$, $S(x)$, $\cosh x$, $\sinh x$ (and $E(x)$ of p. 173) could be found without resort to the algebraic addition theorems.

183. Differentiation of compound functions. By means of the results of this section and the rules of the last section we are now in a position to differentiate any function which is defined by means of a combination of the functions considered in this section, but not to differentiate functions defined as *limits* of such functions except in the special case VII. All functions met naturally in elementary analysis and applications (with the exception of some infinite series which are not power series) can therefore be so dealt with.

To shew the scope of our methods of differentiation let us differentiate the somewhat complicated function

$$\log_e \left[\log_e \left\{ 1 + \sin \left(\frac{x^2 + 1}{x - 1} \right) \right\} \right].$$

The function $= \log_e [\log_e \{1 + \sin w\}]$, where $w = (x^2 + 1)/(x - 1)$,

$= \log_e [\log_e v]$, where $v = 1 + \sin w$,

$= \log_e u$, where $u = \log_e v$,

$= y$, where $y = \log_e u$.

Therefore $\qquad \dfrac{dy}{dx} = \dfrac{dy}{du} \dfrac{du}{dv} \dfrac{dv}{dw} \dfrac{dw}{dx}$

(by repeated application of the rule for the differentiation of a function of a function)

$$= \frac{d \log_e u}{du} \frac{d \log_e v}{dv} \frac{d(1 + \sin w)}{dw} \frac{d\left(\dfrac{x^2 + 1}{x - 1}\right)}{dx}$$

$$= \frac{1}{u} \frac{1}{v} \cos w \frac{(x - 1) D (x^2 + 1) - (x^2 + 1) D (x - 1)}{(x - 1)^2}$$

(by the rule for a quotient)

$$= \frac{1}{u} \frac{1}{v} \cos w \frac{(x - 1) 2x - x^2 - 1}{(x - 1)^2}$$

$$= \frac{\cos \left(\dfrac{x^2 + 1}{x - 1}\right) (x^2 - 2x - 1)}{\log_e \left\{ 1 + \sin \left(\dfrac{x^2 + 1}{x - 1}\right) \right\} \left\{ 1 + \sin \left(\dfrac{x^2 + 1}{x - 1}\right) \right\} (x - 1)^2}.$$

The range of validity of this result will depend on the range of validity of the various steps. It is necessary that $x \neq 1$, that

$$1 + \sin\left(\frac{x^2+1}{x-1}\right) > 0, \text{ and that } \log_e\left\{1 + \sin\left(\frac{x^2+1}{x-1}\right)\right\} > 0;$$

i.e. that $\qquad\qquad \sin\dfrac{x^2+1}{x-1} > 0 \text{ and } x \neq 1.$

These inequations can be solved. Such functions are rare.

EXAMPLES XXII.

1. Differentiate

$(x-1)/(x^2+1)$, $\quad 4x^3 - ax - b$, $\quad 1/\sqrt{[(1-x^2)(1-k^2x^2)]}$, $\quad \sin 2x$,

$2\cos 2x$, $\quad 1/(a+b\cos x)$, $\quad (a+bx)^n$, $\quad x^n \sin bx$,

$\sin x/(a\sin^2 x + b\cos^2 x)$, $\quad 1/\sqrt{(1-2\cos x)}$, $\quad e^{-ax}\cos bx$, $\quad e^{-ax}\sin bx$;

a, b, k, and n being constants.

2. Prove (under appropriate conditions):

   (i) $D\log_e \sin x = \cot x$, $\quad [\sin x > 0]$,

   (ii) $D\log_e \cos x = -\tan x$, $\quad [\cos x > 0]$,

   (iii) $D\log_e \tan(\tfrac{1}{4}\pi + \tfrac{1}{2}x) = D\log_e(\sec x + \tan x) = \sec x$, $\quad [\tan(\tfrac{1}{4}\pi + \tfrac{1}{2}x) > 0]$,

   (iv) $D\log_e \tan(\tfrac{1}{4}\pi - \tfrac{1}{2}x) = D\log_e(\sec x - \tan x) = -\sec x$,

$$[\tan(\tfrac{1}{4}\pi - \tfrac{1}{2}x) > 0],$$

   (v) $D\log_e \log_e x = 1/(x\log_e x)$, $\quad [\log_e x > 0, \text{ i.e. } x > 1]$,

   (vi) $D\log_e \log_e \log_e x = 1/(x\log_e x\log_e \log_e x)$, $\quad [\log_e \log_e x > 0, \text{ i.e. } x > e]$,

   (vii) $D(e^x/x^n) = e^x(x-n)/x^{n+1}$, $\quad [x \neq 0 \text{ if } n \text{ is a positive integer}]$,

   (viii) $D(x^n/\log_e x) = x^{n-1}(n\log_e x - 1)/(\log_e x)^2$, $\quad [x > 0 \text{ and } x \neq 1]$,

   (ix) $Dx^x = D(e^{x\log x}) = x^x(1+\log_e x)$, $\quad [x > 0]$,

   (x) $De^{x^x} = e^{x^x}x^x(1+\log_e x)$, $\quad [x > 0]$.

3. By differentiating the logarithmic series term by term verify that

$$D\log_e x = 1/x \text{ for } 0 < x < 2.$$

4. Most of the inverse functions considered above in the text are capable of extended definition. Thus arc $\sin x$ may be considered to be the multiple-valued function y which is *any* number for which $\sin y = x$. The different values of the function constitute what are called *branches* of the function. The formula for the differentiation of arc $\sin x$ is applicable to any branch, but the sign will depend on the particular branch. Thus for the branch comprising the values of y between $\pi/2$ and $3\pi/2$, $Dy = -1/\sqrt{(1-x^2)}$.

5. By means of the inverse function theorem of p. 220 a function which we may call $\sqrt[n]{x}$ is defined for all values of x, positive and negative, if n is an odd integer. We can similarly build up a power function $x^{p/q}$ for all values of x if q is odd and thus extend the range of definition of the power function in such cases. With this definition $Dx^{p/q} = \dfrac{p}{q}\cdot x^{\frac{p}{q}-1}$ for positive and negative values of x.

[See p. 89 above.]

6. Shew that $\qquad Dx\sin(1/x) = \sin(1/x) - (1/x)\cos(1/x)$

and $\qquad\qquad\quad Dx^2\sin(1/x) = 2x\sin(1/x) - \cos(1/x).$

These functions are not defined for $x=0$. Prove that the first is not differentiable at $x=0$ whatever value be assigned to it to complete its definition, but that the second function is differentiable at $x=0$ if the value 0 be assigned to it there.

7. Establish the series

$$\text{arc}\sin x = x + \frac{1}{2}\frac{x^3}{3} + \frac{1.3}{2.4}\frac{x^5}{5} + \frac{1.3.5}{2.4.6}\frac{x^7}{7} + \cdots$$

for $-1 \leqslant x \leqslant 1$.

By comparing with Ex. 16, p. 189, deduce that $\text{arc}\sin x = \overline{S}(x)$, $\pi = \varpi$, and that $\sin x = S(x)$, firstly for $-\pi/2 \leqslant x \leqslant \pi/2$, and thence for all values of x. The identity of $\cos x$ and $\tan x$ with $C(x)$ and $T(x)$ then follows, or may be proved independently in the same way.

§ 5. APPLICATIONS OF DIFFERENTIATION

184. In this section we shall discuss a few simple problems capable of easy solution by a consideration of differential coefficients. We shall suppose generally that the functions concerned are continuous and differentiable at all points of the ranges concerned. These assumptions will be justified at all ordinary points for the functions hitherto defined in this course. Any exceptional points,—of discontinuity, etc.—can be dealt with separately.

185. Maxima and minima. Let α be a value of x for which a function $y, =f(x)$, has a differential coefficient Dy which is positive. Then, from the definition of a differential coefficient, since the differential coefficient on the right at $\alpha > 0$, it follows that $f(\alpha + h) - f(\alpha)/h > 0$ for all positive values of h sufficiently small; and therefore $f(\alpha + h) > f(\alpha)$ for all positive values of h sufficiently small.

Also, because the differential coefficient on the left > 0,

$$f(\alpha - h) < f(\alpha)$$

for all positive values of h sufficiently small.

That is, there is a neighbourhood to the right of α in which the values of y exceed $f(\alpha)$ and a neighbourhood to the left in which the values of y are less than $f(\alpha)$. We say that $f(x)$ *is increasing at the point.*

Similarly if Dy is negative at α, there is a neighbourhood to the right of α in which $y < f(\alpha)$ and a neighbourhood to the left in which $y > f(\alpha)$; or $f(x)$ *is decreasing at α.*

In these definitions we say nothing as to the magnitude of the neighbourhood in which $y >$ (or $<$) the value at the point α, but only that there is some such neighbourhood,—i.e. that h *can* be chosen small enough to ensure the truth of the inequalities.

If now $f(x)$ is increasing at all points of a range, it is geometrically intuitive that its least and greatest values will occur at the beginning and end of the range respectively; and it is not difficult to prove this strictly analytically *. Similarly (*mutatis mutandis*) if $f(x)$ is decreasing.

Suppose now $Dy > 0$ at all points in a neighbourhood to the left of a point α, and $Dy < 0$ at all points in a neighbourhood to the right of α. The function y increases up to α and decreases beyond α; and the value of the function at α, viz. $f(\alpha)$, exceeds all other values of the function in the neighbourhood. The function $f(x)$ *is then said to have a maximum value at the point α.* At such a point the differential coefficient on the left

$$= \lim_{h \searrow 0} {}^{+} [f(\alpha) - f(\alpha - h)]/h \geqslant 0,$$

and the differential coefficient on the right

$$= \lim_{h \searrow 0} [f(\alpha + h) - f(\alpha)]/h \leqslant 0;$$

and therefore Dy,—the common value of these two differential coefficients,—must $= 0$.

Similarly if $Dy < 0$ at all points in a neighbourhood to the left of α and $Dy > 0$ at all points in a neighbourhood to the right of α, $f(\alpha)$ is the least value of y in the neighbourhood and $f(x)$ has a *minimum* value at α. Again $Dy = 0$.

Thus *if $y, = f(x)$, has a maximum or minimum value at α, $Dy = 0$ at α. If Dy changes from being positive to being negative as x increases through α, y is a maximum there; if from negative to positive, y is a minimum.*

It is possible that $Dy = 0$ at α and is positive at all other points in the neighbourhood of α on both sides. In this case the function has neither a maximum nor a minimum there, and is in fact increasing at the point. Also Dy may be zero at α and negative at all other points in the neighbourhood. It is then decreasing at α. In both these cases it is customary to say that y is *stationary* at the point α. The point on the graph is then a *point of inflexion*.

Other cases may arise in which $Dy = 0$ at a point α but has both positive and negative values on one side (or both sides) of a indefinitely close to a,—e.g. the function $x^2 \sin(1/x)$ of Ex. 6, p. 230. Other cases again arise if we relax the condition that $f(x)$ is differentiable.

---

* This is substantially the theorem of p. 119 above.

† This symbol is self-explanatory.

186. The following are examples:

(i) The function x^2 has a derivative $2x$ for all values of x, which < 0 if $x < 0$, $= 0$ for $x = 0$, and > 0 for $x > 0$.

Therefore x^2 *decreases up to* $x = 0$, *where it has a minimum value, and thereafter increases.*

(ii) The function x^3 has a derivative $3x^2$ for all values of x, which > 0 for all values of x except $x = 0$, for which it $= 0$. Therefore x^3 *is increasing for all values of x though stationary at $x = 0$. The origin is a point of inflexion on the graph* $y = x^3$.

(iii) *The maxima and minima of the polynomial* $x^4 - 3x^2 + 2x + 4$ have already been discussed on p. 158.

(iv) *The rectangle of given perimeter which has the greatest area is a square.*

For, let the given perimeter be $4K$, and let the sides of the rectangle be x and $2K - x$.

The area $= y = x(2K - x)$.

The area y depends on x and is in fact a continuous and differentiable function of x.

$Dy = 2K - 2x$, which $= 0$ for $x = K$, and Dy decreases from positive to negative as x increases through the value K. Therefore y has its only maximum for $x = K$, i.e. when the rectangle is a square of side K.

Note. In a practical problem of this kind it is quite possible for the greatest or least value to occur for the extreme allowable values of the variable x,—in this example $x = 0$ or $x = 2K$. We see in this example that these values correspond to the minimum and not the required maximum.

187. Inequalities. Maxima and minima can be applied to establish inequalities.

(i) The example just considered gives a proof of inequality (i) of p. 29 above,—that *the arithmetic mean of two positive numbers a and $b \geqslant$ the geometric mean.*

For, if $a + b = 2K$, putting $a = x$ and $b = 2K - x$, we have that $ab = x(2K - x)$ and has its maximum value, viz. K^2, when $x = K$; and therefore if $x \neq K$, $x(2K - x)$ is less than K^2.

Hence $ab \leqslant \left(\dfrac{a + b}{2}\right)^2$ and the inequality is established.

(ii) *Similarly we can establish the more general inequality*

$$a^m b^n \leqslant \frac{m^m n^n}{(m+n)^{m+n}} (a+b)^{m+n},$$

where a and b are any positive numbers and m and n any positive integers.

Put $\qquad\qquad a+b=K, \ a=x.$

Then $\qquad\qquad a^m b^n = x^m (K-x)^n = y$ say.

$$Dy = m x^{m-1} (K-x)^n - n x^m (K-x)^{n-1}$$
$$= x^{m-1} (K-x)^{n-1} [m(K-x) - nx],$$

which $= 0$ when $x = 0$, $x = K$, and $x = mK/(m+n)$.

When $x = 0$ or K, y has its minimum value 0.

When $x = mK/(m+n)$, y has its maximum value, because Dy decreases from positive to negative as x increases through this value.

Therefore $\qquad\qquad y \leqslant \left(\frac{mK}{m+n}\right)^m \left(K - \frac{mK}{m+n}\right)^n,$

i.e. $\qquad\qquad a^m b^n \leqslant \frac{m^m n^n}{(m+n)^{m+n}} (a+b)^{m+n}.$ $\qquad$ Q.E.D.

(iii) The *inequality* (iv) *of* p. 29 *may be established thus:*

Consider the function

$$y = n a^n (x^m - a^m) - m a^m (x^n - a^n),$$

a being any positive number and m and n positive integers; and let $m > n$.

$$Dy = n m a^n x^{m-1} - m n a^m x^{n-1}$$
$$= m n a^n x^{n-1} (x^{m-n} - a^{m-n}).$$

Considering only positive values of x, $Dy = 0$ only for $x = a$.

Where $0 < x < a$, $Dy < 0$; and where $x > a$, $Dy > 0$.

Therefore y has its only minimum value (for positive values of x) at $x = a$, where $y = 0$.

Hence if x is positive and $\neq a$, $y > 0$;

i.e. $\qquad \dfrac{x^m - a^m}{x^n - a^n} > \dfrac{m}{n} a^{m-n}$ if $x^n - a^n > 0$, i.e. if $x > a$;

and $\qquad \dfrac{x^m - a^m}{x^n - a^n} < \dfrac{m}{n} a^{m-n}$ if $x < a$.

Hence, if $0 < a < b$ and $m > n$,

$$\frac{m}{n} a^{m-n} < \frac{a^m - b^m}{a^n - b^n} < \frac{m}{n} b^{m-n},$$

and the inequality is proved.

188. Tangents. We have seen that the angle of inclination to the x axis of the tangent at a point (x, y) of a curve given by the equation $y = f(x)$ is ψ, given by $\tan \psi = dy/dx$. The equation of the tangent at any point of such a curve can therefore be found.

(i) *To find the tangent at the point* (1, 1) *on the parabola* $y = x^2$, we have $dy/dx = Dx^2 = 2x = 2$ at the point (1, 1).

Therefore the equation of the tangent is

$$y - 1 = 2(x - 1) \text{ or } y = 2x - 1.$$

(ii) We can verify that *the tangent to a circle* (defined as the limit of the chord) *is the perpendicular to the radius*, thus:

The half above the x axis of the circle whose centre is the origin O and radius R is given by the equation

$$y = + \sqrt{(R^2 - x^2)}.$$

We have

$$dy/dx = \frac{d\sqrt{(R^2 - x^2)}}{d(R^2 - x^2)} \frac{d(R^2 - x^2)}{dx}$$

$$= \frac{1}{2\sqrt{(R^2 - x^2)}} (-2x) = -\frac{x}{\sqrt{(R^2 - x^2)}}.$$

Therefore the tangent at the point $P(x, y)$ makes an angle ψ with OX, where $\tan \psi = -x/\sqrt{(R^2 - x^2)} = -x/y$.

But the radius OP makes the angle θ with OX, where $\tan \theta = y/x$.

These two lines are therefore perpendicular, because

$$\cot(\theta - \psi) = \frac{1 + \tan \theta \tan \psi}{\tan \theta - \tan \psi} = 0.$$

The result is proved.

The equation of the tangent at the point (x_1, y_1) is

$$(y - y_1)/(x - x_1) = \tan \psi = -x_1/y_1$$

or

$$xx_1 + yy_1 = x_1^2 + y_1^2 = R^2.$$

189. Velocity, etc. Let y represent the distance in a given direction moved through in time x by a body moving in a straight line. We have seen that the velocity (v) at any time x is given by dy/dx and the acceleration (f) by dv/dx.

(i) Suppose we know that *a body moves so that* $y = a + bx + cx^2$, a, b and c being known constants. We can at once *deduce the velocity and acceleration at any time during the motion*, for

$$v = dy/dx = b + 2cx, \text{ and } f = dv/dx = 2c.$$

Thus the acceleration is constant and $= 2c$.

(ii) *A body is observed to oscillate on a straight line according to the formula* $y = a \sin nx + b \cos nx$, a, b and n *being constants. The acceleration at any time* x *during the motion is proportional to the distance* y *from the origin.*

For, the velocity

$$= v = dy/dx = na \cos nx - nb \sin nx$$

and the acceleration

$$= f = dv/dx = - n^2 a \sin nx - n^2 b \cos nx = - n^2 y,$$

and the result is proved.

Such motion is called *simple harmonic motion.*

(iii) *A body is observed to oscillate according to the law*

$$y = e^{-mx} (a \sin nx + b \cos nx),$$

where a, b and n are as in (ii) and m is a positive constant.

The acceleration in this case may be divided up into two parts,— one proportional to y as in (ii) *and the other proportional to the velocity v.*

For

$$v = dy/dx = e^{-mx} [n (a \cos nx - b \sin nx) - m (a \sin nx + b \cos nx)]$$
$$= ne^{-mx} (a \cos nx - b \sin nx) - my.$$
$$f = dv/dx = e^{-mx} [- n^2 (a \sin nx + b \cos nx) - mn (a \cos nx - b \sin nx)$$
$$- mn (a \cos nx - b \sin nx) + m^2 (a \sin nx + b \cos nx)]$$
$$= e^{-mx} [(m^2 - n^2) (a \sin nx + b \cos nx) - 2mn (a \cos nx - b \sin nx)]$$
$$= (m^2 - n^2) y - 2m (v + my)$$
$$= - (m^2 + n^2) y - 2mv.$$

190. Rolle's theorem. *If a function $y, = f(x)$, is differentiable at all points of an interval (a, b) and $f(a) = f(b)$, then there is at least one point of the interval, other than a or b, at which the differential coefficient vanishes, i.e. $Dy = 0$.*

The proof of this theorem depends on the following generalisation of the fundamental property of continuous functions:

If a function $f(x)$ is continuous at all points of an interval (a, b) and if M is the upper bound of $f(x)$ in the interval (i.e. the least number not exceeded by any value of $f(x)$ concerned) and m is the corresponding lower bound, then, whatever number k may be, from m to M inclusive, there is at least one point x of the interval (from a to b inclusive) at which $f(x) = k$.

To prove this we first remark that, if, besides being continuous, $f(x)$ is monotone throughout the interval, the theorem follows easily by the fundamental kind of argument of pp. 182—183 above, used in establishing the existence of the number $\varpi/2$.

Secondly, if $f(x)$ is not monotone, suppose for definiteness that $f(a) \leqslant k \leqslant M$. Consider the function $F(x)$, defined as the upper bound of $f(x)$ in the interval (a, x). This function is plainly monotone and continuous for all values of x from a to b and has $f(a)$ and M for its lower and upper bounds. Therefore there are values of x for which $F(x) = k$. Denote by c the lower bound of these values (or the single value if there is only one). We have $F(c) = k$.

It follows now that $f(c) = k$ also. For, if $f(c) \neq k$, say $f(c) = k' < k$, the point c, being a point of continuity of $f(x)$, could be enclosed in an interval throughout which $f(x) < (k + k')/2$ say; and $F(x)$, the upper bound of $f(x)$ in the interval (a, x), being less than k if $x < c$, would then also be less than k for $x = c$,—which is not the case. Hence $f(c) = k$.

The theorem is now established for the case when $f(a) \leqslant k \leqslant M$; it follows immediately for the other possible case,—when $m \leqslant k \leqslant f(a)$.

To deduce Rolle's theorem we remark in the first place that the function $f(x)$, being differentiable, is necessarily continuous throughout the interval. It follows from the theorem just proved that there must be values of x in the interval for which $f(x) = M$ and $f(x) = m$; and it is clear (because $f(a)$ and $f(b)$ are equal and lie between m and M) that at least one of these values must be different from both a and b. For this value the function is a maximum or a minimum and it follows (p. 232) that the differential coefficient there is neither positive nor negative, whence $Dy = 0$ for this value.

It is worth noting that the above theorems are still true if nothing is known about the behaviour of the function $f(x)$ at the end-points a, b of the interval except that $f(x)$ is continuous on the right at a, and continuous on the left at b.

If the function y is a polynomial,

$$y = a_n x^n + a_{n-1} x^{n-1} + \ldots + a_0,$$

and if $x = a$, $x = b$ are two roots of the equation $y = 0$, it follows from Rolle's theorem that $Dy = 0$ for at least one value of x between a and b. Hence *if the equation $Dy = 0$ has no roots in a certain range for x, the original equation $y = 0$ can at most have one root in that range.* Since the equation $Dy = 0$ is of degree one less than that of the equation $y = 0$, it is often possible to discuss the roots of this equation when the equation $y = 0$ cannot be so discussed, and thus obtain indirectly information as to the roots of the equation $y = 0$.

(i) For example, *consider the cubic equation*

$$x^3 + x - 2 = 0.$$

The derivative of $x^3 + x - 2$ equals

$$Dy = 3x^2 + 1.$$

The equation $Dy = 0$ has no real roots. Therefore the original cubic has at most one real root. Being of odd degree, it therefore has one such root (Ex. 5, p. 162). It is evidently $x = 1$.

(ii) Again, *consider the equation*

$$x^4 - 3x^2 + 2x + 4 = 0.$$

Writing y for the left-hand side, we have

$$Dy = 4x^3 - 6x + 2.$$

The roots of the equation $Dy = 0$, i.e. of $2x^3 - 3x + 1 = 0$, are $x = 1$ and the roots of the quadratic

$$\frac{2x^3 - 3x + 1}{x - 1} \equiv 2x^2 + 2x - 1 = 0;$$

i.e. $x = -\tfrac{1}{2} - \tfrac{1}{2}\sqrt{3}$ and $x = -\tfrac{1}{2} + \tfrac{1}{2}\sqrt{3}$.

That is, the equation $Dy = 0$ has roots

$$x = -\tfrac{1}{2} - \tfrac{1}{2}\sqrt{3}, \ x = -\tfrac{1}{2} + \tfrac{1}{2}\sqrt{3}, \text{ and } x = 1.$$

There is at most one root of $y = 0$ between any two of these three roots, or exceeding, or less than all these roots.

By the fundamental property of continuous functions, there is at least one root of $y = 0$ for $x < -\tfrac{1}{2} - \tfrac{1}{2}\sqrt{3}$ and at least one between $-\tfrac{1}{2} - \tfrac{1}{2}\sqrt{3}$ and $-\tfrac{1}{2} + \tfrac{1}{2}\sqrt{3}$ (because the values of y for these critical values of x are $7/4 - (3\sqrt{3})/2 < 0$ and $7/4 + (3\sqrt{3})/2 > 0$). Hence there is precisely one root in each of these ranges.

There is no other real root, for $y > 4$ for all values of x greater than $-\tfrac{1}{2} + \tfrac{1}{2}\sqrt{3}$.

The equation has therefore precisely two real roots, situated as described. Approximations to the actual values of the roots can now be obtained as accurately as desired by considering the value of y for different values of x in the ranges concerned. The graph of the function y is sketched in Fig. 11, p. 158.

191. The mean value theorem (for derivatives). A corollary of Rolle's theorem of very wide applicability is the theorem known as the *mean value theorem:*

If $y, = f(x)$, is a function of x, differentiable throughout an interval (a, b), there is at least one value of x between a and b $(a < x < b)$ for which $Dy = [f(b) - f(a)]/(b - a)$.

To prove this we need only consider the function †

$$F(x) = [f(x) - f(a)] - \left(\frac{x - a}{b - a}\right) [f(b) - f(a)].$$

$F(x)$ is differentiable throughout (a, b) and $F(a) = F(b) = 0$. Rolle's theorem therefore applies to $F(x)$.

The derivative of $F(x) = Dy - [f(b) - f(a)]/(b - a)$; therefore $Dy - [f(b) - f(a)]/(b - a) = 0$ for some value of x between a and b; i.e. $Dy = [f(b) - f(a)]/(b - a)$; and the theorem is proved.

Looked at geometrically, this theorem expresses the evident fact that the tangent at some point P of an arc AB of a curve is parallel to the chord AB‡.

The mean value theorem expresses the fact that, though it is in general not true that the incrementary ratio $[f(x + h) - f(x)]/h$ is equal to $Df(x)$, the limiting value of the incrementary ratio as $h \to 0$, yet there is a point between x and $x + h$ at which the differential coefficient equals the incrementary ratio $[f(x+h) - f(x)]/h$. This is useful in particular in expressing limits (e.g. the length of a curve) in the form of definite integrals, and in many questions involving limits.

Thus, *suppose $f(x)$ and $F(x)$ are two functions which are both continuous and differentiable in the neighbourhood of a point $x = a$, and both $= 0$ when $x = a$. We may require the limit*

$$\lim_{x \to a} f(x)/F(x).$$

We cannot say that this limit $= f(a)/F(a)$ for this apparent fraction is $0/0$, which is meaningless.

But we have $[f(x) - f(a)]/(x - a) = Df(x)$ at some point between

---

* The remarks on p. 237 about the behaviour of $f(x)$ at the end points (a, b) apply also to this theorem.

† If A, P, B are the points on the graph $y = f(x)$ corresponding to the abscissae a, x, b and if Q is the point in which the chord AB cuts the ordinate through P, the function $F(x)$ represents the distance QP.

‡ Cf. geometrical interpretation of $F(x)$, preceding footnote.

a and x, and $[F(x) - F(a)]/(x - a) = DF(x)$ at some (other) point between a and x.

If therefore the derivatives $Df(x)$ and $DF(x)$ are continuous at $x = a$, and if at $x = a$, $Df(x) = f'(a)$ and $DF(x) = F'(a)$, then, if $F'(a) \neq 0$, $\lim\limits_{x \to a} f(x)/F(x) = f'(a)/F'(a)$.

If for example, $f(x) = \sin x$, $F(x) = \sinh x$, $a = 0$; we have:
$Df(x) = \cos x$, $DF(x) = \cosh x$, $f'(0) = \cos 0 = 1$,
$$F'(0) = \cosh 0 = 1;$$
and therefore $\lim\limits_{x \to 0} (\sin x)/(\sinh x) = f'(0)/F'(0) = 1$. To such a case none of the rules for limits previously obtained will apply.

By applying Rolle's theorem to the function
$$\phi(x) \equiv f(x)[F(a) - F(b)] - F(x)[f(a) - f(b)],$$
where $f(x)$ and $F(x)$ are continuous and differentiable throughout the interval (a, b), we prove that $\dfrac{f(a) - f(b)}{F(a) - F(b)} = \dfrac{f'(x)}{F'(x)}$, where $f'(x)$ and $F'(x)$ are the differential coefficients of $f(x)$ and $F(x)$ at some point x of the interval, the *same* for $f'(x)$ and $F'(x)$.

If now, as before, $f(x)$ and $F(x)$ are continuous and differentiable in the neighbourhood of a point $x = a$, and $f(a) = F(a) = 0$, then, if x is any point in this neighbourhood, $\dfrac{f(x)}{F(x)} = \dfrac{f(x) - f(a)}{F(x) - F(a)} = \dfrac{f'(X)}{F'(X)}$, where X lies between a and x. If $\lim\limits_{X \to a} \dfrac{f'(X)}{F'(X)}$ exists and $= L$, $\lim\limits_{x \to a} \dfrac{f(x)}{F(x)}$ will also exist and equal L. We have established the more general theorem that $\lim\limits_{x \to a} \dfrac{f(x)}{F(x)}$ *exists and equals* $\lim\limits_{x \to a} \dfrac{f'(x)}{F'(x)}$ *if this latter limit exists.*

This theorem does not require that $f'(a)$, $F'(a)$, $\lim\limits_{x \to a} f'(x)$, $\lim\limits_{x \to a} F'(x)$ should exist.

EXAMPLES XXIII.

1. Find the vertex of the parabola given by $y = x^2 - 2x - 5$, and the minimum value of y. Find also the tangents to the curve at the points where it cuts the axes.

2. Shew that the tangent at any point P of the parabola $y = x^2$ makes the same angle with SP as it does with OY, S being the point $(0, \frac{1}{4})$.

3. The length (r) of the radius from the centre of an ellipse to a point on the ellipse is given by the formula
$$r^2 (\cos^2 \theta/a^2 + \sin^2 \theta/b^2) = 1,$$
where θ is the angle the radius makes with a fixed line through the centre (the major axis). Shew that the maximum and minimum values of r occur where $\theta = 0$ or π and $\theta = \pm \pi/2$; and those values are a and b.

4. Assuming that the path of a ray of light in a homogeneous medium is a straight line and that in passing from a point A in one medium (air) of refractive index 1 to a point B in another medium (glass) of refractive index μ, separated from the first by a plane face, the path APB is such that $AP + \mu . PB$ is the least possible, verify the usual law of refraction,—that if a and β are the angles (of incidence and refraction) which AP and PB make with the normal to the surface, $\sin a = \mu \sin \beta$.

5. By considering the function $(1+x)^n - nx$ prove $(1+x)^n > 1 + nx$ for all values of x for which $(1+x)^n$ is defined if $n > 1$ and is not an odd integer; and for $x > -1$ in all cases if $n > 1$. If $0 < n < 1$ the inequality is reversed. Consider also the cases when n is negative.

[Cf. inequality (ii) of p. 29; Ex. 1, p. 91; and Ex. 3, p. 132.]

6. Establish the inequality of Ex. 4, p. 36 by the method of this section.

7. Prove that, if x is any positive real number and m and n are positive integers such that $m \leqslant n$, $\dfrac{x^n - 1}{n} \geqslant \dfrac{x^m - 1}{m}$.

8. Find the maxima and minima of the cubic polynomial

$$y = \frac{5}{3} x^3 - 3x^2 + x - 1.$$

Shew (i) that of the three possible real roots of the equation $y = 0$ only one can satisfy each of the inequalities $x < 1/5$, $1/5 < x < 1$, $x > 1$; (ii) that there can be no root satisfying the first of these inequalities; and (iii) that the only real root of the equation is > 1.

Obtain this root correct to within ·1.

9. Use the mean value theorem to verify that if x is positive $\log_e (1+x)$ is positive and less than x. Prove in the same way (i) e^x lies between $1+x$ and $1/(1-x)$ if $0 < x < 1$; (ii) $(1+x)^n$ lies between 1 and $1 + nx$ if $x > 0$ and $n < 1$, or if $-1 < x < 0$ and $n > 1$, but lies outside these limits if $x > 0$ and $n > 1$ or if $-1 < x < 0$ and $n < 1$.

10. The velocity v of a meteor falling vertically to the earth is given approximately in terms of its height (y) above the earth's surface by the relation $v^2 = K/(R+y)$, where R is the radius of the earth (in appropriate units) and K is some constant. Shew that the meteor is subject to a variable acceleration towards the centre of the earth inversely proportional to the square of its distance from the centre of the earth.

11. Determine the limits

(a) $\lim_{x \to 0} (1 - \cos x)/\sin x$,   (b) $\lim_{x \to 0} [\log_e (1+x)]/x$,

(c) $\lim_{x \to 0} (e^x - 1)/x$,   (d) $\lim_{x \to 0} (e^x - 1)/[\log_e (1 - x)]$,

(e) $\lim_{x \to 0} \tanh x / \arcsin (x/2)$.

Deduce from (b) that $\lim_{x \to 0} (1+x)^{\frac{1}{x}} = e$.

12. The generalisation of the fundamental property of continuous functions (p. 236) can be proved alternatively by the method used in the proof of the theorem on derivatives on p. 119 above.

[To prove that there are values of x in the interval (a, b) for which $f(x) = M$,— the upper bound of $f(x)$ in (a, b),—bisect the interval. The upper bound of $f(x)$ in at least one of these half intervals must $= M$. Bisect this half interval, and so on. A sequence of intervals (as on p. 120) is obtained, converging to a limit point, a say. The upper bound of $f(x)$ in any neighbourhood of a is M. It follows from the continuity of $f(x)$ at a that $f(a) = M$, and the result is established. To prove that there are values of x for which $f(x) = k$, where $m \leqslant k \leqslant M$, a sequence of intervals is obtained in which there are values of x for which $f(x) \geqslant k$ and values for which $f(x) \leqslant k$. If a is the limit point $f(a) = k$.]

§ 6. THE DEFINITE INTEGRAL AND ITS EVALUATION

192. Integration in general. In Chapter III, § 1, in considering the question of areas we defined the definite integral. The definition is applicable to other questions. Briefly, where a geometrical, physical, or other quantity can be divided into an indefinitely large number of small portions (or elements) each of which can be measured approximately (or may be regarded as so divided), the measure of the whole quantity is given by a definite integral. Integration is the process of summing the elements to arrive at the total quantity. Areas, volumes, lengths of curves, moments of inertia, centres of mass, etc. provide common practical examples of questions of this kind; as also such questions as the determination of the distance traversed in a given time by a body moving with variable velocity. In mathematical physics most quantities are measured by means of definite integrals. In this section we consider the matter *ab initio*.

193. Function assumed bounded. Let $f(x)$ be a function of x defined for $a \leqslant x \leqslant b$, a and b being two real numbers $(a < b)$. We shall say that $f(x)$ is defined throughout the *interval* (a, b). Let $f(x)$ be *bounded* in this interval,—i.e. $|f(x)| <$ a fixed number K for all values of x in the interval*. Let M and m be the upper and lower bounds of the values of $f(x)$ in the interval.

194. Limits of larger and smaller sums. Suppose now the interval (a, b) is divided into any number of parts,—say by the

---

* Unbounded functions (such as x^{-1} or $x^{-\frac{1}{3}}$ in an interval including $x = 0$) are excluded from our discussion of integration. For such "infinite integrals" see Hardy's *Pure Mathematics*.

points (or numbers) $a, x_1, x_2, \ldots x_{n-1}, b$ (in order). In every one of the sub-intervals so formed, e.g. $(a, x_1), f(x)$ is bounded and has an upper and a lower bound (say M_1 and m_1). Moreover it is clear that $M_1 \leqslant M$ and $m_1 \geqslant m$. If we form the sum

$$S_1 = M_1(x_1 - a) + M_2(x_2 - x_1) + \ldots + M_n(b - x_{n-1}),$$

the terms of which are the products of the lengths of the various sub-intervals* and the upper bounds of $f(x)$ in those sub-intervals, we have $S_1 \leqslant M(b - a)$.

If now the sub-intervals $(a, x_1), (x_1, x_2), \ldots (x_{n-1}, b)$ are themselves divided in the same way, in each of the smaller parts thus formed, the upper bound of $f(x) \leqslant$ the upper bound in the corresponding sub-interval of the first system of division. The sum formed from this second system in the same way as S_1 was formed from the first system,—say

$$S_2 = M_1'(x_1' - a) + M_2'(x_2' - x_1') + \ldots + M_{n'}'(b - x_{n'-1}'),$$

—is clearly $\leqslant S_1$.

By further subdivision we get a third sum, S_3 say, and so on.

We thus obtain a sequence of "larger sums" (as we may call them),

$$S_1, \ S_2, \ S_3, \ \ldots \quad \ldots\ldots\ldots\ldots\ldots\ldots(1).$$

Similarly by taking the lower bounds in each part of the interval we obtain a sequence of "smaller sums,"

$$s_1, \ s_2, \ s_3, \ \ldots \quad \ldots\ldots\ldots\ldots\ldots\ldots(2).$$

Of these sequences we know

$$M(b - a) \geqslant S_1 \geqslant S_2 \geqslant \ldots$$

and

$$m(b - a) \leqslant s_1 \leqslant s_2 \leqslant \ldots$$

and it is evident that every larger sum $\geqslant$ every smaller sum.

We may write

$$m(b - a) \leqslant s_1 \leqslant s_2 \leqslant \ldots \leqslant S_2 \leqslant S_1 \leqslant M(b - a) \quad \ldots\ldots\ldots(3).$$

It follows that the sequence (1) $\searrow$ a unique limit, $\bar{S}$ say, and the sequence (2) $\nearrow$ a unique limit, $\underline{S}$ say; and $\bar{S} \geqslant \underline{S}$.

195. Upper and lower integrals. The definite integral. Definitions. So far this is true whatever system of division has been chosen, but the numbers $\bar{S}$ and $\underline{S}$ will depend on the way the division is carried out. We could make the numbers $\bar{S}$ and $\underline{S}$ definite if we

---

* The *length* of an interval (a, b) is $b - a$.

supposed that in the process of division the greatest of the parts is made to diminish indefinitely to the limit zero. It is not *a priori* evident that, even with this limitation, the numbers $\bar{S}$ and $\underline{S}$ are definite,—the same for all possible modes of division of the above type,—though this is in fact true.

In any case however the set of all possible larger sums is bounded below and has a definite lower bound, $\bar{I}$ say; and the set of smaller sums is bounded above and has an upper bound, $\underline{I}$ say; and $\bar{I} \geqslant \underline{I}$.

To prove that $\bar{I} \geqslant \underline{I}$ we observe that any "larger sum" whatever is greater than or equal to any "smaller sum"; for the upper bound of a (bounded) function in any interval $\geqslant$ the lower bound of the function in any other interval with which the first interval has any part in common, because the upper bound concerned $\geqslant$ the upper bound in the common part, which $\geqslant$ the lower bound in the common part, which $\geqslant$ the lower bound concerned.

$\bar{I}$ is called the *upper integral* of $f(x)$ over the interval and $\underline{I}$ the *lower integral*. Both exist for any bounded function in any bounded interval.

If the upper and lower integrals $\bar{I}$ and $\underline{I}$ coincide, $f(x)$ is said to be integrable over the interval, and the common value $\bar{I} = \underline{I} = I$ say, is called the definite integral of $f(x)$ over the interval (a, b), or between the limits a, b, or from a to b.

a and b are called respectively the *lower* and *upper limits of integration* and the interval (a, b) the *range of integration*.

It is evident that the function will certainly be integrable in this sense if the limits $\bar{S}$ and $\underline{S}$, obtained in the way described above, coincide, because $\underline{S} \leqslant \underline{I} \leqslant \bar{I} \leqslant \bar{S}$. The common value of $\bar{S}$ and $\underline{S}$ will then be the integral $\int_a^b f(x)\,dx$.

It can be proved, conversely, that if a function is integrable in this sense, the limits $\bar{S}$ and $\underline{S}$ are necessarily unique, identical, and equal to the integral I, *provided* the greatest of the parts in the division has been made to diminish indefinitely to the limit zero.

This definition thus agrees with that given in the last chapter (pp. 143—149). We use the notation

$$I = \int_a^b f(x)\,dx.$$

The definition can be modified to cover the case when $b < a$; or we *define* $\int_a^b f(x)\,dx$ when $b < a$ as $-\int_b^a f(x)\,dx$, supposing this latter integral exists.

196. Properties of the definite integral. From the definition we have the following general theorems[*]:

(i) *If $f(x)$ is integrable in an interval (a, b) and also in an interval (b, c), then it is also integrable in the interval (a, c) and*

$$\int_a^c f(x)\,dx = \int_a^b f(x)\,dx + \int_b^c f(x)\,dx.$$

(ii) *If $f_1(x)$ and $f_2(x)$ are two functions which are both integrable in an interval (a, b), then the sum or difference $f_1(x) \pm f_2(x)$ is also integrable and*

$$\int_a^b [f_1(x) \pm f_2(x)]\,dx = \int_a^b f_1(x)\,dx \pm \int_a^b f_2(x)\,dx.$$

(iii) *If $f(x)$ is integrable in (a, b) and k is a constant, then $kf(x)$ is integrable and*

$$\int_a^b kf(x)\,dx = k \int_a^b f(x)\,dx.$$

(iv) *If $f_1(x)$ and $f_2(x)$ are integrable in (a, b) the product $f_1(x) \times f_2(x)$ is also integrable in (a, b).*

The student will find it instructive to write out proofs of these simple fundamental theorems.

To prove (ii) on the basis of the definition of this section, without relying on geometrical intuition or on the unproved statements concerning the limits $\bar{S}$ and $\underline{S}$, we can proceed thus:

The result will follow if it can be proved in general that the upper integral of $f_1(x) + f_2(x)$, say $\bar{I}$, is less than or equal to the sum of the upper integrals of $f_1(x)$ and $f_2(x)$, say $\bar{I}_1 + \bar{I}_2$, and the lower integral of $f_1(x) + f_2(x)$, say $\underline{I}$, is greater than or equal to the sum of the lower integrals of $f_1(x)$ and $f_2(x)$, say $\underline{I}_1 + \underline{I}_2$. For then, $f_1(x)$ and $f_2(x)$ being integrable, we should have $\bar{I}_1 = \underline{I}_1 = I_1$ say, and $\bar{I}_2 = \underline{I}_2 = I_2$ say, and therefore $\bar{I} = \underline{I} = I_1 + I_2$.

To prove the first of the results stated, viz. $\bar{I} \leqslant \bar{I}_1 + \bar{I}_2$, we observe that $\bar{I}_1$ and $\bar{I}_2$ are the lower bounds of the "larger sums" associated with the functions $f_1(x)$ and $f_2(x)$, and therefore systems of division of the interval can be found for which these "larger sums" exceed

---

[*] These theorems are equally true whether $a < b$, etc. or not.

$\bar{I}_1$ and $\bar{I}_2$ respectively by arbitrarily little,—say less than ϵ. By combining the points of division of these two systems a single system of division is obtained for which the "larger sums" exceed their respective lower bounds ($\bar{I}_1$ and $\bar{I}_2$) by less than ϵ. For this system, the "larger sum" associated with the combined function $f_1(x) + f_2(x)$ is less than or equal to the sum of the two "larger sums" and therefore is $\leqslant \bar{I}_1 + \bar{I}_2 + \epsilon$. It follows that the lower bound of such "larger sums," viz. $\bar{I}$, is $\leqslant \bar{I}_1 + \bar{I}_2$.   Q.E.D.

The result for the lower integrals follows similarly.

A proof of theorem (iv) is given in Ex. 12, p. 260 below.

197. Integrability of bounded monotone function and of all elementary functions. So far in this section we have not proved that any functions are integrable. We will now prove theorems establishing the integrability of wide classes of functions,—including all the functions considered in this course, over any interval excluding points of discontinuity.

I. *If $f(x)$ is any (bounded) monotone function in an interval (a, b) then it is integrable in (a, b).*

For the difference between a "larger sum"

$$M_1(x_1 - a) + M_2(x_2 - x_1) + \ldots + M_n(b - x_{n-1})$$

and the corresponding "smaller sum"

$$m_1(x_1 - a) + m_2(x_2 - x_1) + \ldots + m_n(b - x_{n-1})$$

is

$$(M_1 - m_1)(x_1 - a) + (M_2 - m_2)(x_2 - x_1) + \ldots$$
$$+ (M_n - m_n)(b - x_{n-1}) \quad \ldots\ldots(4);$$

and,—supposing for definiteness that $f(x)$ is non-decreasing,—

$$m_1 = f(a), \ M_1 = f(x_1) = m_2, \ M_2 = f(x_2) = m_3, \ldots M_n = f(b).$$

Therefore, if the greatest of the parts,

$$x_1 - a, \ x_2 - x_1, \ldots b - x_{n-1}, \text{ equals } \Delta,$$

the difference (4)

$$\leqslant \Delta \left[(M_1 - m_1) + (M_2 - m_2) + \ldots + (M_n - m_n) \right]$$
$$= \Delta \left[f(x_1) - f(a) + f(x_2) - f(x_1) + \ldots + f(b) - f(x_{n-1}) \right]$$
$$= \Delta \left[f(b) - f(a) \right],$$

which $\to 0$ as $\Delta \to 0$; and it follows that the lower bound of the larger sums and the upper bound of the smaller sums coincide; i.e. $\bar{I} = \underline{I}$.

From theorem (i) above it now follows that any (bounded) function which is monotone in stretches throughout an interval* is integrable in that interval. From theorems (ii) and (iii) any function which is a linear combination of such functions is also integrable. From theorem (iv) and the fact that the reciprocal of an integrable function is integrable over any interval in which $f(x) \neq 0$ it follows that any arithmetical combination of integrable functions is integrable over any interval (excluding points where the function is not defined). *All the particular functions considered in this course are of these types in any interval excluding all points of discontinuity.*

All functions of bounded variation are integrable, in direct consequence of theorems (ii), p. 245, and I, p. 246, and the definition of p. 138 above.

198. The integrability of continuous functions. The proof of the integrability (in appropriate intervals) of the functions of elementary analysis can be made to rest alternatively on the following theorem :

II. *If $f(x)$ is any continuous function in an interval (a, b) then it is integrable in (a, b).*

The proof rests on proving that the difference (4) $\rightarrow 0$ as $\Delta \rightarrow 0$. If the greatest of the differences

$$M_1 - m_1, \ M_2 - m_2, \ \dots \ M_n - m_n$$

is δ,—the difference (4)

$$\leqslant \delta [x_1 - a + x_2 - x_1 + \dots + b - x_{n-1}]$$
$$= \delta (b - a).$$

In virtue of the continuity of $f(x)$ it can be proved that $\delta \rightarrow 0$ as $\Delta \rightarrow 0\dagger$. Hence the difference (4) $\rightarrow 0$ and the theorem follows.

199. It is neither evident nor true that a function defined in a certain range as the sum of a convergent infinite series of monotone or continuous functions is itself monotone or continuous (or is a linear combination of such functions). But it is true that the sum-function of any *power* series is both monotone or expressible as a linear combination of monotone functions,—i.e. of bounded variation,—and continuous, throughout any interval wholly interior to the range of convergence of the series,—though not necessarily if the interval includes the whole range of convergence. These facts are proved above (pp. 171—172 and Ex. 14, p. 195). Any such function is therefore integrable in any such interval.

---

* I.e. such that the interval can be divided into a number of sub-intervals in each of which the function is monotone.

† See Ex. 13, p. 260 below.

200. Functions assumed integrable and continuous. Throughout the remainder of this section we shall suppose the function $f(x)$ under consideration to be integrable *and continuous.*

201. The integral as a function of the upper limit. The fundamental theorem. Let us consider now the definite integral $\int_a^X f(x)\,dx$ where $a < X < b$ and $f(x)$ is integrable and continuous in the range (a, b). Drawing the graph of $f(x)$ (Fig. 23) we see that

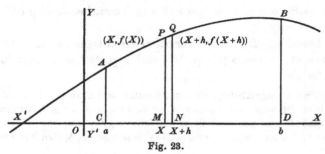

Fig. 23.

$\int_a^X f(x)\,dx$ represents the area $ACMP$. For brevity write

$$I(X) = \int_a^X f(x)\,dx = \text{area } ACMP.$$

This area, or integral $I(X)$, depends on X and is a function of X.

The integral $I(X) = \int_a^X f(x)\,dx$ is also continuous and differentiable with respect to X for any value of X between a and b; and $DI(X) = f(X)$.

For $\quad I(X+h) - I(X) = \text{area } ACNQ - \text{area } ACMP$

$$= \text{area } PMNQ$$

$$\leqslant MN \cdot U = h \cdot U,$$

if U is the upper bound of $f(x)$ in the range $(X, X+h)$,—i.e. over PQ. This $\to 0$ as $h \to 0$ and the continuity is established.

Moreover the incrementary ratio

$$[I(X+h) - I(X)]/h = (\text{area } PMNQ)/h \leqslant U$$

and similarly $\geqslant u$,—the lower bound of $f(x)$ in $(X, X+h)$.

But as $h \to 0$, $U \to f(X)(= MP)$ and $u \to f(X)$.

Therefore the incrementary ratio

$$[I(X+h) - I(X)]/h \to f(X),$$

or the integral $I(X)$ is differentiable with respect to X and $DI(X) = f(X)$.

202. The inverse character of integration and differentiation. This result,—which is an analytical theorem clearly independent of the graphical representation,—establishes what we may call *the inverse character of integration and differentiation*. It is the most important property possessed by integrals and is known as *the fundamental theorem of the integral calculus*. By means of this theorem we are enabled to integrate, i.e. to evaluate the definite integrals of, many of the common functions of mathematics. The theorem is true under wider conditions than those here supposed,— $f(x)$ integrable and continuous,—but, as stated, it is sufficient for all the needs of elementary analysis.

203. Tentative evaluation of definite integrals. To evaluate a definite integral,—say $\int_0^1 x^2 \, dx$,—we might now argue that the integral $I(X) \equiv \int_0^X x^2 \, dx$ is differentiable and $DI(X) = X^2$.

But $$D(X^3/3) = X^2.$$

If therefore $X^3/3$ were the only function whose derivative is X^2, it would follow that $I(X) = X^3/3$ and thence that $\int_0^1 x^2 \, dx = I(1) = 1/3$, —which is in fact the correct result.

There are, however, any number of functions having the same derivative.

Thus $$D(1 + X^3/3) = D(-2 + X^3/3) = D(X^3/3) = X^2.$$

The suggested argument is therefore not sound. Nevertheless we notice that the examples given of functions with the same derivative X^2, though not identical, differ from one another by mere constants; and the student will convince himself that this must be so.

204. Theorem on functions with same derivative. It is in fact true that:

If $f_1(x)$ and $f_2(x)$ are any two functions whose derivatives $Df_1(x)$ and $Df_2(x)$ are equal for all values of x in a given range, then $f_1(x)$ and $f_2(x)$ differ by a constant (or are equal) throughout the range.

This follows at once from the theorem of p. 121 above, viz.

If the derivative $Df(x)$ of a function $f(x)$ is zero throughout a range, then $f(x)$ is constant throughout the range;

for if $Df_1(x) = Df_2(x)$, $D[f_1(x) - f_2(x)] = 0$ and this theorem will shew that $f_1(x) - f_2(x) = $ constant.

This last theorem may be proved alternatively by using the mean value theorem of p. 239 above, thus:

If the values of $f(x)$ at any two points a, b of the range were different, the derivative $Df(x)$ at some point between a and b $= [f(b) - f(a)]/(b - a) \neq 0$, and the theorem follows by *reductio ad absurdum**.

205. We can now remodel the argument by which we sought to determine $\int_0^1 x^2\, dx$ thus:

$$DI(X) = X^2 \text{ and } D(X^3/3) = X^2;$$

therefore $I(X) = X^3/3$ or differs from $X^3/3$ by a constant.

Say $\qquad\qquad\qquad I(X) = X^3/3 + C.$

It is now possible to determine the constant C from other considerations; for we know that as $X \to 0$, $I(X) \to 0$; and therefore in this case $C = 0$.

We have proved therefore that

$$\int_0^X x^2\, dx = X^3/3 \text{ and that } \int_0^1 x^2\, dx = 1/3.$$

206. The fundamental formula. The indefinite integral. This process is evidently general. We have arrived at the standard method of evaluating definite integrals:

To evaluate the definite integral $\int_a^b f(x)\, dx$ of a function $f(x)$ (supposed to be continuous and integrable) we first find any function $F(x)$ whose derivative $DF(x) = f(x)$ for all values of x in the range (a, b). We then have

$$\int_a^b f(x)\, dx = F(b) - F(a).$$

For $\qquad\qquad \int_a^X f(x)\, dx = F(X) + \text{constant}$

$$= F(X) - F(a)$$

because $\qquad \int_a^X f(x)\, dx \to 0$ when $X \to a$.

---

* The theorem of p. 119 may also be proved by this method.

Such a function as $F(x)$ is an *inverse derivative* of $f(x)$. In view of the importance of such functions in integration they are also called *indefinite integrals*. *The indefinite integral of $f(x)$ is any function $F(x)$ which has $f(x)$ for its derivative at all points of a range. The indefinite integral of $f(x)$ is denoted by $\int f(x)\,dx$.* As the name itself implies, the indefinite integral is not a definite unique function.

The formula
$$\int_a^b f(x)\,dx = F(b) - F(a),$$
which is often written
$$\int_a^b f(x)\,dx = \left[F(x)\right]_{x=a}^{x=b} \text{ or } \left[F(x)\right]_a^b,$$
is called the *fundamental formula* of the integral calculus.

In view of the fact (Theorem II, p. 247) that a continuous function is necessarily integrable, it is seen that the fundamental formula will always apply if the function $f(x)$ is known to be continuous throughout the range concerned, $F(x)$ being any function whose derivative is $f(x)$ throughout the range.

207. Standard indefinite integrals. The evaluation of definite integrals depends on the determination of indefinite integrals. We have at once from the list of differential coefficients of § 4 above, a corresponding *list of indefinite integrals*:

I. $\int x^n\,dx = x^{n+1}/(n+1)$ for all values of n other than -1;—valid for all values of x for which x^n is defined.

II. $\int x^{-1}\,dx = \log_e x$;—valid for $x > 0$. If $x < 0$, $\int x^{-1}\,dx = \log_e(-x)$.

Or, for all x except 0, $\int x^{-1}\,dx = \log_e |x|$.

III. $\int e^x\,dx = e^x$;—valid for all x.

IV. (a) $\int \cos x\,dx = \sin x$, $\int \sin x\,dx = -\cos x$;—valid for all x.

$\int \sec^2 x\,dx = \tan x$, $\int \dfrac{\sin x\,dx}{\cos^2 x} = \sec x$;—valid for all values of x except odd multiples of $\pm\,\pi/2$.

$$\int \operatorname{cosec}^2 x\, dx = -\cot x, \quad \int \frac{\cos x\, dx}{\sin^2 x} = -\operatorname{cosec} x;\text{—valid for all}$$

values of x except 0 and multiples of $\pm\, \pi$.

(b) $\displaystyle\int C(x)\, dx = S(x), \int S(x)\, dx = -C(x);$—valid for all values

of x.

$$\int \frac{dx}{[C(x)]^2} = T(x), \int \frac{S(x)\, dx}{[C(x)]^2} = \frac{1}{C(x)};\text{—valid for all values}$$

of x except odd multiples of $\pm\, \varpi/2$.

$$\int \frac{dx}{[S(x)]^2} = -\frac{1}{T(x)}, \int \frac{C(x)\, dx}{[S(x)]^2} = -\frac{1}{S(x)};\text{—valid for all}$$

values of x except 0 and multiples of $\pm\, \varpi$.

V. (a) $\displaystyle\int \frac{dx}{\sqrt{(1-x^2)}} = \arcsin x, \int \frac{dx}{\sqrt{(1-x^2)}} = -\arccos x;$—valid for

$-1 < x < 1$.

$$\int \frac{dx}{1+x^2} = \arctan x, \int \frac{dx}{1+x^2} = -\operatorname{arc cot} x;\text{—valid for all}$$

values of x.

$$\int \frac{dx}{x\sqrt{(x^2-1)}} = \operatorname{arc sec} x, \int \frac{dx}{x\sqrt{(x^2-1)}} = -\operatorname{arc cosec} x;\text{—valid for}$$

$|x| > 1$.

(b) $\displaystyle\int \frac{dx}{\sqrt{(1-x^2)}} = \bar{S}(x), \int \frac{dx}{\sqrt{(1-x^2)}} = -\bar{C}(x);$—valid for $-1 < x < 1$.

$$\int \frac{dx}{1+x^2} = \bar{T}(x);\text{—valid for all values of } x.$$

VI. $\displaystyle\int \cosh x\, dx = \sinh x, \int \sinh x\, dx = \cosh x, \int \operatorname{sech}^2 x\, dx = \tanh x.$

$$\int \frac{\sinh x\, dx}{\cosh^2 x} = -\operatorname{sech} x;\text{—valid for all } x.$$

$$\int \operatorname{cosech}^2 x\, dx = -\coth x, \int \frac{\cosh x\, dx}{\sinh^2 x} = -\operatorname{cosech} x;\text{—valid for } x \neq 0.$$

$$\int \frac{dx}{\sqrt{(x^2+1)}} = \operatorname{arg\, sinh} x;\text{—valid for all } x.$$

$$\int \frac{dx}{\sqrt{(x^2-1)}} = \operatorname{arg\, cosh} x;\text{—valid for } x > 1.$$

$$\int \frac{dx}{1-x^2} = \operatorname{arg\, tanh} x;\text{—valid for } |x| < 1.$$

$$\int \frac{dx}{1 - x^2} = \text{arg coth } x;\text{—valid for } |x| > 1.$$

$$\int \frac{dx}{x \sqrt{(1 - x^2)}} = -\text{arg sech } x;\text{--valid for } 0 < x < 1.$$

$$\int \frac{dx}{x \sqrt{(x^2 + 1)}} = -\text{arg cosech } x;\text{—valid for } x \neq 0.$$

Note: It is usually more convenient to use *alternative forms involving logarithms* in preference to inverse hyperbolic functions; thus

$$\int \frac{dx}{1 - x^2} = -\int \frac{dx}{x^2 - 1} = \tfrac{1}{2} \log_e \frac{1 + x}{1 - x}, \text{ valid for } |x| < 1; \text{ and}$$

$$\int \frac{dx}{1 - x^2} = -\int \frac{dx}{x^2 - 1} = \tfrac{1}{2} \log_e \frac{x + 1}{x - 1}, \text{ valid for } |x| > 1.$$

VII. $\int (a_0 + a_1 x + a_2 x^2 + \dots)\, dx = a_0 x + \tfrac{1}{2} a_1 x^2 + \tfrac{1}{3} a_2 x^3 + \dots;$

—valid for x within the range of convergence of either series.

In this list, in order to have the general form of the indefinite integral, *an arbitrary constant must be added to each result.*

208. Evaluation of definite integrals. The definite integral of any of the above functions over any range within which the function is known to be integrable can now be written down at once.

Thus to evaluate $\int_1^2 x^2\, dx$; the indefinite integral $\int x^2\, dx = x^3/3$ (+ arbitrary constant C); and therefore, from the fundamental formula of p. 251,

$$\int_1^2 x^2\, dx = 2^3/3 - 1^3/3 = 7/3.$$

Or $\qquad \int_{-1}^1 x^2\, dx = 1^3/3 - (-1)^3/3 = 2/3.$

Or again to find the area of a wave of the graph of $y = \sin x$ (cf. Fig. 15, p. 185),—i.e. to evaluate the definite integral

$$\int_0^\pi \sin x\, dx,$$

we have $\qquad \int \sin x\, dx = -\cos x + C$

and therefore $\qquad \int_0^\pi \sin x\, dx = (-\cos \pi) - (-\cos 0) = 2.$

209. Alternative forms for indefinite integrals. In the above table, in V and VI, alternative forms are given for the same integral; for example $\int \dfrac{dx}{\sqrt{(1-x^2)}} = $ arc sin x and also $= -$ arc cos x.

This arises from the essential indefiniteness of the indefinite integral. The two functions arc sin x and $-$ arc cos x differ by a constant, for if arc sin $x = y_1$, and arc cos $x = y_2$, sin $y_1 = x$ and cos $y_2 = x$; and therefore y_1 and y_2 are complementary (we know $-\pi/2 \leqslant y_1 \leqslant \pi/2$ and $0 \leqslant y_2 \leqslant \pi$); i.e. $y_1 + y_2 = \pi/2$; or

$$\text{arc sin } x - (-\text{ arc cos } x) = y_1 + y_2 = \pi/2.$$

Both forms are included in the general forms

$$\text{arc sin } x + C, \ -\text{ arc cos } x + C,$$

C being an arbitrary constant.

Either form will of course give the same result for a definite integral deduced from it.

210. Integration by substitution and by parts. From the above table of standard forms the indefinite integrals of other functions can to some extent be deduced,—by use of theorems similar to those of p. 245 above and by other methods.

There are two methods of frequent applicability by which an integral may be transformed into a simpler one and thence evaluated;—*integration by substitution* and *integration by parts*.

Integration by substitution rests on the law for the differentiation of a function of a function (p. 219) and integration by parts on the law for the differentiation of a product (p. 218).

Thus, if y is a function of x and x may be regarded as a function of another variable u, we have (assuming the existence of all the derivatives and integrals involved)

$$\frac{d}{du}\left(\int y\,dx\right) = \frac{d}{dx}\left(\int y\,dx\right)\frac{dx}{du} = y\frac{dx}{du};$$

i.e. $y\dfrac{dx}{du}$ is the derivative with respect to u of the function $\int y\,dx$ (or any one of these indefinite integrals).

Therefore $\int y\,dx = \int y\dfrac{dx}{du}\,du$,—*the formula for integration by substitution*.

Thus, e.g., if in $\int \dfrac{dx}{\sqrt{(a^2 - x^2)}}$ we put $x = a \sin u$, so that

$$\sqrt{(a^2 - x^2)} = a\sqrt{(1 - \sin^2 u)} = a \cos u \text{ and } dx/du = a \cos u,$$

we have $\int \dfrac{dx}{\sqrt{(a^2 - x^2)}} = \int \dfrac{1}{a \cos u} a \cos u\, du = u = \text{arc } \sin (x/a).$

This result verifies the first result of V above.

Or again $\quad \int e^{ax}\, dx = \int e^{ax} \dfrac{dx}{d(ax)} d(ax) = \dfrac{1}{a} e^{ax};$

or again

$$\int \frac{dx}{x - a} = \int \frac{1}{x - a} \frac{dx}{d(x - a)} d(x - a) = \int \frac{d(x - a)}{x - a} = \log_e (x - a).$$

The formula for the differentiation of a product is

$$\frac{d\,uv}{dx} = u \frac{dv}{dx} + v \frac{du}{dx}$$

which gives at once

$$uv = \int u \frac{dv}{dx}\, dx + \int v \frac{du}{dx}\, dx,$$

whence *the formula for integration by parts*:

$$\int u \frac{dv}{dx}\, dx = uv - \int v \frac{du}{dx}\, dx.$$

For example

$$\int x \cos x\, dx = \int x \frac{d \sin x}{dx}\, dx$$

$$= x \sin x - \int \sin x\, dx,$$

from the formula (putting $u = x$, $v = \sin x$),

$$= x \sin x + \cos x.$$

The use of this formula depends on expressing the function to be integrated in the form $u \dfrac{dv}{dx}$ and on the integration of the function $v \dfrac{du}{dx}.$

In both these formulae we tacitly suppose that the functions concerned are differentiable (and integrable) as required. In the practical work of evaluation of indefinite integrals these assumptions need not be verified. If there is any doubt as to the validity of either formula in any particular case the result is easily verified by direct differentiation.

211. Integration of rational functions. The integration of elementary compound functions is not so systematic as is differentiation. Certain wide classes of functions however can always be integrated. It is also always possible to integrate any function if it can be expressed as a power series, by standard form VII above,—a fact which is often useful in practical work.

If a rational function $R(x)$ can be expressed as a sum of partial fractions of the form

$$R(x) = A_1/(x - a_1) + A_2/(x - a_2) + \ldots + A_n/(x - a_n),$$

we have at once

$$\int R(x)\, dx = A_1 \log_e (x - a_1) + A_2 \log_e (x - a_2) + \ldots + A_n \log_e (x - a_n).$$

Or the function $1/(x^2 + px + q)$, p and q being constants, can be rewritten as $1/[(x + p/2)^2 + (q - p^2/4)]$.

If $q - p^2/4 > 0$ and $= a^2$ say, form V above gives

$$\int \frac{dx}{(x + p/2)^2 + a^2} = \frac{1}{a} \text{arc tan} \, [(x + p/2)/a],$$

and the function is integrated.

If $q - p^2/4 < 0$ and $= -a^2$, form VI above gives

$$\int \frac{dx}{(x + p/2)^2 - a^2} = -\frac{1}{a} \text{arg tanh} \, [(x + p/2)/a]$$

$$\left(\text{or} \ \frac{1}{2a} \log_e \frac{x + \dfrac{p}{2} - a}{x + \dfrac{p}{2} + a} \right),$$

and the function is integrated.

If $\qquad q = p^2/4, \ \displaystyle\int \frac{dx}{(x + p/2)^2} = -1/(x + p/2).$

Rational functions may usually be integrated by these methods.

212. Again the function $1/(\sqrt{(x^2 + px + q)})$ can be integrated similarly and we have

$$\int \frac{dx}{\sqrt{(x^2 + px + q)}} = \text{arg sinh} \, [(x + p/2)/a],$$

where $\qquad\qquad a^2 = q - p^2/4$ if $q > p^2/4$,

or $\quad = \text{arg cosh} \, [(x + p/2)/a]$, where $a^2 = p^2/4 - q$ if $q < p^2/4$,

or $\quad = \log_e (x + p/2)$ if $q = p^2/4$.

Other integrations will be found in the examples.

EXAMPLES XXIV.

[In these examples all letters except x denote constants. The ranges of validity of the results should be supplied.]

1. Verify the following integrations:

(i) $\int \dfrac{dx}{x^2-a^2}=\dfrac{1}{2a}\log_e\dfrac{x-a}{x+a}$;

$\left(\text{Reconcile this with the alternative form } -\dfrac{1}{a}\text{ arg coth }\dfrac{x}{a}\,.\right)$

(ii) $\int \dfrac{dx}{x^2+a^2}=\dfrac{1}{a}\text{ arc tan }\dfrac{x}{a}$ and $=-\dfrac{1}{a}\text{ arc cot }\dfrac{x}{a}$;

(iii) $\int \tan x\,dx=-\log_e\cos x$;

(iv) $\int \cot x\,dx=+\log_e\sin x$;

(v) $\int a^x\,dx=a^x\log_a e=a^x/\log_e a$;

(vi) $\int \dfrac{dx}{a+bx}=\dfrac{1}{b}\log_e(a+bx)$;

(vii) $\int \sqrt{(x+a)}\,dx=\dfrac{2}{3}(x+a)^{\frac{3}{2}}$.

2. Establish:

(i) $\int \cos 2x\,dx=\dfrac{1}{2}\sin 2x$;   (ii) $\int \sin 2x\,dx=-\dfrac{1}{2}\cos 2x$;

(iii) $\int \cos nx\,dx=\dfrac{1}{n}\sin nx$;   (iv) $\int \sin nx\,dx=-\dfrac{1}{n}\cos nx$;

(v) $\int \cos^2 x\,dx=\dfrac{1}{2}\sin x\cos x+\dfrac{1}{2}x$;

(vi) $\int \sin^2 x\,dx=-\dfrac{1}{2}\cos x\sin x+\dfrac{1}{2}x$;

(vii) $\int \cos^3 x\,dx=\dfrac{1}{3}\sin x\cos^2 x+\dfrac{2}{3}\sin x$;

(viii) $\int \sin^3 x\,dx=-\dfrac{1}{3}\cos x\sin^2 x-\dfrac{2}{3}\cos x$;

(ix) $\int \sin^5 x\,dx=-\cos x+\dfrac{2}{3}\cos^3 x-\dfrac{1}{5}\cos^5 x$.

3. Prove:

(i) $\int \tanh x\,dx=\log_e\cosh x$;   (ii) $\int \coth x\,dx=\log_e\sinh x$;

(iii) $\int \text{sech } x\,dx=2\text{ arc tan }e^x$;   (iv) $\int \text{cosech } x\,dx=\log_e\tanh\dfrac{1}{2}x$.

4. By integrating by substitution prove:

(i) $\int \dfrac{x}{x^2+a^2}\,dx = \dfrac{1}{2}\log_e(x^2+a^2)$ [Substitution: $x^2=u$];

(ii) $\int \dfrac{x}{\sqrt{(a^2-x^2)}}\,dx = -\sqrt{(a^2-x^2)}$ [Substitution: $x^2=u$ or $x=a\sin u$];

(iii) $\int \dfrac{dx}{(1-x)^3} = \dfrac{1}{2(1-x)^2}$ [Substitution: $1-x=u$];

(iv) $\int \operatorname{cosec} x\,dx = \log_e \tan(x/2)$ and

$\int \sec x\,dx = \log_e \tan(\pi/4+x/2)$ [Substitution: $\tan x/2 = u$];

(v) $\int \dfrac{dx}{a+b\cos x} = \dfrac{2}{\sqrt{(a^2-b^2)}}\arctan\left(\sqrt{\dfrac{a-b}{a+b}}\tan\dfrac{x}{2}\right)$ if $a>b$ [Substitution: $\tan(x/2)=u$]. Determine the integral also if $a<b$ and if $a=b$.

(vi) $\int \dfrac{dx}{a^2\cos^2 x+b^2\sin^2 x} = \dfrac{1}{ab}\arctan\left(\dfrac{b}{a}\tan x\right)$

[Substitution: $\tan x=u$].

5. By integrating by parts prove:

(i) $\int \arctan x\,dx = x\arctan x - \dfrac{1}{2}\log_e(1+x^2)$;

(ii) $\int \arcsin x\,dx = x\arcsin x + \sqrt{(1-x^2)}$;

(iii) $\int \arccos x\,dx = x\arccos x - \sqrt{(1-x^2)}$;

(iv) $\int \log_e x\,dx = x\log_e x - x$;

(v) $\int xe^{ax}\,dx = \dfrac{1}{a}xe^{ax} - \dfrac{1}{a^2}e^{ax}$;

(vi) $\int x^2 e^{-x}\,dx = -x^2 e^{-x} - 2xe^{-x} - 2e^{-x}$;

(vii) $\int x\sin x\,dx = \sin x - x\cos x$;

(viii) $\int x\cos x\,dx = \cos x + x\sin x$;

(ix) $\int x^2\cos ax\,dx = \dfrac{1}{a}x^2\sin ax + \dfrac{2}{a^2}x\cos ax - \dfrac{2}{a^3}\sin ax$;

(x) $\int x^2\sin ax\,dx = -\dfrac{1}{a}x^2\cos ax + \dfrac{2}{a^2}x\sin ax + \dfrac{2}{a^3}\cos ax$;

(xi) $\int \operatorname{arc\,sec} x\,dx = x\operatorname{arc\,sec} x - \operatorname{arg\,cosh} x$.

6. Prove:

(i) $\int e^{ax} \cos bx \, dx = e^{ax} (b \sin bx + a \cos bx)/(a^2 + b^2)$;

(ii) $\int e^{ax} \sin bx \, dx = e^{ax} (a \sin bx - b \cos bx)/(a^2 + b^2)$;

(iii) $\int \dfrac{dx}{a^2 - b^2 x^2} = \dfrac{1}{2ab} \log_e [(a + bx)/(a - bx)]$;

(iv) $\int \dfrac{x^4 + 4x^3 + 5x^2 + 6x}{(x^2 - 1)^2 (x + 1)} \, dx = \log_e (x - 1) - 2/(x - 1) + 1/[2 (x + 1)^2]$;

[Express the function to be integrated in the form of a sum of partial fractions of the form

$$A/(x - 1) + B/(x - 1)^2 + C/(x + 1) + D/(x + 1)^2 + E/(x + 1)^3.$$

By multiplying up and identifying the numerators by equating the coefficients, $A = 1$, $B = 2$, $C = 0$, $D = 0$, $E = -1$. The integration follows.]

(v) $\int \dfrac{x^2 + 1}{x^3 - 3x^2 + 2x} \, dx = \dfrac{1}{2} \log_e x - 2 \log_e (x - 1) + \dfrac{5}{2} \log_e (x - 2)$;

(vi) $\int \dfrac{1 + x - 3x^2 + 3x^3}{(1 + x^4)(1 - x)^3} \, dx = 1/[2(1 - x)^2] + \dfrac{1}{2} \arctan x^2$;

(vii) $\int \dfrac{dx}{\sqrt{(1 + x^3)}} = x - \dfrac{1}{2} \dfrac{x^4}{4} + \dfrac{1 \cdot 3}{2 \cdot 4} \dfrac{x^7}{7} - \dots$, valid for $|x| < 1$.

7. Any rational function is expressible as a sum of partial fractions of the types

$A_1/(x - a)$, $A_2/(x - a)^2$, $\dots$; $(B_1 x + C_1)/(x^2 + px + q)$, $(B_2 x + C_2)/(x^2 + px + q)^2$, $\dots$.
Of these types all except those with a quadratic denominator raised to a power $\geqslant 2$ can be dealt with by the above methods. Shew how such a function as $(Bx + C)/(x^2 + px + q)^2$ can be integrated.

[The numerator $Bx + C$ can be expressed as $a(x^2 + px + q) + (bx + c)(2x + p)$ and the function then $= a/(x^2 + px + q) - (bx + c) \dfrac{d}{dx} \left(\dfrac{1}{x^2 + px - q} \right)$. The first of these two expressions is integrable as above. The second is reduced to a similar form by integration by parts.

The integration of a rational function thus depends on factorising the denominator into linear and quadratic factors (simple or repeated), for then the function can be expressed in partial fractions of the above type,—as in Ex. 6 (iv). This factorisation can often (but not always) be carried out. Other methods of integrating rational functions are often more convenient (e.g., in particular, by the use of imaginary or complex factors).]

8. Evaluate the definite integrals

$\displaystyle\int_1^2 \dfrac{1}{x^2} \, dx$, $\displaystyle\int_{-1}^1 e^x \, dx$, $\displaystyle\int_0^\pi e^{-x} \sin x \, dx$, $\displaystyle\int_1^{10} \log_e x \, dx$,

$\displaystyle\int_0^1 \dfrac{1}{1 + x^2} \, dx$, $\displaystyle\int_0^\pi \sin (x/n) \, dx$, $\displaystyle\int_0^{\pi/4} \tan x \, dx$, $\displaystyle\int_0^{\pi/2} \sin^2 x \, dx$,

$\displaystyle\int_0^{\pi/2} \sin^2 x \cos^3 x \, dx$, $\displaystyle\int_0^{\pi/2} \sin^4 x \cos^4 x \, dx$.

9. Shew that $\displaystyle\int_0^{\frac{1}{2}} \frac{dx}{\sqrt{(1-x^2)}} = \pi/6 = \varpi/6.$

By expanding $(1-x^2)^{-\frac{1}{2}}$ in a power series and integrating term by term deduce that

$$\pi = \varpi = 6\left[\frac{1}{2} + \frac{1}{2}\frac{1}{3 \cdot 2^3} + \frac{1 \cdot 3}{2 \cdot 4}\frac{1}{5 \cdot 2^5} + \dots\right].$$

10. Establish the identity of the circular and trigonometrical functions $\sin x$ and $S(x)$ (and thence also of the derived functions) from the relations

$$\text{arc } \sin X = \int_0^X \frac{dx}{\sqrt{(1-x^2)}} = \overline{S}(X).$$

11. The function $1/x^2$ is not defined and is discontinuous at $x=0$. It is also not integrable in any range including the point $x=0$. The argument

$$\int_{-1}^1 \frac{dx}{x^2} = \left(-\frac{1}{1}\right) - \left(-\frac{1}{-1}\right) = -2$$

is not valid.

12. Prove that if two (bounded) functions $f_1(x)$ and $f_2(x)$ are integrable and positive throughout a range (a, b), then the product $f_1(x)f_2(x)$ is also integrable. Deduce the same result whether the functions are positive or not.

[If M_1' and M_1'' are the upper bounds of $f_1(x)$ and $f_2(x)$ in the part (a, x_1) of the range (p. 243), the upper bound of $f_1(x)f_2(x)$ in $(a, x_1) \leqslant M_1' M_1''$. With similar notation, the lower bound of $f_1(x)f_2(x)$ in $(a, x_1) \geqslant m_1' m_1''$. Therefore the difference between the upper and lower bounds of $f_1(x)f_2(x)$ in (a, x_1) $\leqslant M_1' M_1'' - m_1' m_1'' = M_1'(M_1'' - m_1'') + m_1''(M_1' - m_1') < K(M_1'' - m_1'' + M_1' - m_1')$, where K is any number exceeding all the values of $f_1(x)$ and $f_2(x)$ in (a, b). Similarly for all the other parts (x_1, x_2) etc. The difference between the larger and smaller sums for the function $f_1(x)f_2(x)$ therefore $<$ the sum of terms such as $K(M_1'' - m_1'')(x_1 - a) + K(M_1' - m_1')(x_1 - a)$, and therefore $\rightarrow 0$ as $\Delta \rightarrow 0$, because of the integrability of $f_1(x)$ and $f_2(x)$.

When $f_1(x)$ and $f_2(x)$ are not restricted to be positive, a constant K can be found so that $K + f_1(x)$ and $K + f_2(x)$ are entirely positive. Theorems (ii) and (iii) of p. 245 complete the proof.]

13. Prove that if a function $f(x)$ is continuous at all points of an interval (a, b), then the difference between the upper and lower bounds of $f(x)$ in any and every part of the interval of width $\leqslant \Delta$ tends to 0 as $\Delta \rightarrow 0$. (*Uniformity of continuity.*)

[Let ϵ be any positive number. Let (a, x_1) be the greatest range starting from a in which the difference between the upper and lower bounds of $f(x) < \epsilon/2$. Because the function is continuous at a, there must be such a range $\leqslant (a, b)$. If $x_1 \neq b$, let (x_1, x_2) be the greatest range starting from x_1 in which the difference between the upper and lower bounds $< \epsilon/2$. Repeat this process. There are two possibilities: (1) sooner or later the points $x_1, x_2, \dots$ arrive at b, or (2) all the points $x_1, x_2, \dots < b$. In case (1) the theorem is conceded, for if Δ is the width of the least of the (finite number of) ranges (a, x_1),

$(x_1, x_2), \dots (x_{n-1}, b)$, any range whatever which $\leqslant \Delta$ overlaps at most two of these ranges, and therefore in any such range the difference between the upper and lower bounds of $f(x) < \epsilon/2 + \epsilon/2 = \epsilon$. In case (2) the sequence $x_1, x_2, \dots$ has an upper bound $= a \leqslant b$. a being a point of continuity of $f(x)$, there is a range $(a - \delta, a + \delta)$ say with a as centre, within which the difference between the upper and lower bounds of $f(x) < \epsilon/2$. This range must overlap some of the points $x_1, x_2, \dots$. If x_m is one such point, the range $(x_m, a + \delta)$ is a range in which the difference between the upper and lower bounds of $f(x) < \epsilon/2$; and the next point to x_m in the sequence of points $x_1, x_2, \dots$ must $\geqslant a + \delta$. This contradicts the fact that a is the upper bound of the unending sequence $x_1, x_2, \dots$. Case (2) therefore cannot arise.

It is interesting to see where this proof breaks down for the function $1/(x - 1)$ in the range $(0, 1)$.]

14. Prove that, if $f(x)$ is any bounded function which is integrable in (a, b), the function $F(X), = \int_a^X f(x)\, dx$, is a continuous function of X.

15. With the notation of Ex. 14, prove that, if $f(x)$ is monotone, or of bounded variation, $F(X)$ is differentiable on the right and on the left for all values of X in the range and that the "semi-differential coefficients" are equal to the limits on the right and on the left of the function $f(X)$.

§ 7. PROPERTIES OF THE DEFINITE INTEGRAL

213. Functions assumed bounded and integrable. The definite integral $\int_a^b f(x)\, dx$ has several important properties, which are almost self-evident when the integral is looked on as representing the area bounded by the graph $y = f(x)$. The student should interpret and verify geometrically all the results of this section. The proofs, based on the analytical definition of the integral, are likewise almost intuitive. *We suppose throughout this section that the functions under consideration are bounded and integrable throughout the ranges concerned.*

214. Obvious properties. We have already, in the last section, stated the theorems expressed symbolically as

I. $\int_a^c f(x)\, dx = \int_a^b f(x)\, dx + \int_b^c f(x)\, dx$, a, b, c being any real numbers;

II. $\int_a^b k f(x)\, dx = k \int_a^b f(x)\, dx$, k being any real constant;

III. $\int_a^b [f_1(x) + f_2(x)]\, dx = \int_a^b f_1(x)\, dx + \int_a^b f_2(x)\, dx.$

Some of the consequences of these theorems have also been developed above.

215. The simple mean value theorem for integrals. No less evident, geometrically or analytically, is the following important theorem:

IV. *If m and M are fixed numbers such that* $m \leqslant f(x) \leqslant M$ *for all values of x in the range* (a, b), *and* $a < b$, *then*

$$m(b-a) \leqslant \int_a^b f(x)\, dx \leqslant M(b-a).$$

For, the undivided interval (a, b) may be regarded as one of the systems of division of the definition (p. 243). For this system, the "smaller" and "larger sums" are $\geqslant m(b-a)$ and $\leqslant M(b-a)$ respectively. The result follows.

If $b < a$, the inequalities will be reversed.

This theorem is often valuable in giving upper and lower approximations to the value of an integral. In the theorem m and M may be taken to be the lower and upper bounds of $f(x)$ in the range (a, b). We may then, *on the assumption that* $f(x)$ *is continuous* throughout the range, express the theorem in the form which will be described as the *simple mean value theorem for integrals*, viz.

$$\int_a^b f(x)\, dx = (b-a) f(X),$$

where X is some number between a and b.

For, by the generalised form of the fundamental property of continuous functions (p. 236), there is at least one value X between a and b for which $f(X) = $ any number between m and M; hence there is a number X for which $f(X) = \left(\int_a^b f(x)\, dx \right) \Big/ (b-a)$; and the theorem is proved*.

As a corollary to Theorem IV we have the following theorem:

---

* Otherwise, by the mean value theorem of p. 239. Denote the indefinite integral $\int f(x)\, dx$ by $F(x)$. Then $F(x)$ is differentiable and $F(b) - F(a) = (b-a) F'(X)$, where X lies between a and b; i.e. $\int_a^b f(x)\, dx = (b-a) f(X)$, by the fundamental formula.

If $f_1(x)$ and $f_2(x)$ are two functions such that $f_1(x) \leqslant f_2(x)$ for every value of x in a range (a, b), and $a < b$, then

$$\int_a^b f_1(x)\, dx \leqslant \int_a^b f_2(x)\, dx.$$

For $f_1(x) - f_2(x) \leqslant 0$ and

$$\int_a^b f_1(x)\, dx - \int_a^b f_2(x)\, dx = \int_a^b [f_1(x) - f_2(x)]\, dx \leqslant 0.$$

We can also deduce at once from Theorem IV that

If the function $f(x)$ is everywhere positive in a range (a, b), then $\int_a^X f(x)\, dx \,(= F(X)$ say$)$ is a positive increasing function of X for X in the range (a, b); and if $f(x) < 0$, $F(X)$ is negative and decreasing.

This follows because, if $a \leqslant X_1 < X_2 \leqslant b$,

then $\qquad F(X_2) - F(X_1) = \int_{X_1}^{X_2} f(x)\, dx > 0;$

and similarly for $f(x) < 0$.

This theorem is in conformity with the fact that the function $F(x)$ is increasing throughout any range in which $DF(x) > 0$ and decreasing in any range in which $DF(x) < 0$ *.

216. The first mean value theorem. Theorem IV may be generalised so as to give upper and lower approximations to the integral of the product of two functions, one of the functions being everywhere positive.

V. *If $f(x)$ and $\phi(x)$ are two functions (both integrable) and $\phi(x) \geqslant 0$ throughout the range (a, b), and $a < b$; and if m and M are any numbers such that $m \leqslant f(x) \leqslant M$ throughout the range, then*

$$m \int_a^b \phi(x)\, dx \leqslant \int_a^b f(x)\, \phi(x)\, dx \leqslant M \int_a^b \phi(x)\, dx.$$

To prove this, we observe that, because $\phi(x)$ is positive (or zero) and $\qquad\qquad m \leqslant f(x) \leqslant M,$
we have $\qquad\qquad m\phi(x) \leqslant f(x)\, \phi(x) \leqslant M\phi(x)$
for all values of x in the range; and the result follows immediately from the corollary to Theorem IV opposite.

The corresponding result when $\phi(x)$ is everywhere negative is

$$M \int_a^b \phi(x)\, dx \leqslant \int_a^b f(x)\, \phi(x)\, dx \leqslant m \int_a^b \phi(x)\, dx.$$

* P. 231 above.

Theorem V can be expressed in the alternative form

$$\int_a^b f(x)\, \phi(x)\, dx = f(X) \int_a^b \phi(x)\, dx,$$

where X is some number between a and b, *provided $f(x)$ is continuous* and $\phi(x)$ everywhere positive. It is then known as the *first mean value theorem*. This theorem and its more general form (Theorem V) are of importance in estimating the magnitude of definite integrals which cannot be evaluated in ordinary terms.

Thus, for example, the graph of the function $\sin x/x$ consists of a succession of waves. The area of the wave between the points $x = 2n\pi$ and $x = (2n+1)\pi$ equals $\int_{2n\pi}^{(2n+1)\pi} \frac{\sin x}{x}\, dx$.

This cannot be evaluated in ordinary terms*; but the function $\sin x$ is positive throughout the range and $1/x$ lies between $1/2n\pi$ and $1/(2n+1)\pi$. Therefore the area lies between

$$\frac{1}{2n\pi} \int_{2n\pi}^{(2n+1)\pi} \sin x\, dx \quad \text{and} \quad \frac{1}{(2n+1)\pi} \int_{2n\pi}^{(2n+1)\pi} \sin x\, dx\,;$$

i.e. between $2/2n\pi$ and $2/[(2n+1)\pi]$. If n is large this gives a good approximation.

217. The *second mean value theorem* is less obvious:

If $f(x)$ and $\phi(x)$ are two integrable functions, and if $f(x)$ is monotonely non-decreasing and positive (or zero) throughout the range of integration (a, b), then

$$\int_a^b f(x)\, \phi(x)\, dx = f(a) \int_a^X \phi(x)\, dx,$$

where X is some number of the range,—i.e. $a \leqslant X \leqslant b$.

To prove this theorem, we observe that the integral of the function $f(x)\,\phi(x)$ is the common value of the lower and upper bounds of such sums as

$$f(a)\, M_1(x_1 - a) + f(x_1)\, M_2(x_2 - x_1) + \ldots + f(x_{n-1})\, M_n(b - x_{n-1})\ldots(1a)$$

and

$$f(a)\, m_1(x_1 - a) + f(x_1)\, m_2(x_2 - x_1) + \ldots + f(x_{n-1})\, m_n(b - x_{n-1}),\ldots(1b)$$

where $M_1, M_2, \ldots M_n$ are the upper bounds of $\phi(x)$ in the intervals (a, x_1), $(x_1, x_2), \ldots (x_{n-1}, b)$ and $m_1, m_2, \ldots m_n$ the corresponding lower bounds.

Taking the numbers

$$f(a), f(x_1), \ldots f(x_{n-1})\,;\ M_1(x_1 - a),\ M_2(x_2 - x_1),\ \ldots M_n(b - x_{n-1})$$

for the numbers $b_1, b_2, \ldots b_n$; $a_1, a_2, \ldots a_n$ of Abel's lemma (p. 29), we have that the sum $(1a) \geqslant f(a)\,\mu$, where μ is the least of all the numbers such as

$$M_1(x_1 - a) + M_2(x_2 - x_1) + \ldots + M_k(x_k - x_{k-1}),\ \ldots\ldots\ldots\ldots\ldots(2)$$

where $1 \leqslant k \leqslant n$.

It is easy to see that the expression $(2) \geqslant \int_a^{x_k} \phi(x)\, dx, = \Phi(x_k)$ say, and

---

* It can be evaluated as the sum of an infinite series.

thence that $\mu \geqslant m$, where m denotes the lower bound of the function $\Phi(X)$ $\left(= \int_a^X \phi(x)\,dx \right)$ for all values of X in the interval (a, b).

Hence the sum $(1\,a) \geqslant f(a)\,m$, and therefore $\int_a^b f(x)\,\phi(x)\,dx \geqslant f(a)\,m$.

Similarly $\int_a^b f(x)\,\phi(x)\,dx \leqslant f(a)\,M$, if M is the upper bound of $\Phi(x)$.

Therefore $\int_a^b f(x)\,\phi(x)\,dx = f(a)\,Y$, where Y is some number between m and M, the lower and upper bounds of $\Phi(X)$, inclusive.

But $\Phi(X)$ is continuous. The theorem follows from the fundamental property (p. 236).

EXAMPLES XXV.

1. Give the geometrical interpretation of Theorem IV, p. 262.

2. Prove that if $f(x)$ is integrable in a range (a, b) then $|f(x)|$ is also integrable in (a, b) and $\left| \int_a^b f(x)\,dx \right| \leqslant \int_a^b |f(x)|\,dx$; a being supposed less than b.

[The difference between the upper and lower bounds of $|f(x)|$ in any range $\leqslant$ the difference between the upper and lower bounds of $f(x)$. The difference between the "larger" and "smaller sums" in the definition of $\int_a^b |f(x)|\,dx \leqslant$ the corresponding difference in the definition of $\int_a^b f(x)\,dx$. The integrability of $|f(x)|$ follows. The corollary to Theorem IV establishes the inequality.]

3. Shew that $\int_\pi^{2\pi} \dfrac{\sin^2 x}{x}\,dx$ lies between $\frac{1}{4}$ and $\frac{1}{2}$.

4. Deduce the theorem of p. 119 above from Theorem IV of p. 262 above in the case where the derivative Dy is supposed continuous and integrable.

5. Rewrite the proof of the logarithmic expansion as given on pp. 122—123 above using Theorem IV of this section in place of the theorem of p. 119.

6. By applying Theorem IV to the integral $\int_a^b Df(x)\,dx$, prove the mean value theorem for derivatives (p. 239) in the case where the derivative $Df(x)$ of $f(x)$ is continuous.

Deduce from this theorem that $\sin x$ lies between $-x$ and x.

7. Apply the first mean value theorem to shew that
$$\int_a^b (b-x)\,D\,[Df(x)]\,dx = \frac{(b-a)^2}{2}\,D\,[Df(X)],$$
where $D\,[Df(X)]$ is the derivative (supposed continuous) of the derivative of $f(x)$ at the point X, X being some number between a and b.

By integrating by parts deduce that
$$f(b) = f(a) + (b-a)\,f'(a) + \frac{(b-a)^2}{2}\,D\,[Df(X)],$$
where $f'(a)$ denotes the value of the derivative $Df(x)$ at the point a, and $D\,[Df(X)]$ is as before.

8. Deduce from Ex. 7 that:

   (i) $\cos x$ lies between 1 and $1 - x^2/2$;

   (ii) if x is positive $\log_e (1+x)$ lies between x and $x - x^2/2$;

   (iii) if x is positive and < 1, $(1-x)^n$ lies between

$$1 - nx \quad \text{and} \quad 1 - nx + \frac{n(n-1)}{2} x^2,$$

   n being any real number;

and (iv) $e^x = 1 + x + x^2 e^{\theta x}/2!,$

where θ is some positive number less than 1.

9. Use Theorem IV to prove successively:

$$\sin x < x; \quad -\cos x < -1 + \frac{x^2}{2!}; \quad -\sin x < -x + x^3/3!; \quad \cos x < 1 - \frac{x^2}{2!} + \frac{x^4}{4!}$$

etc. if x is positive; and similar results when x is negative.

   Deduce $\sin x = x - x^3/3! + x^5/5! - \dots$ and $\cos x = 1 - x^2/2! + x^4/4! - \dots$.

   $\left[\cos x < 1 \text{; therefore } \int_0^X \cos x \, dx < X, \text{ i.e. } \sin x < x. \text{ Again} \right.$

$$\int_0^X \sin x \, dx < \int_0^X x \, dx, \text{ i.e. } 1 - \cos x < x^2/2.$$

We get in general that $\cos x$ differs from $1 - x^2/2! + x^4/4! - \dots \pm x^{2n}/(2n)!$ by less than $x^{2n+2}/(2n+2)!$ and $\sin x$ from $x - x^3/3! + \dots \pm x^{2n+1}/(2n+1)!$ by less than $x^{2n+3}/(2n+3)!$. These $\to 0$ as $n \to \infty$, whatever the value of x.]

10. Use Theorem IV to prove successively that, if x is positive and $< \log_e K$,
$1 + x < e^x < 1 + Kx, \quad 1 + x + x^2/2 < e^x < 1 + x + Kx^2/2, \dots$

$$1 + x + x^2/2! + \dots + x^n/n! < e^x < 1 + x + x^2/2! + \dots + x^{n-1}/(n-1)! + Kx^n/n!$$

and deduce the exponential expansion for $x > 0$.

   Discuss similarly the case where $x < 0$.

11. Shew that if $x > -1$ and n is any real number, $(1+x)^{n+1}$ lies between 1 and $1 + K(n+1)x$, where K is any number exceeding $(1+x)^n$.

   Thence prove successively that

$$(1+x)^{n+2} \text{ lies between } 1 + (n+2)x \text{ and } 1 + (n+2)x + K\frac{(n+2)(n+1)}{2!}x^2,$$

that $(1+x)^{n+3}$ lies between $1 + (n+3)x + \dfrac{(n+3)(n+2)}{2!}x^2$

and $1 + (n+3)x + \dfrac{(n+3)(n+2)}{2!}x^2 + K\dfrac{(n+3)(n+2)(n+1)}{3!}x^3,$

   ....

   $(1+x)^m$ lies between $1 + mx + \dfrac{m(m-1)}{2!}x^2 + \dots + \dfrac{m(m-1)\dots(m-r+1)}{r!}x^r$

and

$$1 + mx + \frac{m(m-1)}{2!}x^2 + \dots + \frac{m(m-1)\dots(m-r+1)}{r!}x^r$$
$$+ K\frac{m(m-1)\dots(m-r+1)(m-r)}{(r+1)!}x^{r+1},$$

where $m = n + r + 1$, and none of the numbers

$$n,\ n+1, \ldots n+r \ (\text{i.e. } m-1,\ m-2, \ldots m-r-1) = -1;$$

(the results being then simpler).

Deduce the general binomial expansion

$$(1+x)^m = 1 + mx + \frac{m(m-1)}{2!}x^2 + \frac{m(m-1)(m-2)}{3!}x^3 + \ldots$$

for $-1 < x < 1$, and m any real number.

[Apply Theorem IV to the integral $\int_0^x (1+x)^n\, dx$ and repeat. The binomial theorem will follow if it can be shewn that the term

$$K\frac{m(m-1)\ldots(m-r)}{(r+1)!}x^{r+1} \to 0 \text{ as } r \to \infty.$$

That this is so if $-1 < x < 1$ is seen most easily by observing that the ratio of the values of this expression for two successive values of r is $(m-r)x/(r+1)$, which numerically $<$ a positive number k, less than 1, for all values of r sufficiently large. The series which has this expression for its rth term is therefore absolutely convergent and (*a fortiori*) this expression $\to 0$.]

12. Deduce the second mean value theorem (p. 264) from the first mean value theorem in the special case when both the functions $f(x)$ and $\phi(x)$ are positive (or zero).

§ 8. VARIOUS APPLICATIONS OF INTEGRATION AND DIFFERENTIATION

218. Areas. In this section we consider a few of the most important simple applications of definite integrals and of differentiation.

The area of any plane region bounded by straight lines and curves which are the graphs of integrable functions is given in terms of definite integrals*. For example *to find the area cut off the parabola $y^2 = 4ax$ by the straight line $x = a$*, we see on drawing the figure that the required area $= 2\int_0^a y\, dx$, where $y = +\sqrt{(4ax)}$.

The area $= 4\sqrt{a}\int_0^a \sqrt{x}\, dx = 4\sqrt{a}\,\frac{2}{3}a^{3/2} = 8a^2/3$.

It is, however, often convenient to suppose the region whose area is required divided up into elemental regions other than the rectangular strips used in defining the area below a graph $y = f(x)$ as $\int_a^b f(x)\, dx$. It is clear that other methods of division will not (in any practical cases) lead to different values for the area of a region.

* See Chapter III, § 1.

To find the area of a sector of an ellipse (or circle).

An ellipse is the closed curve such that the coordinates (x, y) of any point on it, referred to suitably chosen axes, satisfy the equation

$$x^2/a^2 + y^2/b^2 = 1,$$

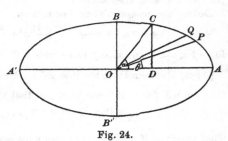

where a and b are positive numbers. We have $y = \pm b \sqrt{(1 - x^2/a^2)}$, and the curve is an oval curve as in Fig. 24, with $OA = a, OB = b$.

Fig. 24.

If $a = b$ it is a circle of centre O and radius a. We wish to find the area of the sector AOC, where the angle $A\hat{O}C = \alpha < \pi/2$ (say).

We could first find the areas of the triangle $ODC (= \frac{1}{2} OD . DC)$ and of the region $ADC \left(= \int_{OD}^{OA} y\, dx, \text{ where } y = b\sqrt{(1 - x^2/a^2)} \right)$.

Let us, however, divide the sector into sub-sectors such as OPQ, where $A\hat{O}P = \theta$, $OP = r$, $P\hat{O}Q = \delta\theta$ say.

The area of such a sub-sector, if $\delta\theta$ is small, will be approximately $\frac{1}{2}r^2\delta\theta$, and we shall have roughly that the required area of the sector AOC is the sum of such terms $\frac{1}{2}r^2\delta\theta$.

Now r^2 is a function of θ given by the equation

$$r^2 \cos^2\theta/a^2 + r^2 \sin^2\theta/b^2 = 1$$

(because $x = r\cos\theta$ and $y = r\sin\theta$);

i.e. $$r^2 = 1/(\cos^2\theta/a^2 + \sin^2\theta/b^2) \quad \dots\dots\dots\dots\dots(1).$$

Hence the sum of the terms such as $\frac{1}{2}r^2\delta\theta \rightarrow \int_0^a \frac{1}{2}r^2 d\theta$ as the greatest $\delta\theta \rightarrow 0$,—from the definition of this definite integral,

i.e. $\rightarrow \int_0^a \dfrac{\frac{1}{2}d\theta}{\cos^2\theta/a^2 + \sin^2\theta/b^2}$.

It is in fact easy to see now that the required area is precisely

$$\int_0^a \frac{\frac{1}{2}d\theta}{\cos^2\theta/a^2 + \sin^2\theta/b^2}.$$

For we have accurately that the area of the elemental sector OPQ lies between $\frac{1}{2}OP^2\delta\theta$ and $\frac{1}{2}OQ^2\delta\theta$, and therefore $= \frac{1}{2}r^2\delta\theta\,(1 + \eta)$, where $\eta \rightarrow 0$ as $\delta\theta \rightarrow 0$; and it follows that the total area is the limit of the sum of the terms $\frac{1}{2}r^2\delta\theta$ as the greatest $\delta\theta \to 0$, as in the definition of the definite integral $\int_a^b f(x)\,dx$;

i.e. $\quad = \int_0^a \frac{1}{2}r^2 d\theta$, where $r^2 = 1/(\cos^2\theta/a^2 + \sin^2\theta/b^2)$.

To evaluate this integral make the substitution $\tan\theta = u$; so that $d\theta/du = \cos^2\theta$.

We then have

$$\int \tfrac{1}{2}r^2\,d\theta = \int \frac{du}{2\,(1/a^2 + u^2/b^2)}$$

$$= \tfrac{1}{2}\int \frac{ab\,dv}{1+v^2}, \text{ putting } v = au/b,$$

$$= \tfrac{1}{2}ab \text{ arc tan } v = \tfrac{1}{2}ab \text{ arc tan } (a\tan\theta/b).$$

The required area of the sector AOC of angle α

$$= \tfrac{1}{2}ab \text{ arc tan } (a\tan\alpha/b) - \tfrac{1}{2}ab \text{ arc tan } 0$$

$$= \tfrac{1}{2} \text{ arc tan } (a\tan\alpha/b).$$

In the special case of a circle,

$a = b$ and the area $= \tfrac{1}{2}a^2 \text{ arc tan tan } \alpha = \tfrac{1}{2}a^2\alpha$.

The area of the complete quadrant OAB

$$= \lim_{a\to\pi/2}(\text{area } OAC) = \lim_{a\to\pi/2} \tfrac{1}{2}ab \text{ arc tan } (a\tan\alpha/b) = \pi ab/4;$$

or we could use the alternative form

$$\tfrac{1}{2}ab\,[\pi/2 - \text{arc tan } (b\cot\alpha/a)] = \tfrac{1}{2}ab\,(\pi/2 - 0).$$

The area of the complete ellipse $= \pi ab$.

The ease with which the solution of a problem of this character can be effected depends very largely on the choice of elemental regions. The rectangular strips of the definition on p. 148 and the sectors are two of the most common types of elements.

In most practical cases *such problems can be simplified by using the method of integration by substitution* or otherwise. Thus in finding the area of the complete ellipse we require

$$2\int_{-a}^{a} y\,dx, \text{ where } y = b\sqrt{1 - \frac{x^2}{a^2}}.$$

To evaluate the indefinite integral $\int b\sqrt{1 - \dfrac{x^2}{a^2}}\,dx$, substitute $\dfrac{x}{a} = u$ and then $u = \cos\theta$, so that

$$\int b \sqrt{1 - \frac{x^2}{a^2}}\, dx = ab \int \sqrt{(1 - u^2)}\, du$$

$$= ab \int \sin\theta\,(-\sin\theta)\, d\theta$$

$$= -\frac{ab}{2} \int (1 - \cos 2\theta)\, d\theta$$

$$= -\frac{ab}{2}\left(\theta - \frac{\sin 2\theta}{2}\right).$$

Instead of substituting for θ in terms of x and then deducing the value of the definite integral from the indefinite integral so found, we notice that as x ranges from $-a$ to a, u ranges from -1 to 1 and θ ranges from π to 0.

Hence

$$2\left[\int b \sqrt{\left(1 - \frac{x^2}{a^2}\right)}\, dx\right]_{x=-a}^{x=a} = 2\left[-\frac{ab}{2}\left(\theta - \frac{\sin 2\theta}{2}\right)\right]_{\theta=\pi}^{\theta=0} = \pi ab,$$

—the required area.

219. Volumes. *To find the volume of a sphere of radius R.*

The sphere may be imagined cut up into slices by a set of parallel planes. Let M and N (Fig. 25) be the points in which the radius OC is cut by two such planes, to which OC is perpendicular. The slice (of thickness $MN = \delta x$) between these two planes will be circular; on one side the radius will be MP, and on the other NQ. The volume of this slice (assuming that the volume of such a region can be satisfactorily defined) is clearly less than $MN \times$ the area of a circle of radius MP and greater than $MN \times$ the area of a circle of radius NQ.

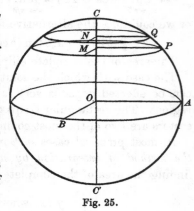

Fig. 25.

Calling $OM = x$, $MP = y$, $MN = \delta x$;

the volume of the slice $= \delta x\,.\,\pi y^2\,(1 + \eta)$ where $\eta \rightarrow 0$ as $\delta x \rightarrow 0$.

Dividing the diameter $C'OC$ up into a number of portions such as MN, and adding up, we see that the required volume is the sum of such terms as $\delta x \,.\, \pi y^2 (1 + \eta)$ and therefore equals the limit of this sum as the greatest $\delta x \to 0$, if this limit exists; i.e. $= \int_{-R}^{R} \pi y^2 \, dx$, where $y^2 = R^2 - x^2$, if this integral exists.

The integral exists and equals

$$\int_{-R}^{R} \pi \, (R^2 - x^2) \, dx = \pi R^2 [R - (- R)] - \pi [R^3/3 - (- R^3/3)] = \frac{4}{3} \pi R^3,$$

—the required volume.

This discussion assumes that a satisfactory definition of the volume of such regions which are not parallelepipeds has been given. It is not difficult to give a definition, applicable to such solids as the sphere, on the lines of the definition of the area under a curve given in Chapter III, § 1. Being concerned here with the practical question of the determination of such volumes, we omit such a definition.

220. Centres of mass. Centres of mass and moments of inertia of continuous bodies may be found similarly.

To find the centre of mass of a semi-circular lamina.

Let O be the centre of the semi-circle $A'BA$ of radius R (Fig. 26). Divide the region up into strips (e.g. $PQQ'P'$) by parallels to the diameter $A'A$. Let $OM = x$, $MN = \delta x$, $MP = y$; so that $y^2 = R^2 - x^2$.

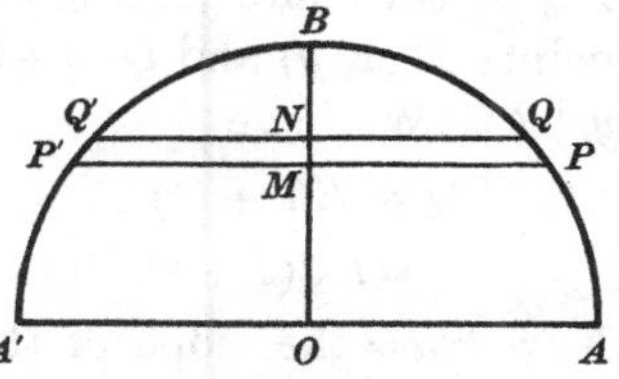

From mechanical principles, the moment of the strip about the line $A'A =$ (area of $PQQ'P'$) $\times x'$, where x' lies between OM and ON,

Fig. 26.

$$= 2y \,.\, \delta x \,.\, x \, (1 + \eta),$$

where $\eta \to 0$ as $\delta x \to 0$.

Hence, as before, the total moment of the whole semi-circle about $A'A$ is equal to

$$\int_{0}^{R} 2xy \, dx = \int_{0}^{R} 2 \sqrt{(R^2 - x^2)} \, x \, dx.$$

Putting $R^2 - x^2 = u$, we have

$$\int 2 \sqrt{(R^2 - x^2)} \, x \, dx = - \int \sqrt{u} \, du = - \frac{2}{3} u^{3/2} = - \frac{2}{3} (R^2 - x^2)^{3/2}.$$

Therefore

$$\int_0^R 2\sqrt{(R^2 - x^2)}\,x\,dx = -\frac{2}{3}(R^2 - R^2)^{3/2} + \frac{2}{3}(R^2 - 0)^{3/2} = \frac{2}{3}R^3.$$

But the area of the semi-circle $= \frac{1}{2}\pi R^2$.

From mechanical principles therefore the centre of mass is on the radius OB, distant $\frac{2}{3}R^3/\frac{1}{2}\pi R^2$, i.e. $4R/3\pi$, from O.

Again we have omitted to give a definition of the moment of an area about a line as the limit of a sum of a type similar to that employed in the definition of an area. The precise definition and justification of the assumptions made "on mechanical principles" are again not difficult.

221. Lengths of curves. The length of an arc of a curve was defined in § 1 of this chapter as the common limit (if it exists) of all sequences of perimeters of inscribed polygons, as the sides are indefinitely diminished in length. (Or as the upper bound of the set of perimeters of all possible inscribed polygons.)

Let the curve be AB (Fig. 27), given by the equation $y = f(x)$. Let the x coordinates of A and B be a and $b\,(a < b)$. Let PQ be the chord joining the points $P(x, y)$ and $Q(x + h, y + k)$. We have

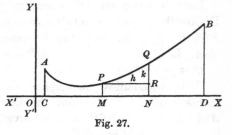

$$PQ = \sqrt{(h^2 + k^2)}$$
$$= h\sqrt{(1 + k^2/h^2)}.$$

Fig. 27.

Now $k/h =$ the slope of the chord PQ, and therefore, by the mean value theorem for derivatives, $k/h = dy/dx$ at some point on the arc PQ,—supposing the function $f(x)$ is differentiable throughout.

Adding up the lengths of all such chords PQ, we get a sum intermediate to the "larger" and "smaller" sums corresponding to the definite integral over the range (a, b) of the function $\sqrt{[1 + (dy/dx)^2]}$. The sum of these chords therefore tends to $\int_a^b \sqrt{[1 + (dy/dx)^2]}\,dx$ as the chords PQ are indefinitely diminished in length,—provided the definite integral exists. *The curve is there-*

fore rectifiable* *and its length is* $\int_a^b \sqrt{[1 + (dy/dx)^2]}\,dx$, *if this integral exists.* This will always be the case if the derivative dy/dx, i.e. $Df(x)$, is monotone or continuous throughout the range (a, b). In particular, any arc of a curve defined by $y = f(x)$, where $f(x)$ is any of the functions considered in this course, will be rectifiable if the arc avoids any exceptional points of discontinuity, etc. As the derivative dy/dx ceases to exist at a point where the graph is parallel to the y axis, the length of an arc which is anywhere

parallel to OY cannot be expressed as $\int_a^b \sqrt{[1 + (dy/dx)^2]}\,dx$. Such a case can be dealt with by using y as independent variable instead of x. The length of the curve is then expressed as $\int_c^d \sqrt{[1 + (dx/dy)^2]}\,dy$, where c and d are the y coordinates of the extremities A and B of the curve.

222. Identity of the numbers ϖ and π. *The circumference of a circle* is easily found in this way†.
Let R be the radius of the circle.
The upper semi-circle $A'BA$ (Fig. 28) is given by the equation

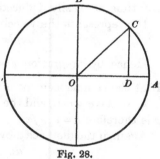

$$y = + \sqrt{(R^2 - x^2)}.$$

Let $A\hat{O}C = \pi/4$ (half a right angle).
Then $\qquad OD = R/\sqrt{2}.$

The arc BC

$$= \int_0^{R/\sqrt{2}} \sqrt{[1 + (dy/dx)^2]}\,dx.$$

Fig. 28.

Now $\quad dy/dx = - x/[\sqrt{(R^2 - x^2)}]$,
and therefore $\qquad 1 + (dy/dx)^2 = R^2/(R^2 - x^2).$
Therefore the circumference $= 8 \times$ arc BC

$$= 8 \int_0^{R/\sqrt{2}} \sqrt{[1 + (dy/dx)^2]}\,dx$$

$$= 8 \int_0^{R/\sqrt{2}} \frac{R}{\sqrt{(R^2 - x^2)}}\,dx.$$

---

* In accordance with either definition, pp. 200, 203.

† We know from the definition of the number π that the circumference of a circle of radius $R = 2\pi R$. Our object is to find the circumference independently of π and thus determine π.

To evaluate this integral use the trigonometrical substitution $x = RS(u)$, so that $dx/du = RC(u)$ and $R^2 - x^2 = R^2 [C(u)]^2$.

Therefore $\int \dfrac{R\,dx}{\sqrt{(R^2 - x^2)}} = \int \dfrac{R^2\,C(u)\,du}{RC(u)} = Ru = R\bar{S}(x\mid R)$.

Therefore the circumference $= 8R\bar{S}(1/\sqrt{2}) = 8R\varpi/4$, by Ex. 4, p. 187, $= 2\varpi R$.

This establishes the identity of the two numbers ϖ and π.

The length of an arc of a parabola may be obtained similarly.

For the parabola $y = x^2$, we have
$$dy/dx = 2x, \quad 1 + (dy/dx)^2 = 1 + 4x^2;$$
and therefore the length of the arc between the points whose abscissae are a and b is
$$\int_a^b \sqrt{(1 + 4x^2)}\,dx.$$

This integral can be evaluated by the substitution $2x = \sinh u$.

223. Motion under constant acceleration. Suppose *a particle moves along a straight line under a constant acceleration, f* say. The formulae giving the velocity and position of the particle at any time are easily deduced thus:

If v represents the velocity and x the time, we have
$$dv/dx = f (= \text{constant}).$$

Hence, by integration, $v = \int \dfrac{dv}{dx}\,dx = xf + C$, where C is a constant.

If at the commencement of the motion (i.e. for $x = 0$) the velocity is u, we have $u = C$, and the formula giving the velocity at any time x is therefore $v = u + fx$.

Again, if the distance moved through is represented by y, we have
$$dy/dx = v = u + fx.$$

Therefore, by integration,
$$y = \int \frac{dy}{dx}\,dx = \int (u + fx)\,dx = ux + fx^2/2 + C',$$

where C' is a constant.

If the distance y is measured from the point where the particle was at time $x = 0$, we have $0 = C'$.

The formula giving the distance traversed is therefore
$$y = ux + \tfrac{1}{2}fx^2.$$

Motion under known variable acceleration can often be similarly discussed.

224. Taylor's series. We know that a power series
$$a_0 + a_1 x + a_2 x^2 + \dots, \qquad \dots\dots\dots\dots\dots\dots(1)$$
if convergent for any value of x (besides $x = 0$), is convergent, and absolutely convergent, for all values of x interior to an interval $(- R, R)$ called the range of convergence*. We know also that the sum function, $f(x)$, is continuous and differentiable for all values of x interior to the interval and that $Df(x) = a_1 + 2a_2 x + 3a_3 x^2 + \dots$ if $- R < x < R,$—this derived series being also absolutely convergent for all such values of x†.

It follows therefore that the derivative $Df(x)$ is a function of x which is continuous and differentiable and
$$D\,[Df(x)] = 2a_2 + 2 \cdot 3a_3 x + 3 \cdot 4a_4 x^2 + \dots$$
for $- R < x < R$, and so on indefinitely.

If we put $x = 0$ in these results we have
$a_0 = f(x)$ when $x = 0, = f(0)$ say; $a_1 = Df(x)$ when $x = 0, = f'(0)$ say;
$2a_2 = D\,[Df(x)]$ when $x = 0, = f''(0)$ say;
$2 \cdot 3 \cdot a_3 = D\,\{D\,[Df(x)]\}$ when $x = 0, = f'''(0)$ say;
and so on.

We can therefore express the function $f(x)$ as
$$f(x) = f(0) + x f'(0) + \frac{x^2}{2!} f''(0) + \frac{x^3}{3!} f'''(0) + \dots$$

for all values of x interior to the range of convergence of this series. This series is known as *Maclaurin's series*. It may be regarded as a special case of *Taylor's series*, which is
$$f(a + h) = f(a) + h f'(a) + \frac{h^2}{2!} f''(a) + \frac{h^3}{3!} f'''(a) + \dots,$$

where $f(a), f'(a)$, etc. denote the values of $f(x), Df(x)$, etc. at the point $x = a$. Taylor's series can itself be deduced from Maclaurin's series thus:

Put $x = a + y$, and write $F(y)$ for $f(a + y)$. By Maclaurin's expansion we have
$$F(y) = F(0) + y F'(0) + y^2 F''(0)/2! + \dots.$$
But $\qquad\qquad F(0) = f(a), F'(0) = f'(a)$, etc.
Hence $\qquad\qquad f(a + h) = f(a) + h f'(a) + \dots,$
the result being valid for all values of h for which the series is convergent, the function $f(a + h)$ being supposed defined for all such values of h by such a series of powers of h.

* P. 193 and Ex. 13, p. 194.   † Ex. 14, p. 195.

225. We have not proved that any function $f(x)$ *can* be expressed as the sum of such a Maclaurin or Taylor series; but only, for example, that the coefficients of such a series as (1) are given by the values of the function and its successive differential coefficients at the point $x = 0$. If we have a function $f(x)$ whose differential properties are known (so that we know for such a function the values of $f(0)$, $f'(0)$, $f''(0)$, etc.) we may desire to express it as the sum of an infinite power series.

For example, take for $f(x)$ the circular function $\sin x$.

We have $f(0) = \sin 0 = 0$; $f'(0) = d\sin x/dx$ when $x = 0$, $= \cos 0 = 1$; $f''(0) = d\cos x/dx$ when $x = 0$, $= -\sin 0 = 0$; $f'''(0) = d(-\sin x)/dx$ when $x = 0$, $= -\cos 0 = -1$; $f^{\mathrm{iv}}(0) = 0$; $f^{\mathrm{v}}(0) = 1$; etc.

The corresponding Maclaurin series

$$f(0) + xf'(0) + \frac{x^2}{2!}f''(0) + \dots \quad \dots\dots\dots\dots(2)$$

is $\qquad\qquad 0 + x + 0 - x^3/3! + 0 + x^5/5! + \dots,$

i.e. $\qquad\qquad x - x^3/3! + x^5/5! - \dots.$

It is tempting in such a case to conclude that the sum of the series is the original function. In practice this is in fact often the case (as it is in the example considered). But we have proved only that *if* the series (2) is convergent for a range of values of x, and if the sum function is called $F(x)$, then $F(0) = f(0)$, $F'(0) = f'(0)$, etc. It is conceivable (i) that the Maclaurin series arising from a given function $f(x)$ may be not convergent for any value of x, and (ii) that there may be two (or any number of) different functions $f(x)$ and $F(x)$ such that $F(0) = f(0)$, $F'(0) = f'(0)$, etc.

These possibilities are indeed very real; and the suggested line of argument is untenable.

226. We may however easily test, after the Maclaurin series is obtained, whether or not it is convergent for any range of values of x. It may also be possible to find some means to test also the second possibility after the series is found. Realising this, we may still use the above process as a suggestion as to what the power series for a given function is likely to be. We will content ourselves with this position;—though it is possible to establish certain general conditions under which the series thus formed necessarily represents the original function*.

227. The identity of the trigonometrical and circular functions. In the case mentioned above of the expansion of $\sin x$ in a Maclaurin series, the series found, viz.

$$x - x^3/3! + x^5/5! - \ldots,$$

is the series considered in Chapter III, § 5 above. This series was there seen to be convergent for all values of x, and we called its sum function $S(x)$. We have now to prove that the two functions $S(x)$ and $\sin x$ are identical.

We will prove also at the same time that the two functions $C(x)$ and $\cos x$ are identical. We shall then have established the Maclaurin expansions of $\sin x$ and $\cos x$ and have identified the trigonometrical functions of Chapter III, § 5 (and their derived functions $T(x)$, etc.) with the circular functions of trigonometry.

228. Write
$$C(x) - \cos x = y,$$
$$S(x) - \sin x = z.$$

Differentiating with regard to x, we have
$$Dy = -S(x) + \sin x = -z,$$
$$Dz = C(x) - \cos x = y.$$

Multiplying these equations by y and z and adding, we get
$$yDy + zDz = 0,$$
i.e.
$$D(y^2 + z^2) = 0.$$

Therefore $y^2 + z^2 = $ constant $(= C$ say$)$, by the theorem at the top of p. 250, and $C = 0$ because, when $x = 0$, $y^2 + z^2 = 0$.

Hence $y^2 + z^2 = 0$ for all values of x;
whence both $y = 0$ and $z = 0$ for all values of x, because y^2 and z^2 cannot be negative.

That is $\cos x = C(x)$ and $\sin x = S(x)$.   Q.E.D.

229. The binomial theorem. Let us try to establish similarly the Maclaurin expansion of the general power of the binomial, viz. $(1 + x)^n$, where n is any real number.

Writing $f(x) = (1 + x)^n$ we have $Df(x) = n(1 + x)^{n-1}$ and therefore $f'(0) = Df(x)$ where $x = 0$, $= n$.

---

* One such case, in which Taylor's expansion is necessarily valid, has been given above, in Ex. 15, p. 195.

Again $\qquad D[Df(x)] = n(n-1)(1+x)^{n-2}$

and therefore $\qquad f''(0) = n(n-1).$

And so on,

$$f'''(0) = n(n-1)(n-2), \quad f^{\text{iv}}(0) = n(n-1)(n-2)(n-3), \text{ etc.}$$

If n is a positive integer, one of the factors n, $n-1$, etc. will be zero, and from that stage the derivatives will all be zero. In all other cases the series thus suggested as the expansion of $(1+x)^n$ is

$$1 + nx + \frac{n(n-1)}{2!}\, x^2 + \frac{n(n-1)(n-2)}{3!}\, x^3 + \dots \quad \dots \dots (3)$$

230. Having found this series we now determine whether it is convergent for any value of x. By the ratio test* or otherwise we see that it is convergent for all values of x between -1 and 1. We have still to prove that the sum function of the series (3) is the function $(1+x)^n$.

This may be proved thus:

Calling the sum function of (3) y, we have

$$Dy = n + n(n-1)x + n\frac{(n-1)(n-2)}{2!}x^2 + \dots$$

$$= n\left[1 + (n-1)x + \frac{(n-1)(n-2)}{2!}x^2 + \dots\right]$$

and $\quad (1+x)Dy = n\left[1 + (n-1)x + \frac{(n-1)(n-2)}{2!}x^2 + \dots \right.$

$$\left. + x + (n-1)x^2 + \dots\right]$$

$$= n\left[1 + nx + \frac{n(n-1)}{2!}x^2 + \dots\right]$$

$$= ny;$$

or $\qquad\qquad (1+x)Dy - ny = 0, \quad \dots \dots \dots \dots \dots (4)$

valid for $-1 < x < 1$.

Multiply the relation (4) by $(1+x)^{-n-1}$.

We have $\qquad (1+x)^{-n}Dy - yn(1+x)^{-n-1} = 0,$

i.e. $\qquad\qquad D[(1+x)^{-n}y] = 0.$

Therefore $(1+x)^{-n}y = \text{constant}$

$$= 1;$$

because when $x = 0$, $(1+x)^{-n} = 1$ and $y = 1$.

* P. 80.

Hence $y = (1 + x)^n$, and the binomial theorem is proved; viz.

$$(1 + x)^n = 1 + nx + \frac{n(n-1)}{2!} x^2 + \frac{n(n-1)(n-2)}{3!} x^3 + \dots,$$

for all real values of the index n and all values of x between -1 and 1. If n is a positive integer the series terminates and the result is true for all values of x.

231. The above proofs of the identity of the functions $\sin x$, $\cos x$, $(1 + x)^n$ with their respective power series have been based on the differential properties of the functions and of the sum functions of the series. The relation (4) in fact, from which the binomial theorem has been deduced, is a "differential equation"; and the above proof consists in determining from this differential equation the nature of the function y involved. Such a process is called solving the differential equation. Differential equations provide a weapon of great power in several branches of mathematics. Their discussion however lies beyond the scope of this course.

EXAMPLES XXVI.

1. Find the area of the segment of the hyperbola $x^2/a^2 - y^2/b^2 = 1$ cut off the branch which lies to the right of the y axis by the line $x = 2a$.

2. Find the volume and centre of mass of a square pyramid of height h, the length of a side of the base being a.
[Divide by planes parallel to the base.]

3. Find the volume and centre of mass of a right circular cone of height h standing on a circular base of radius a.

4. Find the area and centre of mass of the segment of the parabola $y^2 = 4ax$ cut off by the line $x = 4a$.

5. Find the centre of mass of the plane area bounded by the half of the ellipse $x^2/a^2 + y^2/b^2 = 1$ above the x axis and by the x axis.

6. Given that the acceleration of a body moving along a straight line is $f = e^{-x} \sin nx$, where x represents the time and n is a constant, and that the body started at time $x = 0$ at a point O with velocity u in a given direction, find the velocity and position of the body at any time x.

$$\left[dv/dx = f = e^{-x} \sin nx \text{ gives } v = \int f \, dx = \int e^{-x} \sin nx \, dx. \text{ Integrate by parts} \right.$$

(twice). Similarly for the distance y from O, which $= \int v \, dx. \Big]$

7. Given that the acceleration of a body moving along a straight line is $-kv$, where v represents the velocity and k is a positive constant, find the velocity and position of the body after a time x in terms of x, k and the initial velocity (u).

$$\left[v = ue^{-kx}; \quad y = \frac{u}{k}(1 - e^{-kx}). \right]$$

8. Find the lengths of the arcs of the curves $y = \cosh x$ and $y = \sinh x$ between the points $(0, 1)$, $(1, \cosh 1)$ and $(0, 0)$, $(1, \sinh 1)$. [The curve $y = \cosh x$ is the catenary.]

9. Establish the identity of $\sin x$ and $S(x)$ with $\cos x$ and $C(x)$ by the method of Ex. 5, p. 132.

10. Prove the binomial theorem by the method of Ex. 5, p. 132.

11. Assuming the derivatives of $\sin x$, $\cos x$, $S(x)$, and $C(x)$, prove that
$$S(x)\sin x + C(x)\cos x = 1 \text{ and } S(x)\cos x - C(x)\sin x = 0,$$
and thence that $S(x) = \sin x$ and $C(x) = \cos x$.

12. Prove that if y_1 and y_2 are two functions of x such that $Dy_1 = y_1$ and $Dy_2 = y_2$ then $y_1/y_2 = $ constant. Hence, assuming $De^x = e^x$ and $DE(x) = E(x)$, where $E(x)$ denotes the sum function of the series $1 + x + x^2/2! + x^3/3! + ...$, prove that $e^x = E(x)$.

13. Given that the derivative Dy_1 of the sum function y_1 of the binomial series (3) of p. 278 satisfies the relation (4) of that page, and that the function $y_2 = (1+x)^n$ satisfies the same relation, use the formula for the differentiation of a quotient to shew that $y_1/y_2 = $ constant.

Deduce an independent proof of the binomial theorem.

14. By writing $y = 1 - x^2/2! + x^4/4! - ... \pm x^{2n}/2n! - \cos x$, prove that $\cos x = 1 - x^2/2! + x^4/4! - ...$ without assuming any knowledge of the functions $S(x)$, $C(x)$.

15. Obtain the Maclaurin expansions for the functions
$$e^x, \sinh x, \cosh x, \log_e(1+x), e^{-x}\sin x, e^{-x}\cos x, \cos x \cosh x.$$

16. Establish the expansions of Ex. 14 by the method of Ex. 12 or by the method of Ex. 5, p. 132.

APPENDIX
COMPLEX NUMBERS

232. Inadequacy of real numbers for purposes of algebra.
In Chapter I, beginning with the notions of whole number and of the arithmetical operations on whole numbers, we defined the new class of rational numbers so as to render possible, in every case, certain arithmetical operations (of division and subtraction) which are, in some cases, impossible so long as we are restricted to whole numbers. We then introduced the wider class of real numbers, in order, primarily, to overcome the deficiencies of the system of rational numbers for the purposes of measurement. In doing so we had occasion to prove that certain operations, such as the extraction of the square root of a positive non-square number such as 2, which were not possible when our number-system was restricted to that of the rational numbers, were then possible. But we noticed (p. 19) that there were still operations of a similar character that are not possible (for example, extracting the square root of the negative number -2). The question arises whether we can define a new class of numbers so that all such operations are possible.

The answer to the question is that we can define a new system of "numbers," called *complex numbers* (and laws of operation on these numbers), such that all algebraical operations, applied either to real numbers or to these complex numbers, are possible and preserve the validity of the fundamental laws (p. 2). Such operations, however, will not always lead to *unique* results.

The algebraical operations contemplated include all those used in defining a general polynomial and in the solution of polynomial equations. Thus, for example, if z is any real or complex number, z^3, $\sqrt{z}$, and $z^{2/5}$ are definable as complex numbers*.

233. Definitions of complex numbers. The definitions required may be developed naturally by assuming tentatively the existence of such "imaginary numbers" as $\sqrt{(-6)}$ and the truth of the laws of algebra when applied to them. Thus, for example, we should see that $\sqrt{(-6)}$ may be expressed as $\sqrt{(6)} \times \sqrt{(-1)}$ and that any imaginary number like $\sqrt{(-6)}$ can be expressed in terms of the standard imaginary number $\sqrt{(-1)}$; also

$$\{x_1 + y_1\sqrt{(-1)}\} + \{x_2 + y_2\sqrt{(-1)}\} = (x_1 + x_2) + (y_1 + y_2)\sqrt{(-1)}$$

and $\quad \{x_1 + y_1\sqrt{(-1)}\}\{x_2 + y_2\sqrt{(-1)}\} = (x_1x_2 - y_1y_2) + (x_1y_2 + y_1x_2)\sqrt{(-1)}$.

In this way the definitions are suggested. They may be expressed thus:

I. Any (ordered) pair of real numbers x, y defines a *complex number* $[x, y]$.

II. One complex number $[x_1, y_1]$ equals a second complex number $[x_2, y_2]$ if and only if $x_1 = x_2$ and $y_1 = y_2$.

III. The complex number $[x, 0] =$ the real number x.

* Cf. the restrictions on pp. 19 and 89.

IV. The sum and product of two complex numbers are defined by the equalities

$$[x_1, y_1] + [x_2, y_2] = [x_1 + x_2, y_1 + y_2]$$
$$[x_1, y_1] \times [x_2, y_2] = [x_1 x_2 - y_1 y_2, x_1 y_2 + y_1 x_2].$$

Differences and quotients are deduced from sums and products.

The two numbers x, y of a pair $[x, y]$ are called respectively the *real* and *imaginary parts* of the complex number. If $y = 0$ the complex number is said to be *wholly real*; if $x = 0$, *wholly imaginary*.

234. With these definitions we have at once that the fundamental laws of p. 2 hold; for, e.g.,

$$[x_1, y_1] \times [x_2, y_2] = [x_1 x_2 - y_1 y_2, x_1 y_2 + y_1 x_2]$$
$$= [x_2 x_1 - y_2 y_1, x_2 y_1 + y_2 x_1]$$
$$= [x_2, y_2] \times [x_1, y_1],$$

establishing the commutative law for multiplication.

From the equality $[x, 0] = x$ we have $[1, 0] = 1$ and $[x, 0] = [1, 0] \times x$.

Corresponding to this,

$$[0, y] = [0, 1] \times [y, 0]$$
$$= [0, 1] \times y.$$

Also
$$[0, y]^2 = [0, y] \times [0, y] = [-y^2, 0] = -y^2$$

and, in particular,
$$[0, 1]^2 = -1.$$

In view of this last relation, the fixed complex number $[0, 1]$ may be regarded as a *square root of* -1. It is usually denoted by the letter i.

We notice that another number, viz. $[0, -1]$, is also a square root of -1. It clearly equals $[0, 0] - [0, 1]$, i.e. $0 - i$, which is appropriately written $-i$.

That ordinary algebra can be applied to complex numbers follows from the validity of the fundamental laws. If z is any complex number, the polynomial

$$a_n z^n + a_{n-1} z^{n-1} + \ldots + a_1 z + a_0$$

represents a unique complex number.

We may now, if we wish, change the notation and write the complex number $[x, y]$ as $x + iy$ and treat $x + iy$ as an ordinary algebraical expression, replacing i^2 by -1 wherever possible in any algebraical combination. This is always done in using complex numbers.

235. The roots of real and complex numbers. We can shew that every real number a (except 0) has two and only two square roots, i.e. complex numbers z such that $z^2 = a$.

Thus let $z = x + iy$, so that $z^2 = (x^2 - y^2) + i2xy$; then $z^2 = a$ if and only if $x^2 - y^2 = a$ and $2xy = 0$. This will be the case if and only if $x = 0$ and $y^2 = -a$, or $y = 0$ and $x^2 = a$.

If $a > 0$, this will be the case if and only if $y = 0$ and $x = \pm \sqrt{a}$; and then $z = \pm \sqrt{a}$.

If $a < 0$, it will be the case if and only if $x = 0$ and $y = \pm \sqrt{(-a)}$; and then $z = \pm i \sqrt{(-a)}$.

The square roots $\sqrt{a}$, $\sqrt{(-a)}$ occurring here are the ordinary unique (positive) square roots of the positive number a or $-a$.

With the help of the elementary properties of the trigonometrical (or circular) functions it can be proved that every real (or complex) number (other than 0) has precisely n roots of the nth degree,—n being any positive integer. The n nth roots of unity (i.e. 1) are

$$1, \cos 2\pi/n + i\sin 2\pi/n, \cos 4\pi/n + i\sin 4\pi/n, \ldots$$
$$\cos(n-1)2\pi/n + i\sin(n-1)2\pi/n.$$

It can also be proved* that every equation of the nth degree in one unknown has precisely n roots†. This fact establishes—with added precision—the statement of p. 281 as to the possibility of algebraic operations applied to all real or complex numbers.

236. Geometrical representation. Modulus. Corresponding to the use of real numbers for the purposes of measurement along a straight line, complex numbers may be used to represent points in a *plane*. The complex number $x + iy$ is taken to correspond with the point whose rectangular Cartesian coordinates (referred to a chosen pair of axes) are x, y. The resulting diagram is called the *Argand diagram*.

If O is the origin of the Argand diagram, OX, OY the axes, and P the point corresponding to the complex number z ($= x + iy$), the real numbers which measure the distance OP and the angle $X\hat{O}P$ (in radians) are called the *modulus* and the *argument* or *amplitude* of the complex number z. The modulus is taken to be essentially positive. It is alternatively defined purely analytically as $+\sqrt{'}(x^2 + y^2)$. It is written $|z|$. The amplitude will here be denoted by am z.

The amplitude of z would be unique if we agreed to consider only angles from (say) 0 to 2π radians. It is inconvenient as a rule to do this and the amplitude is defined (again analytically) to be any one of those angles (or numbers) θ for which $\cos\theta = x/|z|$ and $\sin\theta = y/|z|$ (or $C(\theta) = x/|z|$ and $S(\theta) = y/|z|$). The value θ for which $-\pi < \theta \leqslant \pi$ is called the *principal value* of the amplitude. If $|z| = r$ and am $z = \theta$, z can be expressed in terms of r and θ as

$$z = r\cos\theta + ir\sin\theta = r(\cos\theta + i\sin\theta).$$

The modulus of a wholly real number x is the modulus in the sense of p. 44.

237. Limits. The notions of greater and less do not apply to complex numbers, but, with the introduction of the modulus, the definition of the unique limit of a sequence of real numbers, as stated on p. 45 (that $|s_m - L| < \epsilon$ under stated conditions), becomes applicable without change to the case of a sequence of complex numbers. Convergent sequences and series of complex numbers can therefore be used. In particular, functions of the complex variable z can be defined by means of power series as in Chapter III, §§ 4, 5, and 6. The notions of continuity, differentiability, and integrability can also be extended to functions of a complex variable.

* See Hardy's *Pure Mathematics*, Appendix I.

† On p. 238 above we have examples of cubic and quartic equations with only one and two real roots respectively. They have of course respectively three and four real or complex roots. Compare also Ex. 4, p. 161.

These extensions in some cases (e.g. integration) involve the consideration of multiple-valued functions, resembling the n-valued nth roots of a real (or complex) number. The extension of the term logarithm which will apply to the case of the logarithm of a negative number to a positive base involves such considerations.

For detailed information on complex numbers the student is referred to Hardy's *Pure Mathematics*, Whittaker and Watson's *Modern Analysis* and books on the theory of functions of a complex variable.

EXAMPLES XXVII.

1. Shew, by direct multiplication, that if x and y are any real numbers,
$$(\cos x + i \sin x)(\cos y + i \sin y) = \cos(x+y) + i \sin(x+y),$$
and deduce de Moivre's theorem for a positive integral index, viz.
$$(\cos x + i \sin x)^n = \cos nx + i \sin nx,$$
where n is any positive integer.

2. Deduce from Ex. 1 that the nth powers of every one of the numbers $\cos(2k\pi/n) + i \sin(2k\pi/n)$, where k has every integral (or zero) value, is 1.

3. Prove that for any positive integral value of n there are precisely n different nth roots of unity.

4. Prove that, if p and q are positive integers having no common factor and z any complex (or real) number, there are precisely q values of $z^{p/q}$, i.e. complex numbers Z such that $Z^q = z^p$.

5. Prove that if z_1 and z_2 are any complex numbers
   (i) $||z_1| - |z_2|| \leqslant |z_1 \pm z_2| \leqslant |z_1| + |z_2|$;
   (ii) $|z_1 z_2| = |z_1| \cdot |z_2|$;
   (iii) $|z_1/z_2| = |z_1|/|z_2|$;
   (iv) $\operatorname{am}(z_1 z_2) = \operatorname{am} z_1 + \operatorname{am} z_2$;
   (v) $\operatorname{am}(z_1/z_2) = \operatorname{am} z_1 - \operatorname{am} z_2$;
where in (iv) and (v) the appropriate values of the amplitude are chosen, not necessarily the principal values on both sides.
Interpret (i) geometrically on the Argand diagram.

6. Prove that a sequence of complex numbers is convergent if and only if the two sequences formed respectively of the real and imaginary parts of the terms are separately convergent; and that, in the case of convergence, if the limits of the separate sequences of real and imaginary parts are s and t, the limit of the complex sequence is $s + it$.

[Denoting the typical term of the sequence by $s_n + it_n$, if $s_n \to s$ and $t_n \to t$, $|(s+it) - (s_n + it_n)| \leqslant |s - s_n| + |t - t_n| < \frac{\epsilon}{2} + \frac{\epsilon}{2} = \epsilon$, for all values of n sufficiently large, whatever positive number the arbitrary ϵ may be; whence $s_n + it_n \to s + it$.
Conversely if $s_n + it_n \to s + it$, $|s - s_n|$ and $|t - t_n|$, each of which $\leqslant |(s+it) - (s_n + it_n)|$, are clearly both less than ϵ whenever $|(s+it) - (s_n + it_n)| < \epsilon$; and the result follows.]

7. Prove that if the real series of positive terms whose typical term is $|u_n + iv_n|$ converges, the complex series whose typical term is $u_n + iv_n$ also converges. (Such a series is said to be *absolutely convergent*.)

[From the hypothesis, the sum of any number of terms of the first series is bounded (i.e. $< K$); whence (*a fortiori*) the sum of any number of terms of each of the series whose typical terms are respectively $|u_n|$ and $|v_n|$ is bounded, and these series are convergent; whence it follows (p. 78) that the series whose typical terms are u_n and v_n are convergent. Hence the result.]

8. Prove that the series

(i) $1 + z + z^2/2! + z^3/3! + ...,$

(ii) . $1 - z^2/2! + z^4/4! - ...,$

(iii) $z - z^3/3! + z^5/5! - ...$

are convergent (and absolutely convergent) for all (complex) values of z.

9. Denoting the sum of series (i) of Ex. 8 by $E(z)$, shew that, if $z = x + iy$,

(i) $E(z) = e^x(\cos y + i \sin y)$, where e^x denotes the (unique) exponential function of Chapters II and III;

(ii) $\cos y = [E(iy) + E(-iy)]/2$, $\sin y = [E(iy) - E(-iy)]/2i$.

(Euler's equalities.)

10. Denoting the sums of series (ii) and (iii) of Ex. 8 by $C(z)$ and $S(z)$, establish the addition theorems and other properties of the "trigonometrical functions" of the complex variable z.

11. Shew that, if Z is any complex number for which $E(Z) = z$, all the numbers $\zeta = Z + 2n\pi i$ (where n is any positive or negative integer or zero) satisfy the equation $E(\zeta) = z$; so that, if the *generalised logarithm* $\log z$ is defined to be any number ζ for which $E(\zeta) = z$, $\log z$ has an indefinite number of values, viz. $Z + 2n\pi i$.

12. If $|z| = r$ and am $z = \theta$, the generalised logarithm (Ex. 11) has the values $\log z = \log_e r + i\theta + 2n\pi i$, where $\log_e r$ is the (real) Napierian logarithm of the positive number r.

In particular $\log(-1) = i\pi + 2n\pi i$.

13. The *generalised exponential e^z* is defined as $E(z \log e)$, where $\log e$ is the generalised logarithm of e, i.e. $\log e = \log_e e + 2n\pi i = 1 + 2n\pi i$, where n is any integer or zero. [This definition is equivalent to $e^z = E(z) E(2n\pi i z)$.]

Shew (i) If z is an integer (or zero) this definition defines e^z uniquely and agrees with the definition of Chapter II, § 1.

(ii) If z is a fraction p/q, p and q being positive integers with no common factor, e^z has q values, which are those numbers Z for which $Z^q = e^p$ (Ex. 4); in particular $e^{1/2}$ has the two values $\sqrt{e}, -\sqrt{e}$.

(iii) If z is an irrational number, e^z has an indefinite number of values, one of which is the positive real number e^z as defined in Chapter II, § 1.

14. The generalised exponential a^z, where a is any real or complex number except 0, is defined similarly:

$a^z = E(z \log a)$, where $\log a$ is the generalised logarithm.

15. The generalised logarithm $\log_a z$, where a is any real or complex number except 0 and 1, is defined as any number Z such that the generalised exponential $a^Z = z$.

BIOGRAPHICAL NOTES
ON MATHEMATICIANS MENTIONED *

EUCLID, famous as the author of the *Elements of Geometry*,—which has been the basis of geometrical teaching for 2000 years,—was born about 330 B.C. and died about 275 B.C. He taught in the Greek University at Alexandria. The *Elements* is a systematic text-book on geometry and arithmetic, as then known to the Greeks, and includes the work of previous geometers such as Pythagoras.

ARCHIMEDES, perhaps the greatest mathematician of antiquity, was born at the Greek city of Syracuse in Sicily in 287 B.C., studied at Alexandria and returned to Syracuse, where he was killed by a Roman soldier at the fall of the city in 212 B.C. His work covered most branches of mathematical knowledge, including the mensuration of the sphere, etc. and work on mechanics, well-known to this day.

JOHN NAPIER, a Scotsman, was born in 1550 and died in 1617. Besides the invention of logarithms, several trigonometrical formulae are due to him.

RENÉ DESCARTES, born in France in 1596, was perhaps even more famous as a philosopher than as a mathematician. His invention of Cartesian geometry, named after him, was epoch-making and may be considered to mark the commencement of modern mathematics. He lived chiefly in Holland and died in Sweden in 1650.

JAMES GREGORY, born in Scotland in 1638, was one of the various forerunners of Newton. He died in 1675.

ISAAC NEWTON, the son of a Lincolnshire farmer, born in 1642, has been described as the greatest mathematician of all time. He was frail as a child, but his mental powers were already phenomenal when an undergraduate at Cambridge. In fact, most of his brilliant discoveries may be said to have been born in his mind about that time or soon afterwards, though not elaborated until many years later. His most important work was embodied in his *Principia*, which contains his results in the theory of gravitation, many of them discovered by the help of his method of "fluxions" (i.e. differential and integral calculus) but expressed in the traditional geometrical form. His results in calculus and geometry were to some extent developments of the work of his predecessors,—notably his tutor Isaac Barrow,—and the time was undoubtedly ripe for Newton's work ; but this work alone would suffice to put Newton well in the first rank. Newton was of an extremely modest disposition and of high moral character. He died in London in 1727, having been a professor at Cambridge since 1669, Master of the Mint since 1699 and a knight since 1705.

GOTTFRIED WILHELM LEIBNIZ, generally accepted as the co-discoverer with Newton of the methods of the infinitesimal calculus, was born at Leipzig in

* Taken from W. W. R. Ball's *Short Account of the History of Mathematics*.

1646 and died at Hanover in 1716. The differential notation which he used has proved more convenient than Newton's fluxional notation and is still retained. He took much interest in politics and visited England in that connection. He was also a philosopher of the first rank.

MICHEL ROLLE, born in 1652 and died in 1719, was a French professor who wrote on algebra and the theory of equations.

ABRAHAM DE MOIVRE was of French birth (born 1667) but was brought up and lived in England. He was one of the founders of that part of analysis (developed later by Euler) which deals with the connection between complex numbers and trigonometry. He died in 1754.

GEORGE BERKELEY, Bishop of Cloyne in Ireland, was born in 1685 and died in 1753. He became famous as an idealist metaphysical philosopher, particularly for his classical "proof" of the existence of God.

BROOK TAYLOR, born near London in 1685 and educated at Cambridge, was one of Newton's admirers. He is chiefly known to mathematicians by his theorem on expansions. He died in 1731.

COLIN MACLAURIN, Professor of Mathematics at Edinburgh, was born in 1698 and died as a result of military privations in opposing the Jacobites in 1745. He wrote a treatise on *Fluxions*, upholding the Newtonian ideas against Berkeley's attacks.

LEONARD EULER was born in Switzerland in 1707 and died at St Petersburg in 1783. He was the best-known mathematician of his time, his mathematical studies covered a wide range, and he left very little without his impress. His work showed very clearly the need for a new critical development of analysis and in this sense he may be regarded as the father of modern analysis.

JEAN LE ROND D'ALEMBERT, born at Paris in 1717, is noted chiefly for his solution of the problem of vibrating strings, which was the starting point of much of modern mathematics. Much of his time was spent on the French encyclopaedia. He died in 1783.

JOSEPH LOUIS LAGRANGE, the greatest mathematician of the eighteenth century, was born at Turin in 1736 and died at Paris in 1813. For twenty years he was at the court of Frederick the Great at Berlin. He was an analytical purist and in his hands analysis proved by no means a less beautiful instrument than geometry had been by tradition. The modern subjects of differential equations, calculus of variations, and theoretical dynamics are almost entirely due to his genius; and his work in the theory of numbers and in the calculus of finite differences is still of active importance. His classical *Mécanique Analytique* has been described as a scientific poem. His *Fonctions Analytiques*, published in Paris at the time of the French Revolution, contains his attempt to found infinitesimal calculus on pure algebraic analysis. He was of a modest disposition and was liked by all,—princes and revolutionaries.

JEAN ROBERT ARGAND, a Swiss, born in 1768 and died in 1822, dealt with the geometrical representation of complex numbers before their more systematic development in the hands of the greater mathematicians Gauss and Cauchy.

JEAN BAPTISTE JOSEPH FOURIER, famous for his *Théorie analytique de la chaleur*, in which,—discussing questions of Heat,—he uses the type of series now known by his name, was born in France in 1768 and died in 1830. Much of modern analysis—connected with the general notion of a function of a real variable—has arisen directly out of a study of the properties associated with Fourier series.

BERNARD BOLZANO, born in Austria in 1781 and died in 1848, was not well-known, but he anticipated in some points the great founders of modern analysis.

AUGUSTIN LOUIS CAUCHY, the chief founder of modern analysis and the theory of functions, was born at Paris in 1789. Though not as precise and acute in mind as Abel and Weierstrass, he was remarkably active and fertile in ideas. The subject of the theory of functions of a complex variable depends fundamentally on Cauchy's well-known theorem in that subject. Cauchy is typical of the French school of analysis of the nineteenth century. He died in 1857.

NIELS HENRIK ABEL, a Norwegian, was born in 1802 and died in 1829. In his short life he shewed remarkable brilliance, and his work on elliptic functions, Abelian integrals, and infinite series has been a model for later mathematicians.

KARL WEIERSTRASS, born in Westphalia in 1815 and later a Professor at Berlin, where he died in 1897, is the chief German representative of the nine-teenth century growth of modern analysis. In his development of the theory of functions he insists on concrete algebraic definitions and hard logic, and in this contrasts to some extent with the more transcendental methods of the French school. Weierstrass's work, together with that of his shorter-lived versatile and brilliant contemporary Riemann (1826—1866), forms the ideal complement to the far-reaching ideas of the French analysts.

JULIUS WILHELM RICHARD DEDEKIND was born at Brunswick in 1831 and died in 1916. He followed up Dirichlet's work on the theory of numbers. Of his several fertile ideas his definition of irrational numbers by Dedekindian sections is the best-known.

INDEX OF SYMBOLS

The numbers denote pages.

The symbols "$\neq$," "$\nrightarrow$," etc. denote the negation of "$=$," "$\rightarrow$," etc.

GENERAL INDEX

The numbers denote pages; those enclosed in brackets refer to examples.

Printed in the United States
By Bookmasters